Control Theory for Practical Applications

Hao Li

Control Theory for Practical Applications

With MATLAB Demonstration Programs

上海交通大學出版社
SHANGHAI JIAO TONG UNIVERSITY PRESS

Springer

Hao Li
SJTU Paris Elite Institute of Technology
Shanghai Jiao Tong University
Shanghai, China

ISBN 978-981-97-5007-8   ISBN 978-981-97-5008-5 (eBook)
https://doi.org/10.1007/978-981-97-5008-5

Jointly published with Shanghai Jiao Tong University Press
The print edition is not for sale in China Mainland. Customers from China Mainland please order the print book from: Shanghai Jiao Tong University Press.

This Springer imprint is published by the registered company Springer Nature Singapore Pte Ltd.
The registered company address is: 152 Beach Road, #21-01/04 Gateway East, Singapore 189721, Singapore

À ma chérie Fangqi AN:
Quand je passe devant le champs de fleurs,
je n'ai pas envie de tourner ma tête;
grâce à la poursuite de la vérité d'une part
et à vous d'autre part.

Preface

Control science is a core representative of the third industrial revolution, together with other core representatives such as computer science and atomic science. Its influence is ubiquitous in engineering activities as well as human daily life, and it contributes enormously to modern civilization. For example, let us consider domestic appliances such as tea machines, washing machines, and refrigerators, consider office equipments such as computers, printers, and scanners, consider transportation facilities such as private cars, public buses, high-speed trains, ships, and aeroplanes, consider industrial equipments such as assemblies, manipulators, and numerical machines, and consider space systems such as rockets, space stations, satellites, and extra-terrestrial explorers, and consider advanced products such as special robots, intelligent vehicles, and unmanned aerial vehicles. Besides, we can also consider various kinds of automation such as agricultural automation, medical automation, warehouse automation, mining automation, and military automation. Control science plays an indispensable role for the prosperity of all these modern fruits.

Since control science is so important to modern civilization, it has already been a core discipline of engineering education in universities for past decades. Writing a textbook on this core discipline is motivated directly by the author's years of teaching activities in the course, *Automation and Control Systems*, at Shanghai Jiao Tong University (SJTU).

Control systems, especially *automatic control systems*, are the main subject of control science. A control system may involve many aspects of consideration, such as hardware consideration, software consideration, operation consideration, maintenance consideration, economy consideration, and society consideration. However, besides all such aspects of consideration, one aspect of consideration, that is most essential to the control system is methodology consideration in mathematical sense, knowledge on which is what we refer to as *control theory*. In other words, control theory focuses on abstracting mathematical essence from engineering aspects of consideration for the control system, analyzing mathematical essence of control phenomena, establishing objective control principles, and providing control methods with guaranteed performance.

Control theory is like a field of mathematics, full of abstract thinking and analysis. It is itself meaningful and interesting enough if purely regarded as a field of mathematics; in fact, a lot of mathematical knowledge does originate and develop in the context of research on control theory. Besides its importance from the mathematical perspective, control theory is even more charming as it is deeply rooted in practical applications, many examples have already been mentioned above.

One charm of control theory for practical applications consists in *know-why*. There are these or those kinds of control phenomena in practical applications. It is true that we may resort to simulation experiments and field experiments to acquire intuitive understanding of control phenomena, yet simulation experiments and field experiments never enable us to really reach essential reasons behind control phenomena. In contrast, control theory which aims for mathematical essence of control systems enables us to reach essential reasons behind control phenomena.

Another charm of control theory for practical applications consists in *know-how*. Control methods are the soul of control systems. Control theory enables us to know how to properly design control methods. It is true that some control methods can be designed by empirical tuning for simple control problems. However, empirical tuning is not applicable to complicated control problems. Control theory can help us avoid blind trials with control systems, and enables us to design guaranteed control methods that essentially possess certain expected or desired properties.

It is the fusion of control theory and practical applications that highlights charms of control theory. According to the author's years of relevant teaching activities, it is neither control theory alone nor practical applications alone, but right the fusion of control theory and practical applications that attracts and motivates students most. So the author's basic pedagogy is naturally to fuse control theory and practical applications.

The author will clarify the inherent close relationship between control theory and practical applications, such as why and how each point of control theory covered by this textbook has been established according to concrete needs in practical applications. The author will also provide many examples of MATLAB simulation code to demonstrate how control theory is applied to handle concrete control problems and for readers to reinforce their understanding of control theory via practical application oriented practice. It is worth noting that some sub-code in given examples can be replaced by MATLAB built-in functions, yet the author insists on providing complete implementation code so that even if MATLAB is unavailable, readers can still easily follow the provided code and adapt it to implementation by other commonly adopted programming languages such as C/C++ and Python.

Prerequisites for this textbook include the standard sequence of undergraduate analysis courses, some knowledge of advanced analysis (including matrix analysis, functional analysis, and numerical computation), and some knowledge of optimization theory. For readers who are not familiar with such mathematical knowledge, this textbook is still readable and self-contained enough so that readers would go through the textbook smoothly on the whole without resorting to references.

Control theory forms a magnificent kingdom that can by no means be covered by a single book. So the author focuses on fundamentals of control theory for practical applications. Even in terms of fundamentals, the author cannot claim that this textbook covers most fundamentals of control theory for practical applications. But, according to the author's field experiences as well as teaching experiences, the fundamentals of control theory covered by this textbook can still help students and field engineers develop a working knowledge of control theory to solve many concrete problems in practical applications and can arouse their interest in acquiring more advanced knowledge of control theory.

The author would like to dedicate this textbook to his mother *Lianhua Zhan* for her priceless contribution to family education and family building during the author's childhood. The author would also like to dedicate this textbook to his father *Jinhong Li* for his mechanical engineering expertise which directly attracted the author into the wonderful kingdom of engineering, especially on automatic systems and robotics. The author would like to express his sincere gratitude and appreciation to his colleagues, especially Shaoyuan Li, Yue Gao, Ming Yang, Cailian Chen, and Jialiang Lu, for their encouragement and help to his teaching activities at Shanghai Jiao Tong University (SJTU), to his former colleagues, especially Arnaud de la Fortelle, Fawzi Nashashibi, Michel Parent, and Christian Laugier at Institut National de Recherche en Informatique et en Automatique (INRIA) for their help to his academic life, to his dear friends in Sophia-Antipolis and Côte d'Azur for their precious friendship in his life.

Shanghai, China Hao Li
2022

Contents

Chapter 1
Introduction

Control science is a core representative of the third industrial revolution, together with other core representatives such as computer science and atomic science. Its influence is ubiquitous in engineering activities as well as our daily life and it contributes enormously to modern civilization.

To demonstrate ubiquitousness of its influence, we may simply take daily-life activities in a normal day as examples. In the morning, we wake up and may turn on a tea machine that prepares a cup of tasty hot tea automatically (namely without human interference or interaction). How does the tea machine do this? After enjoying tea, we leave our house and drive a car towards the working place. On the road, we may occasionally use a driving-assistance module that keeps the car speed at our expected stable value automatically, regardless of road conditions. How does the speed driving-assistance module do this? We may also occasionally use another driving-assistance module that steers the car automatically for safe lane keeping. How does the steering driving-assistance module do this?

We arrive at the working place, park the car, and may take an elevator to go upstairs to the office floor. The elevator ascends to the correct position of the office floor automatically. How does the elevator do this? We arrive at the office and turn on a computer for working. The computer's disk drive system can always rotate quickly and automatically to correct positions for reliable data accessing. How does the disk drive system do this? We may feel hot or cold and turn on an air conditioner that adjusts the room temperature automatically to our desired value. How does the air conditioner do this? When there is a package sent to us, an unmanned vehicle carrier follows a pre-registered map automatically to deliver the package to our office. How does the unmanned vehicle carrier do this?

After working, we may take some cash from an automated teller machine that delivers the correct amount of cash automatically. How does the automated teller machine do this? We may visit a museum where a guidance robot can navigate automatically and accompany us to our destination. How does the guidance robot do this? After visiting the museum, we may feel hungry and resort to a special vending

H. Li, *Control Theory for Practical Applications*,
https://doi.org/10.1007/978-981-97-5008-5_1

machine that prepares hot food automatically. How does the vending machine do this? We finally return home and put our clothes into a washing machine that operates automatically to wash the clothes. How does the washing machine do this?

The "how does" question after each example is intended to arouse readers' reflection on the role that control science plays in the example. Besides above daily-life examples, a large amount of examples can be easily found in various domains. Let us consider transportation facilities such as public buses, high-speed trains, ships, and aeroplanes, consider industrial equipments such as assemblies, manipulators, and numerical machines, consider space systems such as rockets, space stations, satellites, and extra-terrestrial explorers, and consider advanced products such as special robots, intelligent vehicles, and unmanned aerial vehicles. Moreover, we can consider various kinds of automation such as agricultural automation, medical automation, warehouse automation, mining automation, and military automation. Control science plays an indispensable role for the prosperity of all these modern fruits.

1.1 Control Systems

Control systems especially **automatic control systems** are the main subject of control science. We may have intuitive understanding of control systems from above examples. A question arises naturally: what is a control system?

A **system** is the environment segment that concerns an activity of our interest (Dorf and Bishop 2008). For control science, we refer to an activity of our interest as a **process**. A process without humans' intentional control or influence is called a **natural process**. For example, the absolutely most phenomena in the universe are natural processes, such as rotation and revolution of the earth.

In contrast, a process under humans' intentional control is called a **control-targeted process**. In the context of control science, we rarely consider natural processes. Instead, we normally focus on control-targeted processes, which are indispensable part of control systems. For expression conciseness yet without causing confusion throughout this book, we use the term **process** to mean **control-targeted process** only, except rarely when explicit distinction between control-targeted processes and natural processes is necessary.

Normally, a process (control-targeted process) is controlled by certain **input** on one hand and results in certain **output** on the other hand, as illustrated in Fig. 1.1. For example, a process can be the rotating movement of a disk load, where the input is the torque exerted on the disk load and the output is the rotating speed of the disk load.

It is worth noting that the input characteristic is inherent part of the holistic process characteristic. Some processes (control-targeted processes), which seem to be the same process, are actually different if they have input of different characteristics. For example, although the motion of a football only under horizontal kicks, the motion of a football only under vertical kicks, and the motion of a football under both kinds of horizontal and vertical kicks commonly involve motion of a football,

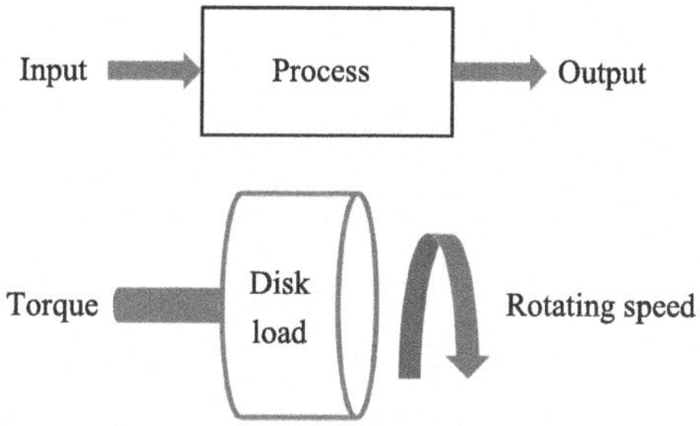

Fig. 1.1 Process (control-targeted process)

they are actually three different processes (control-targeted processes) because they have input of different characteristics namely different kinds of kicks.

It is also worth noting that any output of the process exists objectively,[1] regardless of how we treat it or even whether we treat it or not. Take the rotating movement of a disk load illustrated in Fig. 1.1 as example, although the rotating position of the disk load is not our concern for rotating disk speed control, this by no means implies that the output of the rotating position of the disk load does not exist objectively.

A **control system** involves not only a **process** (control-targeted process) but also a **controller** and an **actuator** that control the process. In the control system, the controller generates and sends control commands to the actuator and then the actuator controls or influences the process directly via the process input. The controller needs to generate proper control commands by which the actuator can control the process to achieve certain control objective, such as expected process output which is usually set as input to the controller.

For example, as illustrated in Fig. 1.2, the process is still the rotating movement of a disk load illustrated in Fig. 1.1. An electric circuit that serves as the controller generates and sends control commands to a motor that serves as the actuator. The motor provides a torque force to drive the disk load to rotate. The electric circuit needs to generate proper control commands by which the motor can drive the disk load to rotate at certain expected rotating speed.

For a control system, if the controller relies only on a priori knowledge to generate control commands without using feedback, then the control system is an **open-loop control system**. The tea machine, the special vending machine, and the washing machine previously mentioned are open-loop control systems. Taking the tea machine as example, a priori knowledge is programmed into the embedded microchip of the tea machine. The embedded microchip relies on the programmed a priori knowledge to generate control commands by which the tea machine actuator puts a standard amount of tea leaves and hot water to form a cup of hot tea.

To facilitate understanding of open-loop control systems, we may resort to an analogue with archery. In archery, flying of the archery arrow is the process (control-targeted process), whereas our brain is the controller and our arms are the actuator.

---

[1] Process output is determined by the **state** of the process, detailed clarification of which is postponed to Chap. 6. For materials before Chap. 6, it is unnecessary to involve the concept of **state**.

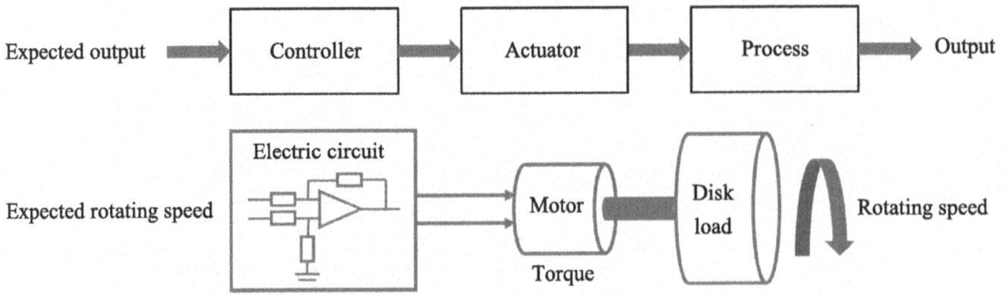

Fig. 1.2 Control system

Fig. 1.3 Archery: **a** brain (controller); **b** arms (actuator); **c** arrow flying (process); **d** expected process output (hitting the small balloon)

Our brain, which stores a priori knowledge namely "feeling of archery" accumulated after a long time of systematic training, relies on such a priori knowledge to control our arms that shoot the archery arrow. Once the archery arrow is shot, no feedback can be used to adjust flying of the archery arrow. The control objective or the expected process output is to hit the target such as a small balloon, as illustrated in Fig. 1.3.

An open-loop control system's performance is largely influenced by a priori knowledge on which the controller relies. If the a priori knowledge is of high quality, then expected process output may be achieved; otherwise, unexpected process output is incurred. It is like a good and stable "feeling of archery" tends to enable us to hit the archery target, whereas a bad "feeling or archery" causes us to miss the archery target.

Open-loop control systems mainly have two limitations. First, a priori knowledge of high quality on which the controller relies is usually unavailable. It is like most people lack a good and stable "feeling of archery" namely a priori knowledge of high quality for archery. Second, the environment where control systems operate may change and a priori knowledge on which the controller relies may no longer be applicable in the changed environment. Consequently, open-loop control systems tend to have non-guaranteed performance in changeable environments. It is like even professional players of archery tend to have poor performance in environments full of wind.

1.2 Closed-Loop Feedback Control Systems

For a control system, if the controller can take advantage of both a priori knowledge and **feedback** (normally process output feedback) to generate control commands, then the control system is a **closed-loop feedback control system**, also known shortly as **closed-loop control system** or **feedback control system**.

Compared with the open-loop control system, the closed-loop feedback control system has two more components, namely the **sensor** (Fraden 2010) and the **estimator** (Li 2022), which provide feedback to the controller. The generic closed-loop feedback control system is illustrated in Fig. 1.4.

The sensor measures process output directly and provides raw output measurements to the estimator which then processes raw output measurements to obtain better output estimates. The obtained estimates are right the feedback provided to the controller so that the controller can take advantage of the estimates besides a priori knowledge to generate control commands.

When the sensor itself is reliable and accurate enough for feedback control, the estimator can be omitted. Even when the estimator cannot be omitted but is indeed needed for feedback control, we may still regard the sensor and the estimator together as a sensor in general sense. Therefore, the generic closed-loop feedback control system can be illustrated as another version in Fig. 1.5.

In contrast with open-loop control systems, closed-loop feedback control systems which take advantage of feedback control can have much more guaranteed performance in changeable environments or when a priori knowledge is not of high quality or even unavailable. Detailed demonstration of utilities of feedback control is postponed to Chap. 2.

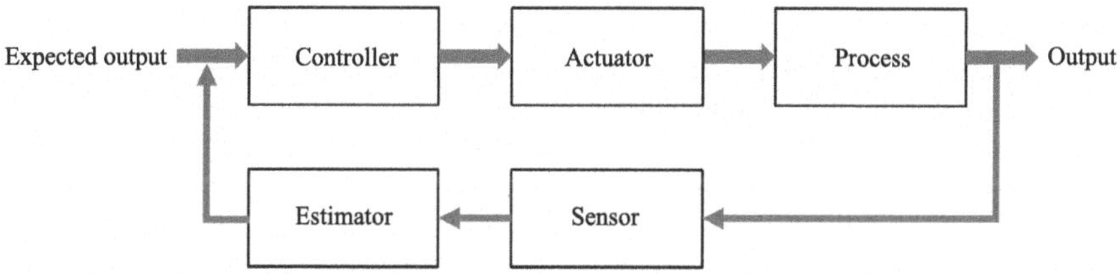

Fig. 1.4 Closed-loop feedback control system

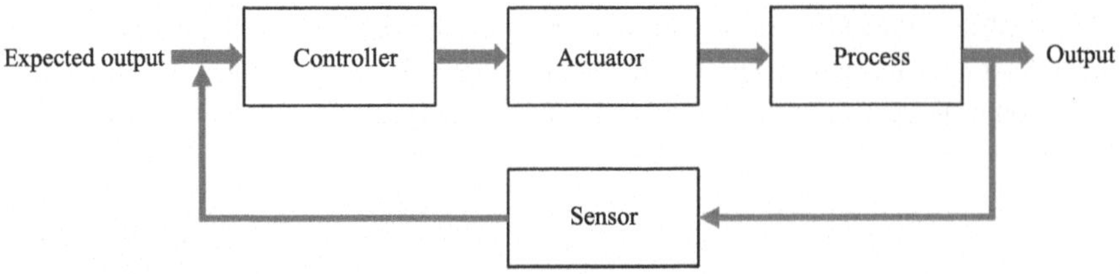

Fig. 1.5 Closed-loop feedback control system (another version)

1.3 Examples of Closed-Loop Feedback Control Systems

Closed-loop feedback control systems play a much more important role than open-loop control systems in control science and engineering. They form the absolutely main part of researches and developments on control theory and relevant practical applications. A number of representative examples of closed-loop feedback control systems are described below.

Rotating disk speed control

Many modern facilities and devices involve rotating disk speed control. For example, the speed driving-assistance module previously mentioned, which appears as car speed control from the driving perspective, is actually based on rotating disk speed control (regard the car driving wheel as the rotating disk). Similarly, the elevator previously mentioned, which needs to keep smooth ascending or descending, also relies on rotating disk speed control.

A closed-loop feedback control system for rotating disk speed control is illustrated in Fig. 1.6. The process is the rotating movement of a disk load and our concerned process output is the rotating disk speed. An electric circuit that serves as the controller generates and sends control commands to a motor that serves as the actuator. The motor provides a torque force to drive the disk load to rotate. A tachometer serves as the sensor that measures the rotating speed of the disk load and provides rotating speed measurements as feedback to the electric circuit controller.

In the rotating disk speed feedback control system, the electric circuit controller can take advantage of not only a priori knowledge but also rotating disk speed feedback to generate control commands. As will be demonstrated in details in Chap. 2, the rotating disk speed feedback control system can succeed in achieving the control objective of rotating disk speed control.

A simple intuitive interpretation for the working mechanism and principle of the rotating disk speed feedback control system is given as follows: When the rotating speed is lower than the expected rotating speed, the electric circuit controller generates and sends control commands to the motor to increase the torque exerted on the disk load. When the rotating speed is higher than the expected rotating speed, the electric circuit controller generates and sends control commands to the motor to decrease the torque exerted on the disk load. The electric circuit controller adjusts its control commands dynamically until the rotating disk speed feedback control system

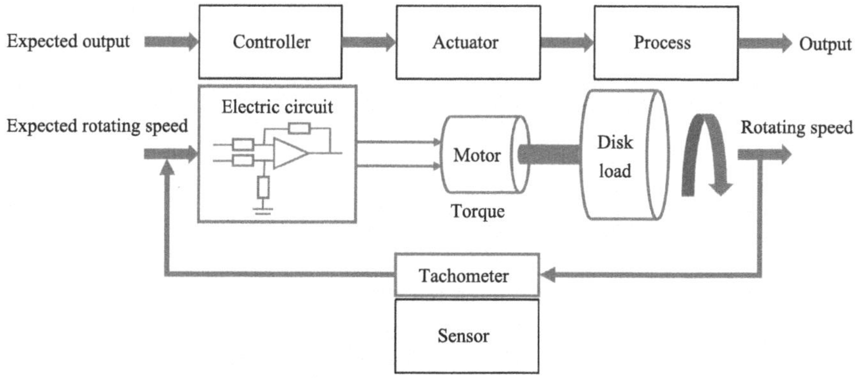

Fig. 1.6 Rotating disk speed control

converges to the equilibrium status that the rotating speed is actually the expected rotating speed. It is worth noting that in such feedback control way, the electric circuit controller takes advantage of only feedback error namely error between the rotating speed feedback and the expected rotating speed, without necessarily relying on *a priori* knowledge of the disk load or the environment.

Rotating disk position control

Many modern facilities and devices also involve rotating disk position control. For example, the elevator previously mentioned not only relies on rotating disk speed control to ascend or descend smoothly but also relies on rotating disk position control to stop at correct floor positions. The computer's disk drive system previously mentioned relies on rotating disk position control to access data at correct positions. A surveillance camera servomechanism relies on rotating disk position control to adjust the camera view correctly towards targets.

A closed-loop feedback control system for rotating disk position control is illustrated in Fig. 1.7. The process is the rotating movement of a disk load and our concerned process output is the rotating disk position. An electric circuit that serves as the controller generates and sends control commands to a motor that serves as the actuator. The motor provides a torque force to drive the disk load to rotate. A goniometer serves as the sensor that measures the rotating position of the disk load and provides rotating position measurements as feedback to the electric circuit controller.

In the rotating disk position feedback control system, the electric circuit controller can take advantage of rotating position feedback to generate control commands. Whenever there is feedback error namely error between the rotating position feedback and the expected rotating position, the electric circuit controller can adjust its control commands dynamically to control the motor to rotate in the direction of reducing such error, until the rotating disk position feedback control system converges to the equilibrium status that the rotating position is actually the expected rotating position.

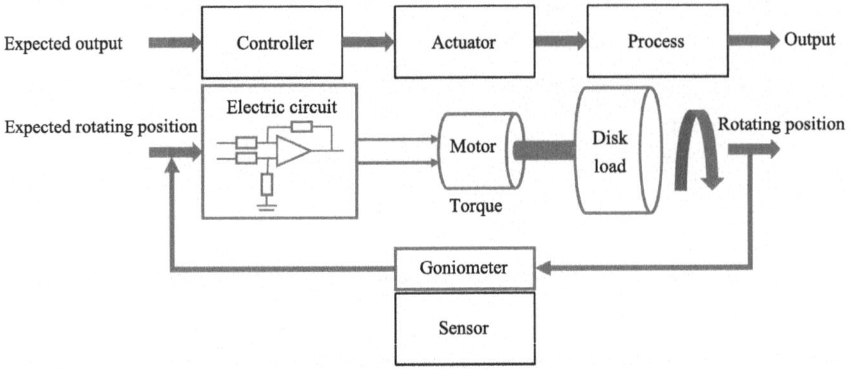

Fig. 1.7 Rotating disk position control

Similar to the rotating disk speed feedback control system, the rotating disk position feedback control system takes advantage of only feedback error, without necessarily relying on a priori knowledge of the disk load or the environment.

Intelligent vehicle steering control

Intelligent vehicles or autonomous vehicles (Laurgeau 2009; Eskandarian 2012) have been developing rapidly in terms of researches and have been attracting more and more public attention as well as investment interests during recent years. Intelligent vehicles are an important representative of high technology. An intelligent vehicle is a comprehensive intelligent system and involves a wide range of research domains and functionalities. Among all functionalities of an intelligent vehicle, the functionality that impresses people most directly may be intelligent vehicle steering control (also known as lateral control) for lane keeping, namely the functionality that enables the intelligent vehicle to steer itself automatically during driving on a target lane.

A closed-loop feedback control system for intelligent vehicle steering control is illustrated in Fig. 1.8. The process is vehicle motion and our concerned process output is the vehicle lateral position on the target lane. A microchip that serves as the controller generates and sends control commands to a steering mechanism that serves as the actuator. A module of lane detection (Li and Nashashibi 2013) serves as the sensor that estimates the vehicle lateral position on the target lane and provides vehicle lateral position measurements as feedback to the microchip controller. The expected process output is the middle of the target lane.

In the intelligent vehicle steering feedback control system, the microchip controller can take advantage of vehicle lateral position feedback to generate control commands. Simply speaking, when the intelligent vehicle is to the left of the target lane middle, the microchip controller generates and sends control commands to the steering mechanism to turn right. When the intelligent vehicle is to the right of the target lane middle, the microchip controller generates and sends control commands to the steering mechanism to turn left. The microchip controller adjusts its control

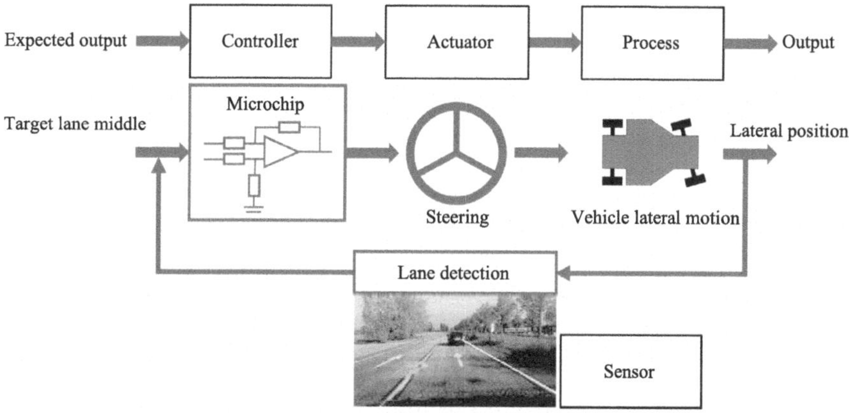

Fig. 1.8 Intelligent vehicle steering control. Lane detection figure from Li and Nashashibi (2013)

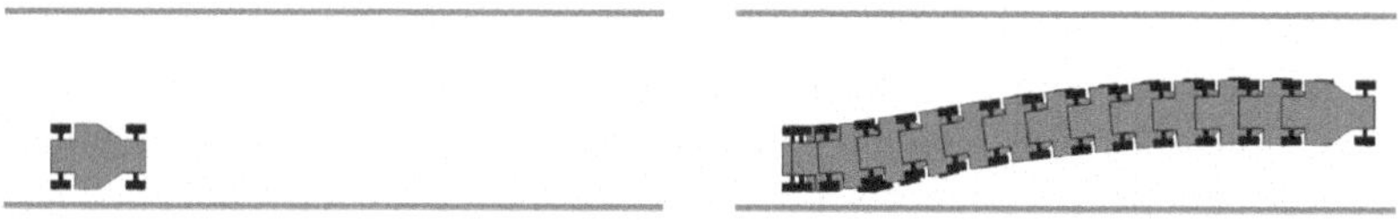

Fig. 1.9 Lane keeping: (left) initial deviation, (right) convergence to the lane middle

commands dynamically until the intelligent vehicle steering feedback control system converges to the equilibrium status that the intelligent vehicle is actually in the middle of the target lane, as illustrated in Fig. 1.9.

Intelligent vehicle parking control

Intelligent vehicle steering control for lane keeping is not the only fascinating functionality of an intelligent vehicle. Intelligent vehicle parking control is another fascinating functionality, as illustrated in Fig. 1.10. Intelligent vehicle parking control aims at driving the intelligent vehicle automatically from an initial pose to the destination pose in a parking slot.

In a closed-loop feedback control system for intelligent vehicle parking control, the process is vehicle motion and our concerned process output is the vehicle pose

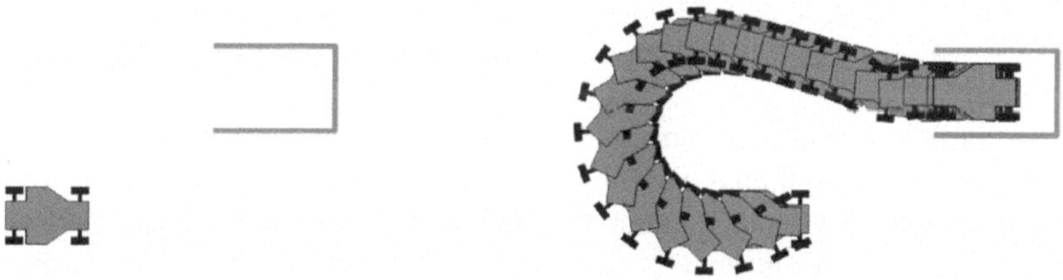

Fig. 1.10 Intelligent vehicle parking control: (left) initial pose, (right) parking

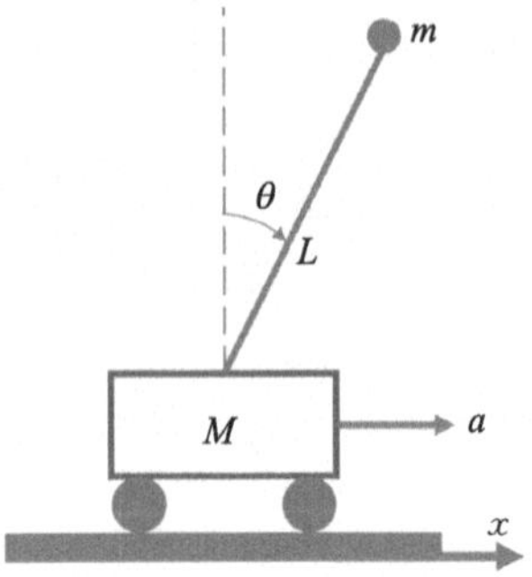

Fig. 1.11 Inverted pendulum control (cart acceleration a, cart position x, and inverted pendulum angle θ)

(including position and orientation). The expected process output is the destination pose in the parking slot. The controller of the intelligent vehicle parking feedback control system can take advantage of vehicle pose feedback to generate control commands.

Compared with intelligent vehicle steering control for lane keeping, intelligent vehicle parking control is more complicated in terms of two aspects. First, intelligent vehicle parking control involves not only steering control (also known as lateral control) but also speed control (also know as longitudinal control). Second, intelligent vehicle parking control usually necessitates certain advanced form of feedback control such as model predictive control, presentation of which is postponed to Chap. 7.

Inverted pendulum control

Inverted pendulum control aims at moving a cart to a target position while balancing an inverted pendulum on the cart, as illustrated in Fig. 1.11. In a closed-loop feedback control system for inverted pendulum control, the process is movement of the cart as well as the inverted pendulum and our concerned process output is the inverted pendulum state that consists of the cart position, the cart speed, the inverted pendulum angle, and the inverted pendulum angular speed. The expected process output is the expected inverted pendulum state that consists of the target cart position, zero cart speed, zero inverted pendulum angle, and zero inverted pendulum angular speed.

Inverted pendulum control is an elegant application example to highlight the charm of control theory. First, inverted pendulum control is interesting enough and resembles a typical game of our childhood, namely to balance a long stick on our hand without holding it. Second, inverted pendulum control is difficult enough and hence offers a performance stage for a number of important control methods to show their power.

Some intuitive explanations hover over the difficulty of inverted pendulum control. First, the inverted pendulum is extremely unstable. Whenever control commands are inappropriate, the inverted pendulum would quickly fall down, leaving little chance for remedy. Second, for inverted pendulum control, the process has control input of only one dimension namely the cart acceleration, whereas it has concerned process output of multiple dimensions namely the cart position, the cart speed, the inverted

Fig. 1.12 Motorcycle control (steering angle β and rolling angle θ)

pendulum angle, and the inverted pendulum angular speed. It is like we use only one hand to do multiple things simultaneously.

Motorcycle control

Motorcycle (or bicycle) control is also an interesting application example close to our daily life, as illustrated in Fig. 1.12. Motorcycle control consists of motorcycle steering control for lane keeping and motorcycle balance control. Motorcycle steering control for lane keeping is similar to intelligent vehicle steering control for lane keeping. Motorcycle balance control, which aims at maintaining the motorcycle rolling angle at zero, is realized by motorcycle steering control indirectly.

Similar to inverted pendulum control, motorcycle control is also an application example where the process has control input of only one dimension but has concerned process output of multiple dimensions. In motorcycle control, only the steering angle is used to simultaneously control the motorcycle lateral position on a target lane and control the motorcycle rolling angle with respect to the vertical line. In fact, motorcycle control is essentially the same to inverted pendulum control. Intuitively speaking, we may imagine controlling the motorcycle lateral position as controlling the cart position and imagine balancing the motorcycle as balancing the inverted pendulum on the cart.

Industrial manipulator control

Control theory has long since been flourishing in modern industry, instantiated as a large variety of industrial manipulators (Murray et al. 1994), as illustrated in Fig. 1.13.

In contrast with inverted pendulum control and motorcycle control where the process has control input of less dimensions but has concerned process output of more dimensions, industrial manipulator control is a representative application example where the process usually has control input of more dimensions but has concerned process output of less dimensions. Take the industrial manipulator illustrated in Fig. 1.13 as example, where the process is manipulator motion and our concerned process output is the end effector position. The process has one-dimensional control

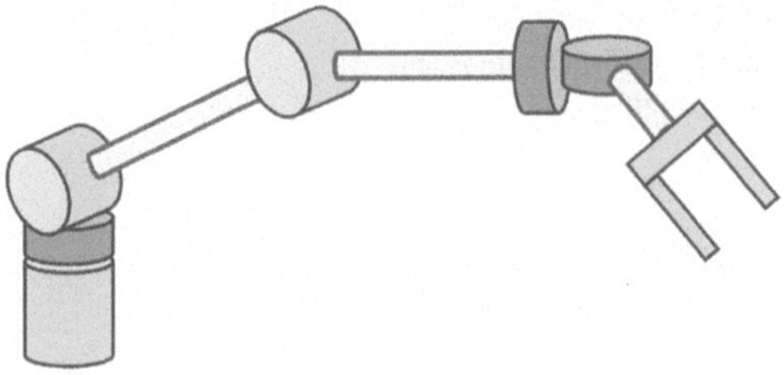

Fig. 1.13 Industrial manipulator control

input at each manipulator joint and hence has control input of totally six dimensions, whereas it has concerned process output of only three dimensions.

The fact that an industrial manipulator control system usually has control input of more degrees of freedom does not imply that industrial manipulator control is less difficult or complicated than inverted pendulum control and motorcycle control. There is an old saying: "One monk has water to drink, whereas three monks have no water to drink".[2] If the various dimensional parts of control input are inappropriate or do not cooperate well, they may even incur worse consequences than control input of less degrees of freedom. On the other hand, if the various dimensional parts of control input are appropriate and cooperate well, they do tend to achieve desirable performance and outperform control input of less degrees of freedom.

Special robot control

A large variety of special robots have been researched and developed; many of them have already been put into practical applications. These special robots include agriculture robots, construction robots, mining robots, underwater robots, search and rescue robots, medical robots, health care robots, domestic robots, education robots, humanoids, social robots, biologically inspired robots, soft-body robots, etc. (Siciliano and Khatib 2016). These special robots represent humans' unlimited imagination for automation and artificial intelligence and provide colourful performance stages for control theory. Fig. 1.14 shows a multiped robot, which is developed by the team in the charge of Professor GAO Yue at SJTU, was performing automatic skiing in Beijing Winter Olympic Games in 2022.

Unmanned aerial vehicle control

Unmanned aerial vehicles become more and more common in daily life, in industry, and in military activities (Valavanis and Vachtsevanos 2015). A Da-Jiang (meaning "broad territory") unmanned aerial vehicle is illustrated in Fig. 1.15. For unmanned aerial vehicle control, the process is motion of the unmanned aerial vehicle and our concerned process output is the unmanned aerial vehicle pose in the three-dimensional sky. The control input is driving (or lifting) force provided by propulsion components of the unmanned aerial vehicle. The controller of the unmanned aerial

[2] The three monks all rely on others to take water and nobody acts to take water, consequently they all have no water to drink.

Fig. 1.14 Multiped robot skiing

Fig. 1.15 Unmanned aerial vehicle control (a Da-Jiang UAV)

vehicle needs to generate control commands that enable the unmanned aerial vehicle to achieve certain expected motion status, such as hovering stably over a target position.

Space vehicle control
Space exploration and exploitation have been providing an enormous impetus for development of control theory in modern history. Space exploration and exploitation involve many advanced control systems that play important roles in high-level economic activities and national defence. Space vehicle control systems are a representative of such advanced control systems. A famous application example of space vehicle control, named after Tsien Hsue-shen, is illustrated in Fig. 1.16.

1.4 Control Theory

A control system especially a closed-loop feedback control system may involve many aspects of consideration:

- Hardware consideration, for example, what kind of actuator (a fuel motor, an electrical motor, a hydraulic device etc.) should be used? What kind of sensor (a mechanical sensor, an electrical sensor, a photonic sensor etc.) should be used? What kind of controller (a computer, a microchip, an *ad hoc* device etc.) should be used? What kind of interface among various components should be used? How to arrange and install the controller, the actuator, the sensor, and peripherals? How to verify hardware quality for control purpose?

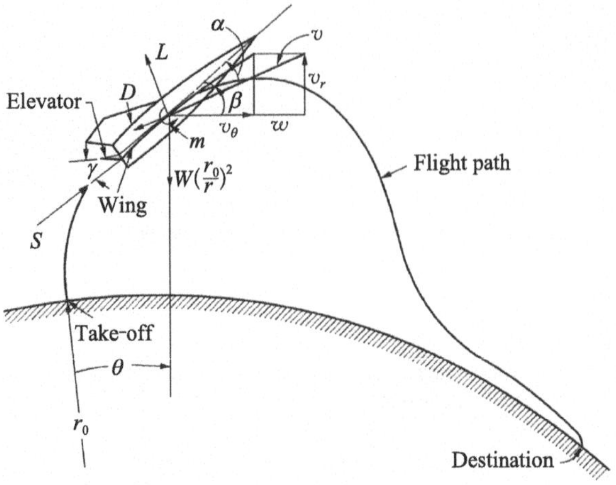

Fig. 1.16 Tsien Hsue-shen ballistics. Figure from Tsien (2015)

- Software consideration, for example, what kind of programming language (assembly, C/C++, *ad hoc* language etc.) should be used? What is the execution efficiency of the programming language (on certain computational hardware)? Is the algorithm implemented correctly and efficiently? Is parallel processing needed? What kind of communication protocols among various components should be used? How to verify if the software is reliable and exempt from bugs? How to guarantee that the software is secured and robust to malicious or adversarial attacks?
- Operation and maintenance consideration, for example, how to provide energy to the controller, the actuator, the sensor, and peripherals? How to verify all hardware can always operate well after integration? How to verify all software can always operate well after integration? How to maintain the control system in good conditions in the long run?
- Economy consideration, for example, is the control system affordable for research purpose? Is the control system economical and competitive for commercial purpose?
- Society consideration, for example, is the control system acceptable to public? Does the control system respect cultures and conventions of a society in which the control system operates? Does the control system violate national laws?

Besides above aspects of consideration, one aspect of consideration that is most essential to the control system is methodology consideration in mathematical sense, knowledge on which is what we refer to as **control theory**. In other words, control theory focuses on *abstracting mathematical essence from engineering aspects of consideration for the control system, analysing mathematical essence of control phenomena, establishing objective control principles, and providing control methods with guaranteed performance*.

Taking rotating disk speed control as example, mathematical essence of the process, the actuator, and the sensor is illustrated in Fig. 1.17. The process dynamics of how the process input namely the torque T influences the process output namely

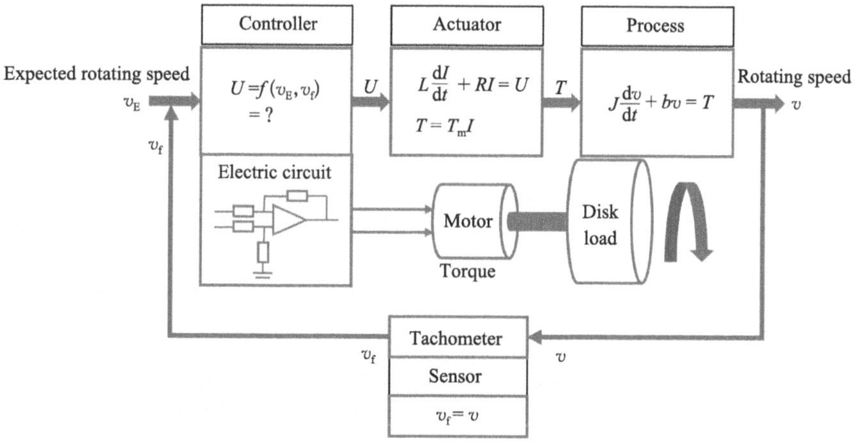

Fig. 1.17 Methodology consideration in mathematical sense

the rotating speed v is described by a first-order differential equation. The actuator dynamics of how the control command namely the voltage U influences the actuator output namely the torque T is described by a first-order differential equation together with a proportional equation. The sensor is abstracted as a bijective mapping between the process output v and the sensor output namely the rotating speed feedback v_f (an identity mapping $v_f = v$ is usually realized in practical applications). For a generic control system, abstracting its mathematical essence is what we refer to as **system modelling**.

In the closed-loop feedback control system for rotating disk speed control, the controller takes the expected rotating speed v_E and the rotating speed feedback v_f as input and generates the control command U as output. The mathematical essence of the controller is the mapping between $\{v_E, v_f\}$ and U. For a generic control system, the mathematical essence of the controller is the mapping between the controller input and the controller output, and this mapping in mathematical sense is what we refer to as **control method**.

Different control methods have different influence on control systems and result in different control phenomena. Why are there these or those kinds of control phenomena? How to analyse control phenomena based on certain methodology of system modelling and reach essential reasons behind them? How to analyse essential influence of different control methods on control systems? How to avoid blind trials with control systems? Can objective control principles instead of empirical tuning rules be established according to theoretical analysis? How to design guaranteed control methods that essentially possess certain expected or desired properties? **Control theory** aims at providing answers to such kinds of methodology consideration in mathematical sense that are essential to control systems.

References

R. Dorf, R. Bishop. *Modern control systems*. Pearson Prentice Hall, 2008.

A. Eskandarian (Ed.). *Handbook of intelligent vehicles*. Springer-Verlag London, 2012.

J. Fraden. *Handbook of modern sensors: physics, designs, and applications*. Springer Science & Business Media, 2010.

C. Laurgeau. *Le siècle de la voiture intelligente*. Presses des MINES, 2009.

H. Li, F. Nashashibi. Lane detection (part i): Mono-vision based method. *INRIA Tech Report*, RT-433, 2013.

H. Li. *Fundamentals and applications of recursive estimation theory*. Shanghai Jiao Tong University Press, 2022.

R. Murray, Z. Li, S. Sastry. *A mathematical introduction to robotic manipulation*. CRC Press, 1994.

B. Siciliano, O. Khatib. *Springer handbook of robotics*. Springer, 2016.

H. Tsien. *Engineering cybernetics*. Shanghai Jiao Tong University Press, 2015.

K. Valavanis, G. Vachtsevanos. *Handbook of unmanned aerial vehicles*. Springer, 2015.

Chapter 2
Feedback Control

In Chap. 1, we have mentioned that closed-loop feedback control systems play a much more important role than open-loop control systems in control science and engineering. We have also presented a variety of representative examples of closed-loop feedback control systems. Questions arise naturally: why cannot we simply use open-loop control systems? Why is feedback control so important to control systems? What are basic utilities of feedback control? In this chapter, we gradually clarify answers to these questions.

2.1 Open-Loop Control

Taking rotating disk speed control as example, mathematical essence of the involved components is illustrated in Fig. 1.17. For the moment, consider only the open-loop control part namely the process, the actuator, and the controller without the sensor feedback loop, as illustrated in Fig. 2.1.

The process dynamics of how the process input namely the torque T influences the process output namely the rotating speed v is described by a first-order differential equation:

$$J\frac{\mathrm{d}v}{\mathrm{d}t} + bv = T,$$

or

$$\frac{J}{b}\frac{\mathrm{d}v}{\mathrm{d}t} + v = \frac{1}{b}T, \tag{2.1}$$

where J denotes the inertia of the rotating disk load, and b denotes the friction coefficient.

H. Li, *Control Theory for Practical Applications*,
https://doi.org/10.1007/978-981-97-5008-5_2

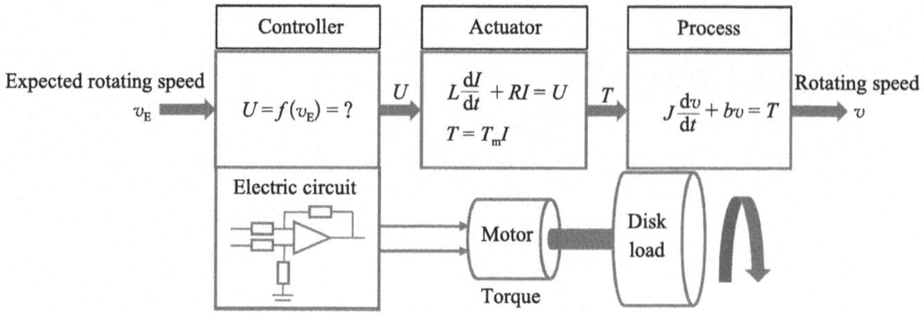

Fig. 2.1 Rotating disk speed open-loop control

The actuator dynamics of how the control command namely the voltage U influences the actuator output namely the torque T is described by a first-order differential equation together with a proportional equation:

$$L\frac{dI}{dt} + RI = U,$$
$$T = T_m I.$$

Above two equations can be combined into one single first-order differential equation as

$$\frac{L}{R}\frac{dT}{dt} + T = \frac{T_m}{R}U, \tag{2.2}$$

where R denotes the motor armature resistance, L denotes the motor armature inductance, and T_m denotes the motor torque constant. An example of rotating disk speed control system parameters J, b, R, L, and T_m is given in Table 2.1.

We assume that readers are familiar with fundamental physics. *So for expression conciseness throughout this book, we omit variable physical units which can be easily completed according to fundamental physics knowledge*. For example, parameters in Table 2.1 can be easily completed with physical units, as listed in Table 2.2.

Table 2.1 Rotating disk speed control system parameters

Parameter	Value
Rotating disk load inertia J	10
Friction coefficient b	5
Motor armature resistance R	1
Motor armature inductance L	0.001
Motor torque constant T_m	5

Table 2.2 Rotating disk speed control system parameters (with physical units)

Parameter	Value
Rotating disk load inertia J	10 N m s^2/rad
Friction coefficient b	5 N m s/rad
Motor armature resistance R	1 Ω
Motor armature inductance L	0.001 H
Motor torque constant T_m	5 N m/A

Since $L/R \ll J/b$ usually holds in practical applications, we neglect L/R. Here, we do not explain why we can fairly neglect L/R if $L/R \ll J/b$. First, readers may naturally understand the reason as they acquire more knowledge of control theory from this book. Second, negligence of the trivial factor L/R has no influence on clarifying knowledge points of this chapter.

Once L/R is neglected, mathematical description of the actuator dynamics is simplified into a bijective mapping between U and T as

$$T = \frac{T_m}{R} U$$

and hence can be incorporated into the controller as a unified controller-actuator, as illustrated in Fig. 2.2.

Mathematical essence of the unified controller-actuator is the mapping between its input v_E and its output T, namely $T = F(v_E)$. Whenever we have defined the unified controller-actuator mapping $T = F(v_E)$, we can uniquely determine the corresponding original controller mapping $U = f(v_E)$ by setting

$$f(v_E) = \frac{R}{T_m} F(v_E).$$

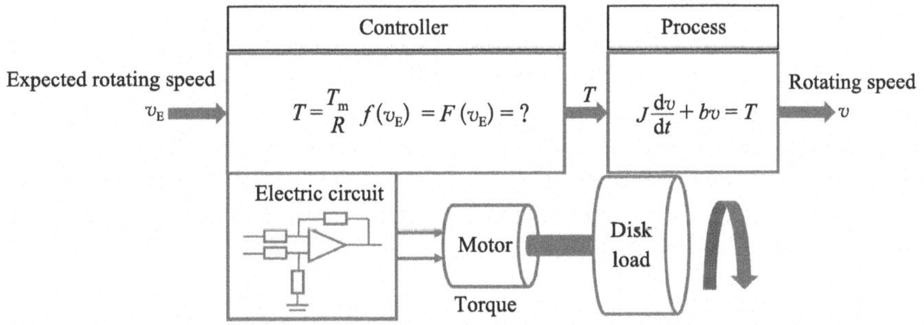

Fig. 2.2 Rotating disk speed open-loop control (unified controller-actuator)

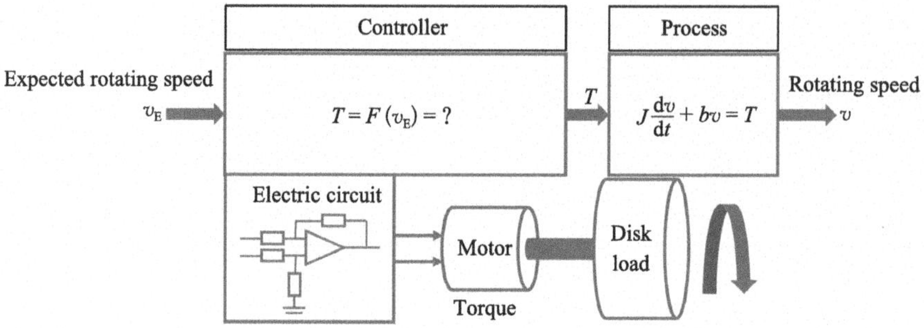

Fig. 2.3 Rotating disk speed open-loop control (simplified equivalence)

Vice versa, whenever we have defined the original controller mapping $U = f(v_E)$, we can also uniquely determine the corresponding unified controller-actuator mapping $T = F(v_E)$ by setting

$$F(v_E) = \frac{T_m}{R} f(v_E).$$

Therefore, if we regard the unified controller-actuator as a controller in general sense, then the rotating disk speed open-loop control system illustrated in Fig. 2.1 is transformed into a simplified mathematical equivalence, as illustrated in Fig. 2.3.

2.2 Control System Simulation

Consider a task of rotating disk speed control: *Design a control method namely a controller mapping $T = F(v_E)$ such that the expected rotating speed v_E can be achieved.* For example, given rotating disk speed control system parameters J and b specified in Table 2.1, suppose the expected rotating speed $v_E = 1$. How to design the control method $T = F(v_E)$?

Readers may have no idea of answering above question. For the moment, instead of just putting forward a qualified control method abruptly, we resort to **simulation** for readers to acquire more intuitive understanding of the control problem. Simulation of a control system is a set of synthetic experiments that are realized via numerical computation according to mathematical essence of the control system. We can use simulation to test tentative control methods for observation and analysis. Matlab code for simulation of rotating disk speed control is given as follows.

RotatingDiskSpeedOpenLoopControl.m

```
%% Rotating disk speed control system parameters
J = 10; % Rotating disk load inertia
b = 5; % Friction coefficient

%% Simulation preliminary configuration
dt = 0.001; % Numerical computation step
tSpan = 0:dt:10; % Simulation time span
% Record rotating disk speeds during simulation
vAll = zeros(size(tSpan)); k = 1;
v = 0; % Initial rotating disk speed

%% Simulation of rotating disk speed control
for t = tSpan
    %% Control method
    T = 1;

    %% Rotating disk dynamics
    dv = (T-b*v)/J;
    v = v + dv*dt;
    vAll(k) = v; k = k+1;
end
plot(tSpan, vAll, 'LineWidth', 2); ylim([0, 2]);
xlabel('Time'); ylabel('Rotating Disk Speed'); grid on;
```

We suggest that readers try above Matlab simulation code[1] to acquire intuitive understanding of rotating disk speed control. For example, set the control method as $T = 1$. The simulation result is demonstrated in Fig. 2.4, where the horizontal axis represents the simulation time span and the vertical axis represents the rotating speed during simulation.

As demonstrated in Fig. 2.4, when the control method $T = 1$ is adopted, the rotating speed rises steadily until arriving at a stable value $v(\infty) = 0.2$. This is not the expected rotating speed $v_{\mathrm{E}} = 1$. We may try another control method $T = 2$ and the simulation result is demonstrated in Fig. 2.5. When the control method $T = 2$ is adopted, the rotating speed rises steadily until arriving at another stable value $v(\infty) = 0.4$. Unfortunately, this is still not the expected rotating speed $v_{\mathrm{E}} = 1$.

Although above two tentative control methods cannot achieve the expected rotating speed $v_{\mathrm{E}} = 1$, the simulation experiments convey some regularity about performance of the control methods. It seems that the stable rotating speed achieved by a control method is proportional to the **control law** namely the controller output T. So we may try one more control method $T = 5$ such that v_{E}/T equals the ratio $0.4/2 = 0.2/1$. The simulation result is demonstrated in Fig. 2.6.

When the control method $T = 5$ is adopted, the rotating speed rises steadily and converges to a stable value $v(\infty) = 1$ which is exactly the expected rotating speed $v_{\mathrm{E}} = 1$. Therefore, we have designed a control method $T = 5$ that can achieve

[1] If Matlab is unavailable, readers can adapt above Matlab code to implementation by other programming languages such as C/C++ and Python.

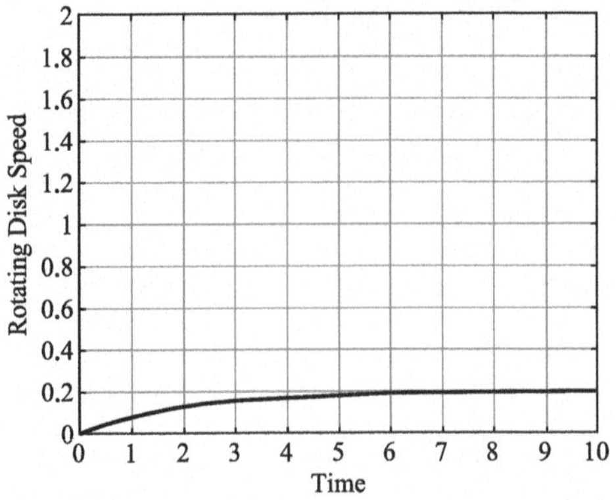

Fig. 2.4 Rotating disk speed open-loop control ($T = 1$)

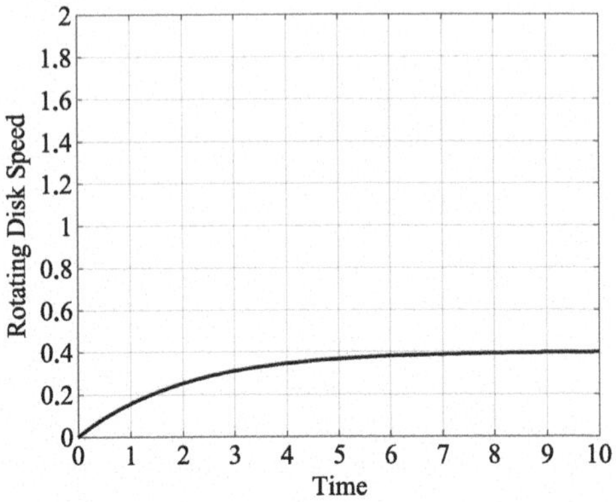

Fig. 2.5 Rotating disk speed open-loop control ($T = 2$)

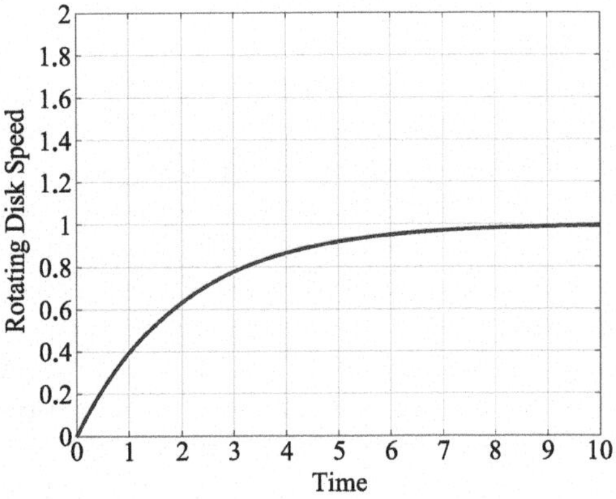

Fig. 2.6 Rotating disk speed open-loop control ($T = 5$)

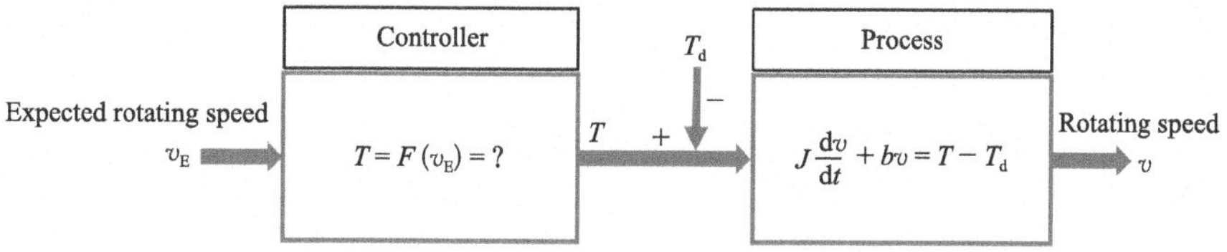

Fig. 2.7 Rotating disk speed control with unknown torque disturbance

the expected rotating speed $v_E = 1$. Besides, given any arbitrary expected rotating speed v_E, we can set the control method as $T = 5v_E$ to achieve the expected rotating speed v_E.

2.3 Feedback Control: First Utility

Simulation experiments presented in Sect. 2.2 may give us an illusion that open-loop control is effective enough for practical applications. If readers do have such illusion, then we may consider a more challenging task of rotating disk speed control: *Given a rotating disk speed control system with uncertainty, design a control method such that the expected rotating speed v_E can be achieved.*

2.3.1 Control System with Uncertainty

In Sect. 2.2, the rotating disk speed control system is completely certain. In other words, all configurations or parameters of the rotating disk speed control system are known. However, in practical applications, a rotating disk speed control system usually faces unknown torque disturbance. For example, an elevator may take more or less people in different rounds of operation. The undetermined total people weight has an influence on the elevator actuator motor like unknown torque disturbance does. Taking the speed driving-assistance module of a vehicle as another example, the vehicle may ascend or descend a slope, which incurs unknown torque disturbance to the speed driving-assistance module during vehicle driving. Unknown torque disturbance is a typical aspect of uncertainty inevitable to a rotating disk speed control system in practical applications.

A rotating disk speed control system with uncertainty caused by unknown torque disturbance is illustrated in Fig. 2.7. The process dynamics is no longer influenced by the controller output torque T only, but influenced by the difference between the controller output torque T and unknown torque disturbance T_d.

As in Sect. 2.2, we can use simulation to test tentative control methods for observation and analysis. Rotating disk speed control system parameters J and b specified in

Table 2.1 are still adopted. Matlab code for simulation of rotating disk speed control with unknown torque disturbance is given as follows.

RotatingDiskSpeedControlWithUnknownTorqueDisturbance.m

```
%% Rotating disk speed control system parameters
J = 10; % Rotating disk load inertia
b = 5; % Friction coefficient
Td = random('Uniform', -5, 5); % Unknown torque disturbance

%% Simulation preliminary configuration
dt = 0.001; % Numerical computation step
tSpan = 0:dt:10; % Simulation time span
% Record rotating disk speeds during simulation
vAll = zeros(size(tSpan)); k = 1;
v = 0; % Initial rotating disk speed
vE = 1; % Expected rotating disk speed

%% Simulation of rotating disk speed control
for t = tSpan
    %% Control method
    T =

    %% Rotating disk dynamics suffering from unknown toruqe
    disturbance
    dv = (T-Td-b*v)/J;
    v = v + dv*dt;
    vAll(k) = v; k = k+1;
end
plot(tSpan, vAll, 'LineWidth', 2); ylim([0, 2]);
xlabel('Time'); ylabel('Rotating Disk Speed'); grid on;
```

Above Matlab simulation code is a little different from the previous Matlab simulation code presented in Sect. 2.2. Different code lines are highlighted in bold font in the Matlab simulation code. Most part of the Matlab simulation code is the same to that presented in Sect. 2.2. The little code difference only involves a newly-added variable T_d that denotes unknown torque disturbance.

In above Matlab simulation code, setting T_d randomly is to simulate changeable torque disturbance that cannot be known a priori for the control system. Therefore, in simulation, we must refrain from using the value of T_d in the control law of T. It is worth noting that setting T_d randomly according to uniform distribution does not imply that unknown torque disturbance in practical applications always appears as a random variable that follows uniform distribution. In simulation, we can fairly set T_d randomly according to other kinds of distribution. Here, the key point of simulation is to set T_d randomly to imply that T_d cannot be known a priori and cannot be used in the control law of T.

The control law of T is intentionally left blank in the Matlab simulation code. We suggest that readers try the Matlab simulation code and try to design a control method such that the expected rotating speed v_E can be achieved even under influence of unknown torque disturbance T_d.

2.3.2 Debut of Feedback Control

After a number of simulation trials, readers would find that the task of rotating disk speed control with unknown torque disturbance presented in Sect. 2.3 is more challenging than the task of rotating disk speed control without torque disturbance presented in Sect. 2.2. The challenge comes from the torque disturbance that is unknown a priori and cannot be used in the control law of T. Otherwise, if torque disturbance T_d could be known in advance, we would easily design a control method as $T = 5v_E + T_d$ which achieves the expected rotating speed v_E. Unfortunately, such utopian control method normally cannot be realized in practical applications.

How to design a control method that suppresses torque disturbance regardless of concrete values of torque disturbance? For this purpose, we can resort to **feedback control**. In Chap. 1, we have briefly introduced an intuitive idea of feedback control: when the rotating speed is lower than the expected rotating speed, the controller increases its output torque T. When the rotating speed is higher than the expected rotating speed, the controller decreases its output torque T. We instantiate this intuitive idea as the following *extreme* feedback control method:

$$T = \begin{cases} T_{\max} & e > 0 \\ -T_{\max} & e < 0 \end{cases}, \tag{2.3}$$

where $e = v_E - v$ is **feedback error**. In other words, when the rotating speed is lower than the expected rotating speed, the controller sets its output torque T to the maximum allowable positive value $T_{\max}$ to increase the rotating speed as quickly as possible. When the rotating speed is higher than the expected rotating speed, the controller sets its output torque T to the minimum allowable negative value $-T_{\max}$ to decrease the rotating speed as quickly as possible. Such intuitive feedback control is illustrated in Fig. 2.8.

Matlab code for simulation of rotating disk speed intuitive feedback control with unknown torque disturbance is given as follows.

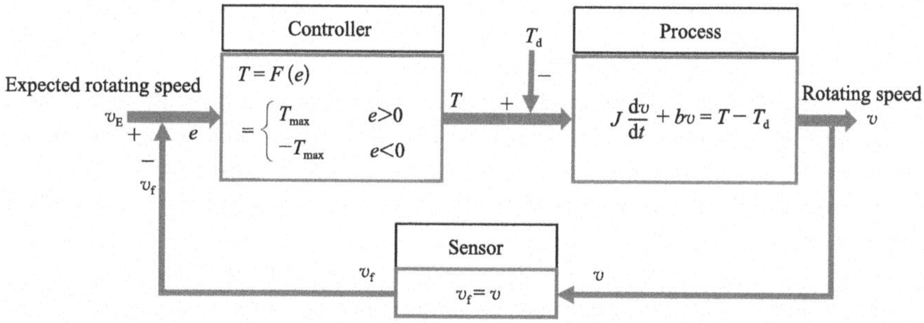

Fig. 2.8 Rotating disk speed intuitive feedback control

RotatingDiskSpeedIntuitiveFeedbackControl.m

```
%% Rotating disk speed control system parameters
J = 10; % Rotating disk load inertia
b = 5; % Friction coefficient
Td = random('Uniform', -5, 5); % Unknown torque disturbance
Tmax = 50; % Maximum torque

%% Simulation preliminary configuration
dt = 0.001; % Numerical computation step
tSpan = 0:dt:10; % Simulation time span
% Record rotating disk speeds during simulation
vAll = zeros(size(tSpan)); k = 1;
v = 0; % Initial rotating disk speed
vE = 1; % Expected rotating disk speed

%% Simulation of rotating disk speed control
for t = tSpan
    %% Control method
    e = vE-v; % Feedback error
    if (e>0)
        T = Tmax;
    elseif (e<0)
        T = -Tmax;
    end

    %% Rotating disk dynamics suffering from unknown toruqe
    disturbance
    dv = (T-Td-b*v)/J;
    v = v + dv*dt;
    vAll(k) = v; k = k+1;
end
plot(tSpan, vAll, 'LineWidth', 2); ylim([0, 2]);
xlabel('Time'); ylabel('Rotating Disk Speed'); grid on;
```

The simulation result is demonstrated in Fig. 2.9, where the horizontal axis represents the simulation time span and the vertical axis represents the rotating speed during simulation. The expected rotating speed $v_E = 1$ is achieved by the intuitive feedback control method.

We suggest that readers try above Matlab simulation code for multiple rounds. In each round of simulation, a different value of torque disturbance T_d is generated randomly. On the other hand, readers will find that simulation results are almost the same to that demonstrated in Fig. 2.9, regardless of concrete values of torque disturbance T_d in different rounds of simulation. This implies that the intuitive feedback control method can always achieve the expected rotating speed v_E, even under influence of unknown torque disturbance T_d.

In practical applications, the intuitive feedback control method, which switches frequently and abruptly between two extreme controller output torques T_{max} and $-T_{max}$, may be too severe and harmful to the control system. To facilitate understanding of the harm, imagine how harmful it is to a television if we turn on and turn

Fig. 2.9 Rotating disk speed intuitive feedback control

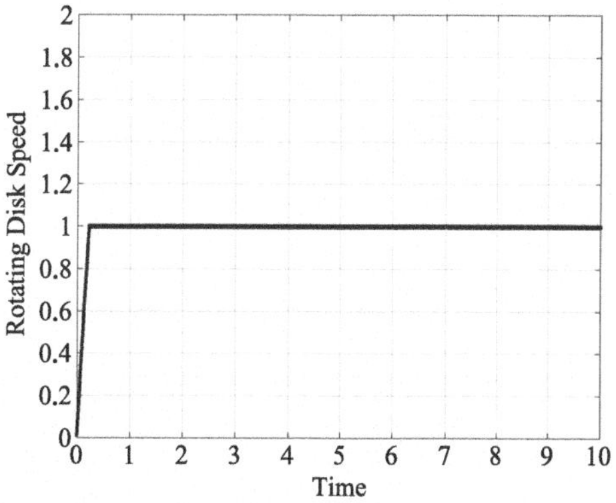

off the television frequently. We can adapt the intuitive feedback control method to a proportional feedback control form constrained by controller output torque limits $[-T_{\max}, T_{\max}]$, as described in (2.4) and illustrated in Fig. 2.10. A large proportional coefficient P is usually used.

$$T = \begin{cases} T_{\max} & e > T_{\max}/P \\ Pe & -T_{\max}/P \leqslant e \leqslant T_{\max}/P \\ -T_{\max} & e < -T_{\max}/P \end{cases} \tag{2.4}$$

The adapted intuitive feedback control method namely the proportional feedback control method described in (2.4) resembles the original intuitive feedback control method described in (2.3), as demonstrated in Fig. 2.11.

The slight difference between the two intuitive feedback control methods exists only about $e = 0$. The controller output torque T associated with the original intuitive feedback control method described in (2.3) switches abruptly from $-T_{\max}$ to $T_{\max}$ at $e = 0$, whereas the controller output torque T associated with the proportional feedback control method described in (2.4) switches continuously from $-T_{\max}$ to

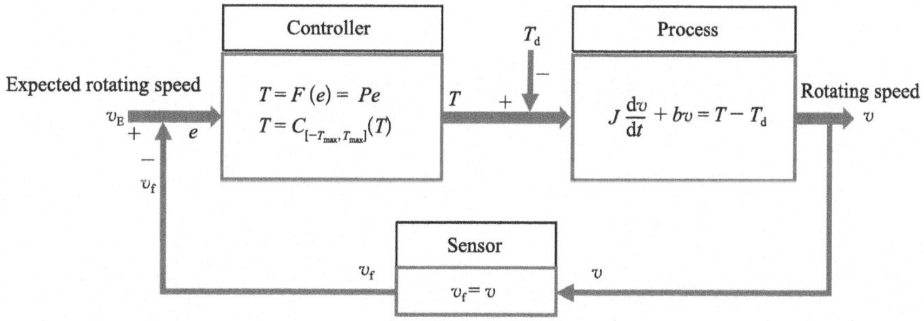

Fig. 2.10 Rotating disk speed proportional feedback control

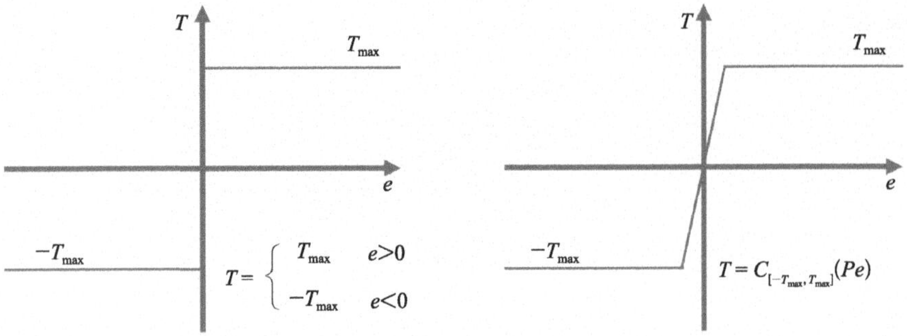

Fig. 2.11 Comparison between two intuitive feedback control methods

T_{max} about $e = 0$. Such continuity of the control law enables the proportional feedback control method described in (2.4) to have smoother control performance than the original intuitive feedback control method described in (2.3). Matlab code for simulation of rotating disk speed proportional feedback control with unknown torque disturbance is given as follows.

RotatingDiskSpeedProportionalFeedbackControl.m

```
%% Rotating disk speed control system parameters
J = 10; % Rotating disk load inertia
b = 5; % Friction coefficient
Td = random('Uniform', -5, 5); % Unknown torque disturbance
Tmax = 50; % Maximum torque

%% Simulation preliminary configuration
dt = 0.001; % Numerical computation step
tSpan = 0:dt:10; % Simulation time span
% Record rotating disk speeds during simulation
vAll = zeros(size(tSpan)); k = 1;
v = 0; % Initial rotating disk speed
vE = 1; % Expected rotating disk speed

%% Simulation of rotating disk speed control
for t = tSpan
    %% Control method
    e = vE-v; % Feedback error
    T = 1000*e; % Proportional feedback control
    T = max(min(T, Tmax), -Tmax); % Control law constraint

    %% Rotating disk dynamics suffering from unknown toruqe
    disturbance
    dv = (T-Td-b*v)/J;
    v = v + dv*dt;
    vAll(k) = v; k = k+1;
end
plot(tSpan, vAll, 'LineWidth', 2); ylim([0, 2]);
xlabel('Time'); ylabel('Rotating Disk Speed'); grid on;
```

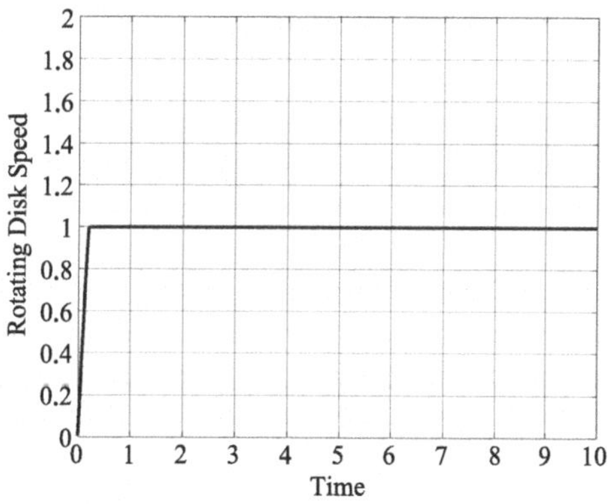

Fig. 2.12 Rotating disk speed proportional feedback control

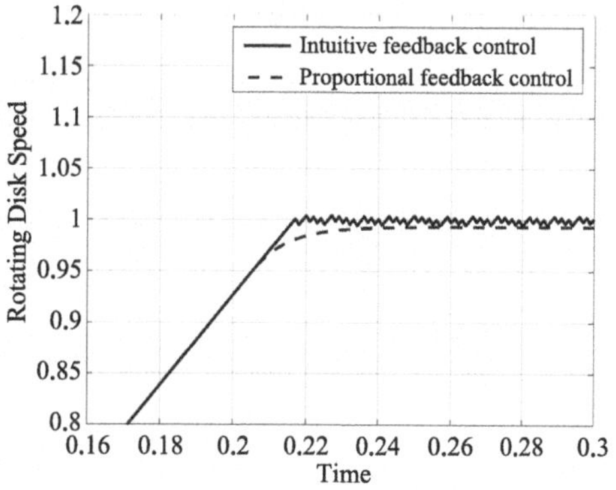

Fig. 2.13 Performance comparison between two intuitive feedback control methods

The simulation result is demonstrated in Fig. 2.12. The expected rotating speed $v_{\mathrm{E}} = 1$ is achieved by the proportional feedback control method—It is worth noting that the expected rotating speed is not exactly achieved by the proportional feedback control method, but achieved with steady error. If a large proportional coefficient P is used, the steady error will be small and even negligible. For the moment, we neglect steady error associated with the proportional feedback control method and skip theoretical analysis of why such steady error exists.

As demonstrated in Figs. 2.9 and 2.12, the original intuitive feedback control method and the proportional feedback control method (with large proportional coefficient) seem to have the same performance. However, the two intuitive feedback control methods have performance differences that become visible when we zoom into figure details, as demonstrated in Fig. 2.13. The proportional feedback control method has apparently smoother control performance than the original intuitive feedback control method.

The simulation experiments demonstrate that both the original intuitive feedback control method described in (2.3) and the adapted intuitive feedback control method namely the proportional feedback control method described in (2.4) can achieve the expected rotating speed v_{E}, even under influence of unknown torque disturbance T_{d}.

2.3.3 Feedback Control's Ability in Handling Uncertainty

Besides adding unknown torque disturbance which is one aspect of uncertainty, we can add more aspects of uncertainty into the rotating disk speed control system in simulation to evaluate feedback control's ability in handling various aspects of uncertainty. For example, we can set the rotating disk load inertia J and the friction coefficient b randomly within certain reasonable ranges. This is to simulate uncertainty of control system parameters that are not well identified or calibrated in practical applications. Matlab code for simulation of rotating disk speed proportional feedback control with various aspects of uncertainty is given as follows.

RotatingDiskSpeedProportionalFeedbackControl2.m

```
%% Rotating disk speed control system parameters
J = random('Uniform', 5, 20); % Unknown rotating disk load
    inertia
b = random('Uniform', 2, 10); % Unknown Friction coefficient
Td = random('Uniform', -5, 5); % Unknown torque disturbance
Tmax = 50; % Maximum torque

%% Simulation preliminary configuration
dt = 0.001; % Numerical computation step
tSpan = 0:dt:10; % Simulation time span
% Record rotating disk speeds during simulation
vAll = zeros(size(tSpan)); k = 1;
v = 0; % Initial rotating disk speed
vE = 1; % Expected rotating disk speed

%% Simulation of rotating disk speed control
for t = tSpan
    %% Control method
    e = vE-v; % Feedback error
    T = 1000*e; % Proportional feedback control
    T = max(min(T, Tmax), -Tmax); % Control law constraint

    %% Rotating disk dynamics suffering from uncertainty
    dv = (T-Td-b*v)/J;
    v = v + dv*dt;
    vAll(k) = v; k = k+1;
end
plot(tSpan, vAll, 'LineWidth', 2); ylim([0, 2]);
xlabel('Time'); ylabel('Rotating Disk Speed'); grid on;
```

The simulation result is demonstrated in Fig. 2.14, where the horizontal axis represents the simulation time span and the vertical axis represents the rotating speed during simulation. The expected rotating speed $v_E = 1$ is achieved by the proportional feedback control method.

We suggest that readers try above Matlab simulation code for multiple rounds. In each round of simulation, a different value of torque disturbance T_d, a different value of rotating disk load inertia J, and a different value of friction coefficient b are generated randomly. On the other hand, readers will find that simulation results are

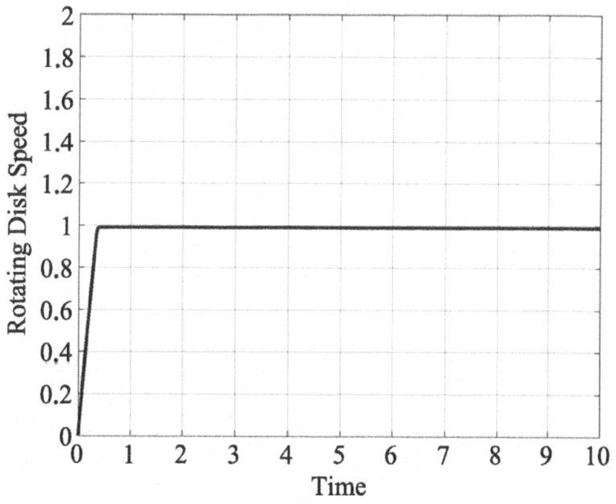

Fig. 2.14 Rotating disk speed proportional feedback control with uncertainty

similar to that demonstrated in Fig. 2.14. The expected rotating speed v_E is always achieved by the proportional feedback control method, regardless of concrete values of these unknown parameters in different rounds of simulation.

With all simulation experiments presented in Sect. 2.3, we have no intention to focus only on how we apply feedback control to a concrete control task namely rotating disk speed control. Instead, we intend to use this concrete application example to highlight an important utility of feedback control, namely that *feedback control enables a control system even with uncertainty to guarantee control accuracy (in terms of achieving certain expected process output).*

> **Feedback control utility I**: *Feedback control enables a control system even with uncertainty to guarantee control accuracy.*

2.4 Feedback Control: Second Utility

For a control system with uncertainty, the advantage of feedback control over open-loop control is obvious, as demonstrated previously in Sect. 2.3. On the other hand, for a control system without uncertainty, can feedback control still outperform open-loop control?

Consider the same task of rotating disk speed control presented in Sect. 2.2. Rotating disk speed control system parameters J and b specified in Table 2.1 are still adopted. Suppose that there is no torque disturbance T_d, no uncertainty about the rotating disk load inertia J and the friction coefficient b, and no other aspects of uncertainty. In one word, suppose that the rotating disk speed control system is completely known.

Both the open-loop control method described in Sect. 2.2 and the proportional feedback control method (or the original intuitive feedback control method) described in Sect. 2.3 can accomplish the control task of achieving the expected rotating speed

$v_E = 1$. However, this does not imply that the proportional feedback control method and the open-loop control method have the same performance. Matlab simulation code for performance comparison between the proportional feedback control method and the open-loop control method is given as follows.

RotatingDiskSpeedProportionalFeedbackControlUtility2.m

```
%% Rotating disk speed control system parameters
J = 10; % Rotating disk load inertia
b = 5; % Friction coefficient
Tmax = 50; % Maximum torque

%% Simulation preliminary configuration
dt = 0.001; % Numerical computation step
tSpan = 0:dt:10; % Simulation time span
% Record rotating disk speeds during simulation
vAll = zeros(size(tSpan)); k = 1;
vE = 1; % Expected rotating disk speed
v = 0; % Initial rotating disk speed
%% Simulation of rotating disk speed control
for t = tSpan
    %% Control method
    T = 5;

    %% Rotating disk dynamics
    dv = (T-b*v)/J;
    v = v + dv*dt;
    vAll(k) = v; k = k+1;
end
vOpenLoop = vAll;

vAll(:) = 0; k = 1; % Record rotating disk speeds during
    simulation
v = 0; % Initial rotating disk speed
%% Simulation of rotating disk speed control
for t = tSpan
    %% Control method
    e = vE-v; % Feedback error
    T = 1000*e; % Proportional feedback control
    T = max(min(T, Tmax), -Tmax); % Control law constraint

    %% Rotating disk dynamics
    dv = (T-b*v)/J;
    v = v + dv*dt;
    vAll(k) = v; k = k+1;
end
vProportional = vAll;

plot(tSpan, vOpenLoop, 'r--', tSpan, vProportional, 'b-',
     'LineWidth', 2);
legend('Open-loop control', 'Proportional feedback control');
ylim([0, 2]);
xlabel('Time'); ylabel('Rotating Disk Speed'); grid on;
```

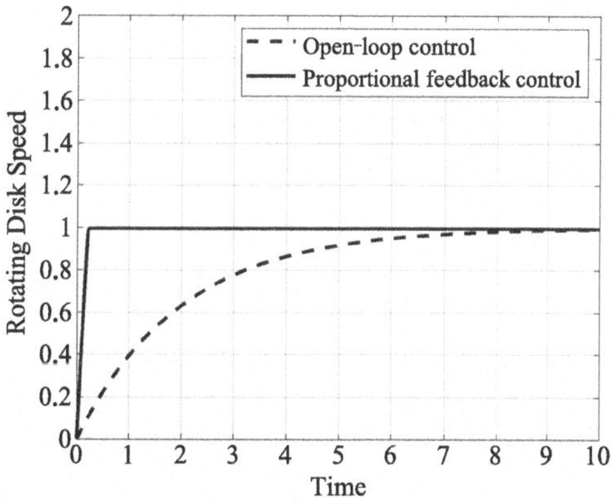

Fig. 2.15 Performance comparison between the proportional feedback control method and the open-loop control method

Simulation results associated with the proportional feedback control method and the open-loop control method are demonstrated together in Fig. 2.15 for comparison. Although both control methods can achieve the expected rotating speed $v_E = 1$, the proportional feedback control method achieves the expected rotating speed much more quickly than the open-loop control method. In other words, the proportional feedback control method is much more *responsive* than the open-loop control method.

By the way, if the original intuitive feedback control method instead of its adapted version namely the proportional feedback control method is applied in above simulation, simulation results will be almost the same to those demonstrated in Fig. 2.15. The original intuitive feedback control method is also much more responsive than the open-loop control method.

As already mentioned in Sect. 2.3, we have no intention to focus only on how we apply feedback control to the concrete task of rotating disk speed control. By performance comparison associated with this concrete application example, we intend to highlight another important utility of feedback control, namely that *feedback control enables a control system to enhance control responsiveness.*

Feedback control utility II: *Feedback control enables a control system to enhance control responsiveness.*

2.5 Limitations of Intuitive Feedback Control

Simulation experiments presented in Sects. 2.3 and 2.4 may give readers an illusionary confidence that we have found an omnipotent strategy of feedback control, namely the intuitive feedback control methods, which would accomplish any potential control task. If readers do have such illusionary confidence, we may consider another control task closely related to rotating disk speed control, namely the task of

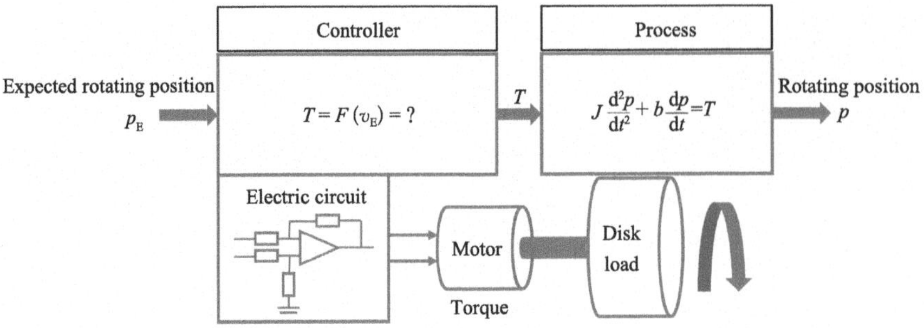

Fig. 2.16 Rotating disk position open-loop control

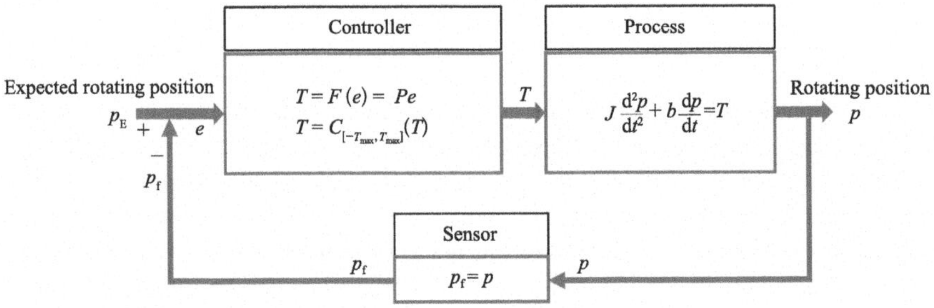

Fig. 2.17 Rotating disk position proportional feedback control

rotating disk position control: *Given a rotating disk position control system, design a control method such that the expected rotating position can be achieved.*

Similar to the simplified mathematical equivalence of a rotating disk speed open-loop control system illustrated in Fig. 2.3, the simplified mathematical equivalence of a rotating disk position open-loop control system is illustrated in Fig. 2.16. Note that the rotating speed is the first-order derivative of the rotating position, i.e. $v = \mathrm{d}p/\mathrm{d}t$. Substitute this relationship into the process dynamics illustrated in Fig. 2.3 and hence obtain the rotating disk position open-loop control system illustrated in Fig. 2.16.

For the moment, we neglect any control system uncertainty and suppose that the rotating disk position control system is completely known. To accomplish the task of achieving the expected rotating position p_E, we resort to the adapted intuitive feedback control method namely the proportional feedback control method described in (2.5) and illustrated in Fig. 2.17 (where $e = p_E - p$ is feedback error):

$$T = \begin{cases} T_{\max} & e > T_{\max}/P \\ Pe & -T_{\max}/P \leqslant e \leqslant T_{\max}/P \\ -T_{\max} & e < -T_{\max}/P \end{cases} . \tag{2.5}$$

Matlab code for simulation of rotating disk position proportional feedback control is given as follows.

RotatingDiskPositionProportionalFeedbackControl.m

```
%% Rotating disk position control system parameters
J = 10; % Rotating disk load inertia
b = 5; % Friction coefficient
Tmax = 50; % Maximum torque

%% Simulation preliminary configuration
dt = 0.001; % Numerical computation step
tSpan = 0:dt:10; % Simulation time span
% Record rotating disk positions during simulation
pAll = zeros(size(tSpan)); k = 1;
pE = 1; % Expected rotating disk position
p = 0; v = 0; % Initial rotating disk position and speed

%% Simulation of rotating disk position control
for t = tSpan
    %% Control method
    e = pE-p; % Feedback error
    T = 1000*e; % Proportional feedback control
    T = max(min(T, Tmax), -Tmax); % Control law constraint

    %% Rotating disk dynamics
    dv = (T-b*v)/J;
    v = v + dv*dt;
    p = p + v*dt;
    pAll(k) = p; k = k+1;
end
plot(tSpan, pAll, 'LineWidth', 2); ylim([0, 2]);
xlabel('Time'); ylabel('Rotating Disk Position'); grid on;
```

The simulation result is demonstrated in Fig. 2.18, where the horizontal axis represents the simulation time span and the vertical axis represents the rotating position during simulation. Although the rotating position has a tendency to converge to the expected rotating position $p_E = 1$, it has a long transition time span with severe oscillation (note that it has not converged yet even at time $t = 10$ s). Severe transitional oscillation is undoubtedly undesirable to a control system in practical applications.

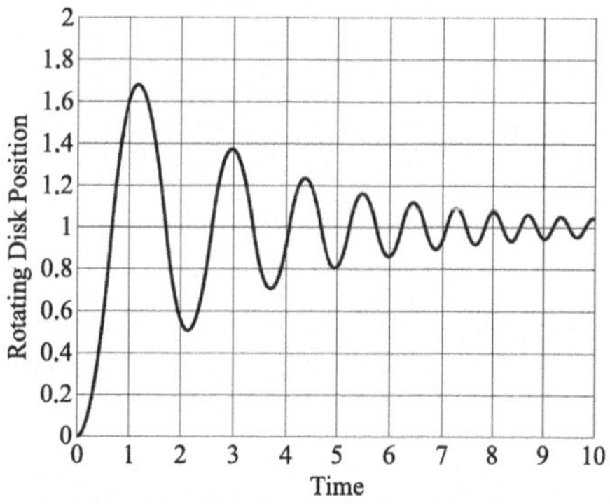

Fig. 2.18 Rotating disk position proportional feedback control ($b = 5$)

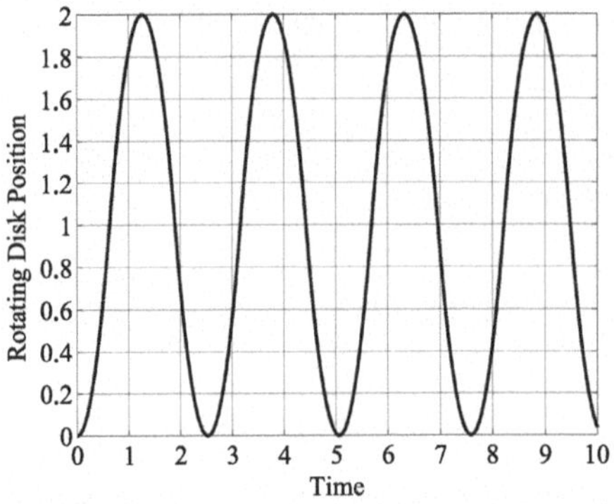

Fig. 2.19 Rotating disk position proportional feedback control ($b = 0$)

Why does the proportional feedback control method work desirably for rotating disk speed control but undesirably for rotating disk position control? An even more tricky control phenomenon takes place if we set the friction coefficient b to zero in simulation. Setting the friction coefficient b to zero implies that the rotating disk suffers from no friction at all, which is an ideal condition. So we may intuitively imagine that the proportional feedback control method would have better performance if the friction coefficient is zero. However, the simulation result demonstrated in Fig. 2.19 contradicts our intuition. The proportional feedback control method has much worse performance: the rotating position oscillates more severely and has no tendency to converge at all. Why does this tricky control phenomenon take place?

For the moment, we skip theoretical analysis for above "why" questions. Although the "why" questions are put aside, simulation results demonstrated in Figs. 2.18 and 2.19 clearly reflect limitations of the adapted intuitive feedback control method namely the proportional feedback control method. Besides, the original intuitive feedback control method also has such limitations.

2.6 Necessity of Control Theory

Simulation experiments presented in Sect. 2.5 do not imply that feedback control is no longer effective for rotating disk position control. In fact, we can replace the intuitive feedback control methods by another feedback control method namely the proportional-derivative feedback control method described in (2.6) and illustrated in Fig. 2.20:

$$T = \begin{cases} T_{\max} & Pe + D\frac{de}{dt} > T_{\max} \\ Pe + D\frac{de}{dt} & -T_{\max} \leqslant Pe + D\frac{de}{dt} \leqslant T_{\max} \\ -T_{\max} & Pe + D\frac{de}{dt} < -T_{\max} \end{cases} . \tag{2.6}$$

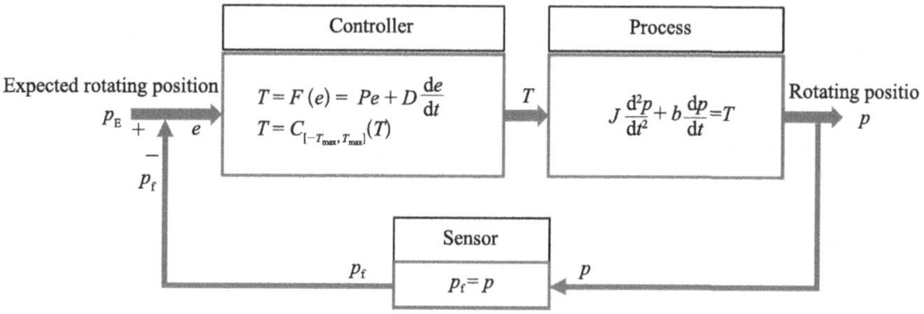

Fig. 2.20 Rotating disk position proportional-derivative feedback control

Compared with proportional feedback control, proportional-derivative feedback control involves one more term namely the derivative term $D\frac{de}{dt}$ where D is the derivative coefficient. Matlab code for simulation of rotating disk position proportional-derivative feedback control is given as follows (note that $\mathrm{d}e/\mathrm{d}t = \mathrm{d}(p_E - p)/\mathrm{d}t = -\mathrm{d}p/\mathrm{d}t = -v$).

RotatingDiskPositionPDFeedbackControl.m

```
%% Rotating disk position control system parameters
J = 10; % Rotating disk load inertia
b = 5; % Friction coefficient
Tmax = 50; % Maximum torque

%% Simulation preliminary configuration
dt = 0.001; % Numerical computation step
tSpan = 0:dt:10; % Simulation time span
% Record rotating disk positions during simulation
pAll = zeros(size(tSpan)); k = 1;
pE = 1; % Expected rotating disk position
p = 0; v = 0; % Initial rotating disk position and speed

%% Simulation of rotating disk position control
for t = tSpan
    %% Control method
    e = pE-p; % Feedback error
    de = -v; % Feedback error derivative
    T = 1000*e + 200*de; % proportional-derivative feedback
    control
    T = max(min(T, Tmax), -Tmax); % Control law constraint

    %% Rotating disk dynamics
    dv = (T b*v)/J;
    v = v + dv*dt;
    p = p + v*dt;
    pAll(k) = p; k = k+1;
end
plot(tSpan, pAll, 'LineWidth', 2); ylim([0, 2]);
xlabel('Time'); ylabel('Rotating Disk Position'); grid on;
```

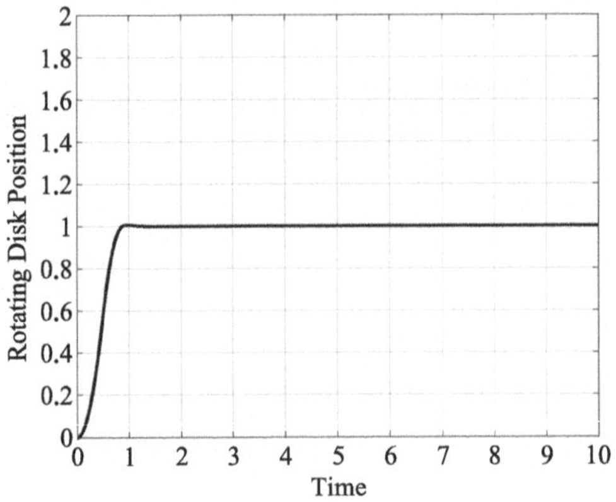

Fig. 2.21 Rotating disk position proportional-derivative feedback control ($b = 5$)

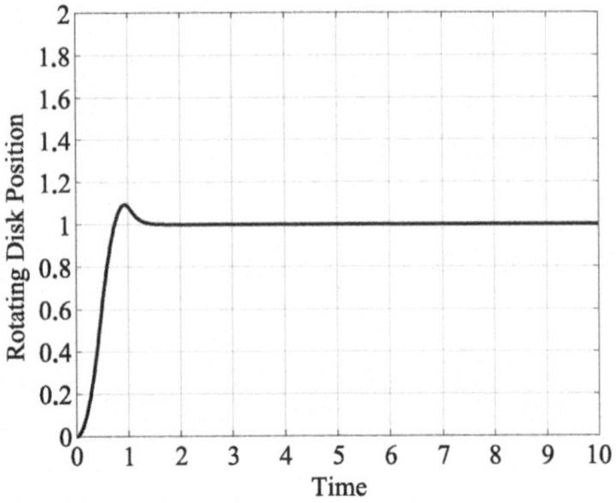

Fig. 2.22 Rotating disk position proportional-derivative feedback control ($b = 0$)

Simulation results are demonstrated in Fig. 2.21 ($b = 5$ in simulation) and Fig. 2.22 ($b = 0$ in simulation), where the horizontal axis represents the simulation time span and the vertical axis represents the rotating position during simulation. In both simulation experiments, the rotating position converges rapidly to the expected rotating position $p_E = 1$, which reflects effectiveness of the rotating disk position proportional-derivative feedback control method.

Here, our focus is neither on a specific control method namely the proportional-derivative feedback control method nor on how we apply it to accomplish a concrete control task namely rotating disk position control. In retrospect of this chapter our focus is not on any specific control method or any concrete control task. Instead, we would like to guide readers to reflect on a number of control phenomena and essential reasons behind them.

- Why do the intuitive feedback control methods enable the rotating disk speed control system to guarantee control accuracy?
- Why does the adapted intuitive feedback control method namely the proportional feedback control method enable the rotating disk speed control system to achieve the expected rotating speed yet with a small enough but existing steady error?

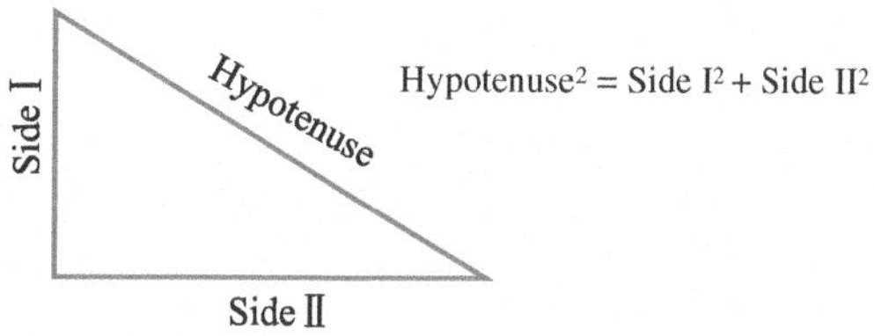

Fig. 2.23 Orthogonal triangle

- Why do the intuitive feedback control methods enable the rotating disk speed control system to enhance control responsiveness?
- Why do the intuitive feedback control methods not work for the task of rotating disk position control?
- Why does the proportional feedback control method perform even worse for the task of rotating disk position control when control system conditions become better and even ideal?
- Why does the proportional-derivative feedback control method work for the task of rotating disk position control?

We can continue to raise more questions like those above for reflection, yet it is unnecessary to do so for the moment. For these questions, we may resort to simulation experiments and field experiments to acquire intuitive understanding of control phenomena. However, simulation experiments and field experiments never enable us to really reach essential reasons behind control phenomena. It is like by experiments we can fabricate and measure a billion of orthogonal triangles (also known as right-angled triangles) to demonstrate the phenomenon that an orthogonal triangle's hypotenuse square equals the sum of its two side squares (see Fig. 2.23), but all these experiments themselves can neither tell us whether this phenomenon is really a general truth nor tell us any essential reason behind this phenomenon.

To really reach essential reasons behind the phenomenon that an orthogonal triangle's hypotenuse square equals the sum of its two side squares, we need mathematics on geometry. By analogue, to really reach essential reasons behind control phenomena, we need **control theory** which aims for mathematical essence of control systems and essential reasons behind their control phenomena. Since Chap. 3, we will formally enter the kingdom of control theory.

Chapter 3
System Modelling via Laplace Transform

A control system consists of various physical systems, including at least a process, an actuator, and a controller. A sensor is also included if feedback control is involved. A process may further consist of various sub-processes, an actuator may further consist of various sub-actuators, a sensor may further consist of various sub-sensors, and a controller may further consist of various sub-controllers. All these sub-processes, sub-actuators, sub-sensors, and sub-controllers are also physical systems. A fundamental part of control theory is **system modelling** that abstracts mathematical essence of a control system, including mathematical essence of physical systems that compose the control system.

3.1 Differential Equations of Physical Systems

For control purpose, we need to study dynamics of a control system. **Differential equations** are good mathematical tools for modelling dynamics of physical systems that compose a control system. For example, translational motion dynamics of an object is modelled by a differential equation as

$$\frac{\mathrm{d}^2 \boldsymbol{p}}{\mathrm{d}t^2} = \frac{\mathrm{d}\boldsymbol{v}}{\mathrm{d}t} = \frac{\boldsymbol{F}}{m},$$

where $\boldsymbol{p}$ denotes the object position, $\boldsymbol{v}$ denotes the object velocity, m denotes the object mass, and $\boldsymbol{F}$ denotes the force exerted on the object. Similarly, one-dimensional rotational motion dynamics of an object is modelled by a differential equation as

$$\frac{\mathrm{d}^2 \theta}{\mathrm{d}t^2} = \frac{\mathrm{d}\omega}{\mathrm{d}t} = \frac{T}{J},$$

H. Li, *Control Theory for Practical Applications*,
https://doi.org/10.1007/978-981-97-5008-5_3

where θ denotes the rotating angle of the object around the rotating axis, ω denotes the rotating speed, J denotes the rotating inertia, and T denotes the torque exerted on the object around the rotating axis.

Besides mechanical dynamics, electrical and magnetic dynamics can also be modelled by differential equations. For example, given an electrical inductance L, dynamics of how the current I influences the voltage U is modelled by a differential equation as

$$U = L\frac{\mathrm{d}I}{\mathrm{d}t}.$$

Given an electrical capacitance C, dynamics of how the voltage U influences the current I is modelled by a differential equation as

$$I = C\frac{\mathrm{d}U}{\mathrm{d}t}.$$

The relationship between motion of a particle with electrical charge q in a magnetic field with magnetic field vector $\boldsymbol{B}$ and the magnetic force $\boldsymbol{F}$ on the particle is modelled by a differential equation as

$$q\left(\frac{\mathrm{d}\boldsymbol{p}}{\mathrm{d}t} \times \boldsymbol{B}\right) = q(\boldsymbol{v} \times \boldsymbol{B}) = \boldsymbol{F}.$$

From above relationship, the magnetic force on a current-carrying wire is derived as

$$\frac{\mathrm{d}\boldsymbol{F}}{\mathrm{d}l} = \boldsymbol{I} \times \boldsymbol{B},$$

where $\boldsymbol{I}$ is the vector current in the wire and l is the wire length.

Other physical phenomena such as thermal dynamics can also be modelled by differential equations. We refrain from giving more examples here because how to establish differential equations of physical systems is beyond the scope of control theory in this book. For a comprehensive review of physics fundamentals, a classical book worth recommendation is *The Feynman Lectures on Physics* (Feynman, 2004).

Given a control system with proper differential equations established for its composing physical systems and given a specific control method, we can deduce performance of the control method by solving differential equations of the control system. For example, consider the rotating disk speed open-loop control system illustrated in Fig. 2.3 in Chap. 2, where the process dynamics is modelled by a first-order differential equation as

$$J\frac{\mathrm{d}v}{\mathrm{d}t} + bv = T. \tag{3.1}$$

To solve the differential equation described in (3.1), compute its characteristic equation namely

$$J\alpha + b = 0$$

and obtain $\alpha = -b/J$. Apply the following function transformation

$$v = \mathrm{e}^{\alpha t} u = \mathrm{e}^{-\frac{b}{J}t} u \tag{3.2}$$

to (3.1) and obtain

$$\frac{\mathrm{d}u}{\mathrm{d}t} = \frac{T}{J}\mathrm{e}^{\frac{b}{J}t},$$
$$u = u_0 + \int_0^t \frac{T}{J}\mathrm{e}^{\frac{b}{J}\tau}\mathrm{d}\tau.$$

Substitute above relationship into (3.2) and obtain

$$v = \mathrm{e}^{-\frac{b}{J}t}\left(u_0 + \int_0^t \frac{T}{J}\mathrm{e}^{\frac{b}{J}\tau}\mathrm{d}\tau\right) = \mathrm{e}^{-\frac{b}{J}t}u_0 + \int_0^t \frac{T}{J}\mathrm{e}^{-\frac{b}{J}(t-\tau)}\mathrm{d}\tau.$$

Consider the initial condition

$$v_0 = v(t=0) = u_0,$$

and finally obtain the differential equation solution as

$$v = \mathrm{e}^{-\frac{b}{J}t}v_0 + \int_0^t \frac{T}{J}\mathrm{e}^{-\frac{b}{J}(t-\tau)}\mathrm{d}\tau. \tag{3.3}$$

Suppose that the controller output torque $T = T_0$ is constant, and the solution of v described in (3.3) is further derived as

$$v = \frac{T_0}{b} - \left(\frac{T_0}{b} - v_0\right)\mathrm{e}^{-\frac{b}{J}t}. \tag{3.4}$$

If the initial rotating speed $v_0 = 0$, then the solution of v is given as

$$v = \frac{T_0}{b} - \frac{T_0}{b}\mathrm{e}^{-\frac{b}{J}t}. \tag{3.5}$$

As $t \to \infty$ implies $\mathrm{e}^{-\frac{b}{J}t} \to 0$, so by both (3.4) and (3.5) we have

$$v_s \equiv v(\infty) = \frac{T_0}{b}, \tag{3.6}$$

where the subscript "s" of v described in (3.6) means the **steady state** or **final value** of v. To achieve the expected rotating speed v_E, we can set a control law as $T = T_0$ such that $T_0/b = v_E$, namely to set $T_0 = bv_E$. If the parameter b specified in Table 2.1 is adopted, then above theoretical analysis is exactly consistent with simulation results presented in Sect. 2.2 in Chap. 2.

The rotating disk speed open-loop control system illustrated in Fig. 2.3 with process dynamics modelled by (3.1) is a rather simple application example, so to analyse it by solving (3.1) directly is not difficult. However, for a much more complicated control system, to analyse it by solving its associated differential equations directly is usually tedious and even difficult.

On the other hand, for control systems, we normally do not need to study all details of their dynamics. Instead, we only need to study details of their dynamics related to important concerns for control purpose. For example, an important concern is **control system stability** that will be introduced in Chap. 4, and other important concerns include **control system accuracy** and **control system responsiveness** that will be introduced in Chap. 5. For control systems, we have an effective mathematical tool that is much more convenient than solving control system differential equations directly. This mathematical tool is the **Laplace transform**.

3.2 Laplace Transform

3.2.1 Common Laplace Transform Pairs

We review mathematical knowledge on the **Laplace transform** in this section and present system modelling via Laplace transform in following sections.

Given a time-domain function $f(t)$ which satisfies

$$\int_{0^-}^{\infty} |f(t)|e^{-\sigma t}dt < \infty \tag{3.7}$$

for some positive real σ. The **Laplace transform** is defined as

$$F(s) \equiv L\{f(t)\} = \int_{0^-}^{\infty} f(t)e^{-st}dt. \tag{3.8}$$

In other words, the s-domain function $F(s)$ is obtained via Laplace transform of the original time-domain function $f(t)$. Equation (3.7) is sufficient for $f(t)$ to be Laplace transformable and the s-domain $\text{Re}(s) > \sigma$ is a domain of absolute convergence for the Laplace transform described in (3.8). The **inverse Laplace transform** from the s-domain function $F(s)$ to the original time-domain function $f(t)$ is given as

$$f(t) \equiv L^{-1}\{F(s)\} = \frac{1}{2\pi \mathrm{j}} \int_{\sigma-\mathrm{j}\infty}^{\sigma+\mathrm{j}\infty} F(s)\mathrm{e}^{+st}\mathrm{d}s. \tag{3.9}$$

We review the Laplace transform for common functions. In following derivations, the Laplace variable s is assumed to be located in the s-domain of absolute convergence by default.

(1) *Impulse function* $\delta(t)$:

$$L\{\delta(t)\} = \int_{0^-}^{\infty} \delta(t)\mathrm{e}^{-st}\mathrm{d}t = \mathrm{e}^{-st}|_{t=0} = 1.$$

(2) *Step function* $u(t)$:

$$L\{u(t)\} = \int_{0^-}^{\infty} u(t)\mathrm{e}^{-st}\mathrm{d}t = -\frac{1}{s}\mathrm{e}^{-st}|_{0^-}^{\infty} = \frac{1}{s}.$$

(3) *Power function* t^n:

$$\begin{aligned} L\{t^n\} &= \int_{0^-}^{\infty} t^n \mathrm{e}^{-st}\mathrm{d}t = -\frac{1}{s}\int_{0^-}^{\infty} t^n \mathrm{d}\mathrm{e}^{-st} \\ &= -\frac{t^n \mathrm{e}^{-st}}{s}\bigg|_{0^-}^{\infty} + \frac{1}{s}\int_{0^-}^{\infty} \mathrm{e}^{-st}\mathrm{d}t^n \\ &= \frac{n}{s}\int_{0^-}^{\infty} t^{n-1}\mathrm{e}^{-st}\mathrm{d}t = \frac{n}{s}L\{t^{n-1}\} \\ &\quad \text{(recursive relationship)} \\ &= \frac{n}{s}\frac{n-1}{s}L\{t^{n-2}\} = \cdots = \frac{n!}{s^n}L\{u(t)\} = \frac{n!}{s^{n+1}}. \end{aligned}$$

(4) *Exponential function* e^{-at}:

$$\begin{aligned} L\{\mathrm{e}^{-at}\} &= \int_{0^-}^{\infty} \mathrm{e}^{-at}\mathrm{e}^{-st}\mathrm{d}t = \int_{0^-}^{\infty} \mathrm{e}^{-(s+a)t}\mathrm{d}t \\ &= -\frac{1}{s+a}\mathrm{e}^{-(s+a)t}\bigg|_{0^-}^{\infty} = \frac{1}{s+a}. \end{aligned}$$

(5) *Sine function* $\sin(\omega t)$:

$$\begin{aligned} L\{\sin(\omega t)\} &= \int_{0^-}^{\infty} \sin(\omega t)\mathrm{e}^{-st}\mathrm{d}t = -\frac{1}{\omega}\int_{0^-}^{\infty} \mathrm{e}^{-st}\mathrm{d}\cos(\omega t) \\ &= -\frac{1}{\omega}\mathrm{e}^{-st}\cos(\omega t)\bigg|_{0^-}^{\infty} + \frac{1}{\omega}\int_{0^-}^{\infty} \cos(\omega t)\mathrm{d}\mathrm{e}^{-st} \\ &= \frac{1}{\omega} - \frac{s}{\omega}\int_{0^-}^{\infty} \cos(\omega t)\mathrm{e}^{-st}\mathrm{d}t = \frac{1}{\omega} - \frac{s}{\omega^2}\int_{0^-}^{\infty} \mathrm{e}^{-st}\mathrm{d}\sin(\omega t) \end{aligned}$$

$$= \frac{1}{\omega} - \frac{s}{\omega^2}\left(\mathrm{e}^{-st}\sin(\omega t)\Big|_{0^-}^{\infty} - \int_{0^-}^{\infty}\sin(\omega t)\mathrm{d}\mathrm{e}^{-st}\right)$$

$$= \frac{1}{\omega} - \frac{s^2}{\omega^2}\int_{0^-}^{\infty}\sin(\omega t)\mathrm{e}^{-st}\mathrm{d}t = \frac{1}{\omega} - \frac{s^2}{\omega^2}L\{\sin(\omega t)\}.$$

Above derivation leads to an equation in terms of $L\{\sin(\omega t)\}$, from which $L\{\sin(\omega t)\}$ is computed as

$$L\{\sin(\omega t)\} = \frac{1}{\omega}/\left(1 + \frac{s^2}{\omega^2}\right) = \frac{\omega}{s^2 + \omega^2}.$$

(6) *Cosine function* $\cos(\omega t)$:

$$L\{\cos(\omega t)\} = \int_{0^-}^{\infty}\cos(\omega t)\mathrm{e}^{-st}\mathrm{d}t = \frac{1}{\omega}\int_{0^-}^{\infty}\mathrm{e}^{-st}\mathrm{d}\sin(\omega t)$$

$$= \frac{1}{\omega}\mathrm{e}^{-st}\sin(\omega t)\Big|_{0^-}^{\infty} - \frac{1}{\omega}\int_{0^-}^{\infty}\sin(\omega t)\mathrm{d}\mathrm{e}^{-st}$$

$$= \frac{s}{\omega}\int_{0^-}^{\infty}\sin(\omega t)\mathrm{e}^{-st}\mathrm{d}t = -\frac{s}{\omega^2}\int_{0^-}^{\infty}\mathrm{e}^{-st}\mathrm{d}\cos(\omega t)$$

$$= -\frac{s}{\omega^2}\left(\mathrm{e}^{-st}\cos(\omega t)\Big|_{0^-}^{\infty} - \int_{0^-}^{\infty}\cos(\omega t)\mathrm{d}\mathrm{e}^{-st}\right)$$

$$= \frac{s}{\omega^2} - \frac{s^2}{\omega^2}\int_{0^-}^{\infty}\cos(\omega t)\mathrm{e}^{-st}\mathrm{d}t = \frac{s}{\omega^2} - \frac{s^2}{\omega^2}L\{\cos(\omega t)\}.$$

Above derivation leads to an equation in terms of $L\{\cos(\omega t)\}$, from which $L\{\cos(\omega t)\}$ is computed as

$$L\{\cos(\omega t)\} = \frac{s}{\omega^2}/\left(1 + \frac{s^2}{\omega^2}\right) = \frac{s}{s^2 + \omega^2}.$$

(7) *Exponentially-decaying sine function* $\mathrm{e}^{-at}\sin(\omega t)$:

$$L\{\mathrm{e}^{-at}\sin(\omega t)\} = \int_{0^-}^{\infty}\mathrm{e}^{-at}\sin(\omega t)\mathrm{e}^{-st}\mathrm{d}t = \int_{0^-}^{\infty}\sin(\omega t)\mathrm{e}^{-(s+a)t}\mathrm{d}t$$

(via Laplace transform of the sine function)

$$= \frac{\omega}{(s+a)^2 + \omega^2}.$$

(8) *Exponentially-decaying cosine function* $e^{-at}\cos(\omega t)$:

$$L\{e^{-at}\cos(\omega t)\} = \int_{0^-}^{\infty} e^{-at}\cos(\omega t)e^{-st}dt = \int_{0^-}^{\infty} \cos(\omega t)e^{-(s+a)t}dt$$

(via Laplace transform of the cosine function)

$$= \frac{s+a}{(s+a)^2+\omega^2}.$$

(9) *Exponentially-decaying generic function* $e^{-at}f(t)$ where $L\{f(t)\} = F(s)$:

$$L\{e^{-at}f(t)\} = \int_{0^-}^{\infty} e^{-at}f(t)e^{-st}dt = \int_{0^-}^{\infty} f(t)e^{-(s+a)t}dt = F(s+a).$$

For example, suppose $f(t) = t^n$, then

$$L\{e^{-at}t^n\} = \frac{n!}{(s+a)^{n+1}}.$$

(10) *Differential operator* $f^{(k)}(t) \equiv \frac{d^k f(t)}{dt^k}$:

$$\begin{aligned}
L\{f^{(k)}(t)\} &= \int_{0^-}^{\infty} \frac{d^k f(t)}{dt^k}e^{-st}dt = \int_{0^-}^{\infty} e^{-st}d\frac{d^{k-1}f(t)}{dt^{k-1}} \\
&= e^{-st}\frac{d^{k-1}f(t)}{dt^{k-1}}\Bigg|_{0^-}^{\infty} - \int_{0^-}^{\infty}\frac{d^{k-1}f(t)}{dt^{k-1}}de^{-st} \\
&= -f^{(k-1)}(0^-) + s\int_{0^-}^{\infty}\frac{d^{k-1}f(t)}{dt^{k-1}}e^{-st}dt \\
&= sL\{f^{(k-1)}(t)\} - f^{(k-1)}(0^-) \\
&\quad \text{(recursive relationship)} \\
&= s(sL\{f^{(k-2)}(t)\} - f^{(k-2)}(0^-)) - f^{(k-1)}(0^-) \\
&= s^2 L\{f^{(k-2)}(t)\} - sf^{(k-2)}(0^-) - f^{(k-1)}(0^-) = \cdots \\
&= s^k L\{f(t)\} - s^{k-1}f(0^-) - \cdots - sf^{(k-2)}(0^-) - f^{(k-1)}(0^-).
\end{aligned}$$

If the initial condition of $f(t)$ namely $\{f(0^-), \ldots, f^{(k-1)}(0^-)\}$ is neglected, above relationship is reduced to the following relationship that highlights the essential correspondence between the Laplace variable s and the differential operator $\frac{d}{dt}$:

$$L\{f^{(k)}(t)\} = s^k L\{f(t)\}.$$

In other words, multiplying the s-domain function $F(s)$ by s once corresponds to performing the differential operator on the associated time-domain function $f(t)$ once. So the Laplace variable s can be regarded as the differential operator:

$$s \Leftrightarrow \frac{\mathrm{d}}{\mathrm{d}t}.$$

(11) *Integral operator* $\int_{-\infty}^{t} f(t)\mathrm{d}t$:

$$\begin{aligned}
L\left\{\int_{-\infty}^{t} f(t)\mathrm{d}t\right\} &= \int_{0^-}^{\infty}\left(\int_{-\infty}^{t} f(\tau)\mathrm{d}\tau\right)\mathrm{e}^{-st}\mathrm{d}t = -\frac{1}{s}\int_{0^-}^{\infty}\left(\int_{-\infty}^{t} f(\tau)\mathrm{d}\tau\right)\mathrm{d}\mathrm{e}^{-st} \\
&= -\frac{1}{s}\left(\mathrm{e}^{-st}\int_{-\infty}^{t} f(\tau)\mathrm{d}\tau\Big|_{0^-}^{\infty} \cdot - \int_{0^-}^{\infty}\mathrm{e}^{-st}\mathrm{d}\int_{-\infty}^{t} f(\tau)\mathrm{d}\tau\right) \\
&= \frac{1}{s}\int_{-\infty}^{0^-} f(t)\mathrm{d}t + \frac{1}{s}\int_{0^-}^{\infty} f(t)\mathrm{e}^{-st}\mathrm{d}t \\
&= \frac{1}{s}L\{f(t)\} + \frac{1}{s}\int_{-\infty}^{0^-} f(t)\mathrm{d}t.
\end{aligned}$$

If the historical condition of $f(t)$ namely $\int_{-\infty}^{0^-} f(t)\mathrm{d}t$ is neglected, above relationship is reduced to the following relationship that highlights the essential correspondence between the inverse of the Laplace variable s and the integral operator $\int_{0^-}^{t}\mathrm{d}t$:

$$L\left\{\int_{0^-}^{t} f(t)\mathrm{d}t\right\} = \frac{1}{s}L\{f(t)\}.$$

In other words, the inverse of the Laplace variable s can be regarded as the integral operator:

$$\frac{1}{s} \Leftrightarrow \int_{0^-}^{t}\mathrm{d}t.$$

(12) *Convolution operator* $f(t) * g(t) \equiv \int_{0^-}^{\infty} f(\tau)g(t-\tau)\mathrm{d}\tau$ ($f(t) = g(t) = 0$ for $t < 0$):

$$\begin{aligned}
&L\left\{\int_{0^-}^{\infty} f(\tau)g(t-\tau)\mathrm{d}\tau\right\} \\
&= \int_{0^-}^{\infty}\left(\int_{0^-}^{\infty} f(\tau)g(t-\tau)\mathrm{d}\tau\right)\mathrm{e}^{-st}\mathrm{d}t = \int_{0^-}^{\infty}\left(\int_{0^-}^{\infty} f(\tau)g(t-\tau)\mathrm{e}^{-st}\mathrm{d}t\right)\mathrm{d}\tau \\
&= \int_{0^-}^{\infty}\left(\int_{0^-}^{\infty} g(t-\tau)\mathrm{e}^{-s(t-\tau)}\mathrm{d}t\right) f(\tau)\mathrm{e}^{-s\tau}\mathrm{d}\tau \\
&= \int_{0^-}^{\infty}\left(\int_{0^-}^{\infty} g(t)\mathrm{e}^{-st}\mathrm{d}t\right) f(\tau)\mathrm{e}^{-s\tau}\mathrm{d}\tau \\
&\quad (\text{note } g(t) = 0 \text{ for } t < 0) \\
&= \int_{0^-}^{\infty} g(t)\mathrm{e}^{-st}\mathrm{d}t\int_{0^-}^{\infty} f(\tau)\mathrm{e}^{-s\tau}\mathrm{d}\tau \quad \text{or} \quad \int_{0^-}^{\infty} f(\tau)\mathrm{e}^{-s\tau}\mathrm{d}\tau\int_{0^-}^{\infty} g(t)\mathrm{e}^{-st}\mathrm{d}t \\
&= L\{g(t)\}L\{f(t)\} \quad \text{or} \quad L\{f(t)\}L\{g(t)\}.
\end{aligned}$$

In other words, the s-domain function associated with the convolution of two time-domain functions (without history) is the product of the s-domain functions associated with the two time-domain functions. It is worth noting that the convolution operator may also be defined as $\int_{-\infty}^{\infty} f(\tau)g(t-\tau)\mathrm{d}\tau$ or $\int_{0^-}^{t^+} f(\tau)g(t-\tau)\mathrm{d}\tau$. However, with the *no history* condition $f(t) = g(t) = 0$ (for $t < 0$) which is commonly-adopted for practical applications, above three definitions of the convolution operator have no essential difference.

Above common Laplace transform pairs are summarized in Table 3.1.

3.2.2 *Solve Differential Equations via Laplace Transform*

The Laplace transform and the inverse Laplace transform are both bijective and are mutually the inverse mapping of each other. However, we rarely use the inverse Laplace transform described in (3.9) to directly compute the original time-domain function $f(t)$ associated with a s-domain function $F(s)$. To compute $f(t)$ from $F(s)$, a more practical way is to decompose $F(s)$ into partial fractions that belong to common Laplace transform pairs listed in Table 3.1 and to check Table 3.1 for the time-domain functions associated with these partial fractions.

Table 3.1 Common Laplace transform pairs

Time-domain function $f(t)$	s-domain function $F(s)$
$\delta(t)$	1
$u(t)$	$\frac{1}{s}$
t^n	$\frac{n!}{s^{n+1}}$
e^{-at}	$\frac{1}{s+a}$
$\sin(\omega t)$	$\frac{\omega}{s^2+\omega^2}$
$\cos(\omega t)$	$\frac{s}{s^2+\omega^2}$
$\mathrm{e}^{-at}\sin(\omega t)$	$\frac{\omega}{(s+a)^2+\omega^2}$
$\mathrm{e}^{-at}\cos(\omega t)$	$\frac{s+a}{(s+a)^2+\omega^2}$
$\mathrm{e}^{-at}f(t)$ where $L\{f(t)\} = F(s)$	$F(s+a)$
e.g. $L\{\mathrm{e}^{-at}t^n\}$	$\frac{n!}{(s+a)^{n+1}}$
$f^{(k)}(t) \equiv \frac{\mathrm{d}^k f(t)}{\mathrm{d}t^k}$	$s^k L\{f(t)\} - s^{k-1}f(0^-) - \cdots - f^{(k-1)}(0^-)$
$\int_{-\infty}^{t} f(t)\mathrm{d}t$	$\frac{1}{s}L\{f(t)\} + \frac{1}{s}\int_{-\infty}^{0^-} f(t)\mathrm{d}t$
$f(t) * g(t) \equiv \int_{0^-}^{\infty} f(\tau)g(t-\tau)\mathrm{d}\tau$	$L\{g(t)\}L\{f(t)\}$ or $L\{f(t)\}L\{g(t)\}$

We still take the rotating disk speed control system as example and demonstrate how to use above method to conveniently solve the differential equation described in (3.1). Perform the Laplace transform on both sides of (3.1) as

$$J(sv(s) - v_0) + bv(s) = T(s).$$

We have

$$v(s) = \frac{T(s) + Jv_0}{Js + b} = \frac{1}{J}\frac{1}{s + b/J}T(s) + \frac{v_0}{s + b/J} \tag{3.10}$$

(partial fraction decomposition)

and

$$\begin{aligned} v(t) &= \frac{1}{J}L^{-1}\left\{\frac{1}{s + b/J}\right\} * T(t) + v_0 L^{-1}\left\{\frac{1}{s + b/J}\right\} \\ &\text{(check Table 3.1)} \\ &= \frac{1}{J}e^{-\frac{b}{J}t} * T(t) + v_0 e^{-\frac{b}{J}t} = \int_0^t \frac{T(\tau)}{J}e^{-\frac{b}{J}(t-\tau)}d\tau + v_0 e^{-\frac{b}{J}t}, \end{aligned}$$

which is the same to the solution described in (3.3). Suppose that the controller output torque $T = T_0$ is constant. Then, $T(s) = T_0/s$. Substitute it into (3.10) and obtain

$$v(s) = \frac{T_0}{J}\frac{1}{s(s + b/J)} + \frac{v_0}{s + b/J} = \frac{T_0}{b}\left(\frac{1}{s} - \frac{1}{s + b/J}\right) + \frac{v_0}{s + b/J} \tag{3.11}$$

(partial fraction decomposition)

and

$$\begin{aligned} v(t) &= \frac{T_0}{b}\left(L^{-1}\left\{\frac{1}{s}\right\} - L^{-1}\left\{\frac{1}{s + b/J}\right\}\right) + v_0 L^{-1}\left\{\frac{1}{s + b/J}\right\} \\ &\text{(check Table 3.1)} \\ &= \frac{T_0}{b} - \frac{T_0}{b}e^{-\frac{b}{J}t} + v_0 e^{-\frac{b}{J}t}, \end{aligned}$$

which is the same to the solution described in (3.4).

3.2.3 *Partial Fraction Decomposition*

From above example of rotating disk speed control we can see effectiveness and convenience of solving differential equations via Laplace transform, yet readers may wonder how to do partial fraction decomposition. For an s-domain function (usually in fractional-polynomial form), if it is simple, we may do partial fraction decomposition directly by observation of its form. However, if it is complex and its partial fraction decomposition is not obvious, then how do we decompose it in a systematic way?

We take the following fractional-polynomial s-domain function described in (3.12) as example to demonstrate a systematic method of partial fraction decomposition.

$$F(s) = \frac{N(s)}{D(s)} = \frac{12\,s^3 + 38\,s^2 + 40\,s + 15}{4\,s^4 + 16\,s^3 + 21\,s^2 + 11\,s + 2}. \tag{3.12}$$

Step 1: Find roots of the $F(s)$ denominator namely $D(s)$ which is normally the characteristic polynomial of a differential equation. Check whether $D(s)$ contains a *repeated root* or *multiple root* by computing the *greatest common divisor* of $D(s)$ and its derivative $D(s)' \equiv \frac{\mathrm{d}D(s)}{\mathrm{d}s}$:

$$\begin{aligned}
(D(s), D(s)') &= (4\,s^4 + 16\,s^3 + 21\,s^2 + 11\,s + 2,\ 16\,s^3 + 48\,s^2 + 42\,s + 11) \\
&\quad \text{(by Euclid method)} \\
&= (2\,s^2 + 3\,s + 1,\ 16\,s^3 + 48\,s^2 + 42\,s + 11) \\
&= (2\,s^2 + 3\,s + 1,\ 2\,s + 1) = 2\,s + 1.
\end{aligned}$$

So $D(s)$ has a repeated root $s = -1/2$ and a factor $(2s + 1)^2$. It is worth noting that the greatest common divisor obtained in above way is not necessarily a first-order polynomial. For example, consider $\bar{D}(s) = (s + 1)^2(s + 2)^2$. Then, $(\bar{D}(s), \bar{D}(s)') = (s + 1)(s + 2)$ which is a second-order polynomial.

Divide $D(s)$ by $(2s + 1)^2$ and obtain $D_1(s) = D(s)/(2s + 1)^2 = s^2 + 3s + 2$ which has no more repeated root. Compute remaining roots of $D_1(s)$, which are obviously -1 and -2 in above example. When a more complicated polynomial is encountered, we may resort to numerical computation for root finding. After finding all roots of the denominator $D(s)$, we can factor it as

$$D(s) = 4\,s^4 + 16\,s^3 + 21\,s^2 + 11\,s + 2 = 4\left(s + \frac{1}{2}\right)^2 (s + 1)(s + 2).$$

Checking the greatest common divisor $(D(s), D'(s))$ is not necessary for finding the polynomial roots, yet it can considerably enhance computational efficiency when $D(s)$ has repeated roots. Even when $D(s)$ has no repeated root, a preliminary step of checking the greatest common divisor $(D(s), D'(s))$ has almost no computational cost in comparison with cost of numerical computation for finding polynomial roots. By the way, the computational cost of finding characteristic polynomial roots should not be counted only on the method of solving differential equations via Laplace Transform, because such computational cost is after all inevitable for differential equation solving.

Step 2: Express the s-domain function $F(s)$ as the sum of partial fractions with coefficients to be determined:

$$F(s) = \frac{c_1}{s+1} + \frac{c_2}{s+2} + \frac{c_3}{(s+1/2)^2} + \frac{c_4}{s+1/2}. \tag{3.13}$$

Step 3: Compute coefficients associated with partial fractions. Compute c_1 as

$$\begin{aligned} c_1 &= \lim_{s\to -1}(s+1)F(s) = \lim_{s\to -1}(s+1)\frac{N(s)}{D(s)} = \lim_{s\to -1}\frac{N(s)}{D'(s)} \\ &= \frac{12\,s^3+38\,s^2+40\,s+15}{16\,s^3+48\,s^2+42\,s+11}\Big|_{s=-1} = 1. \end{aligned}$$

Compute c_2 similarly as

$$c_2 = \lim_{s\to -2}\frac{N(s)}{D'(s)} = \frac{12\,s^3+38\,s^2+40\,s+15}{16\,s^3+48\,s^2+42\,s+11}\bigg|_{s=-2} = 1.$$

Generally for a partial fraction $c_\alpha/(s+\alpha)$ where $-\alpha$ is a single root (be $-\alpha$ a real root or a complex root), compute c_α as

$$c_\alpha = \frac{N(s)}{D'(s)}\bigg|_{s=-\alpha}.$$

Compute c_3 as

$$c_3 = \lim_{s\to -1/2}\left(s+\frac{1}{2}\right)^2 F(s) = \lim_{s\to -1/2}\frac{12\,s^3+38\,s^2+40\,s+15}{4(s^2+3\,s+2)} = 1.$$

Compute c_4 as

$$\begin{aligned} c_4 &= \lim_{s\to -1/2}\left[\left(s+\frac{1}{2}\right)^2 F(s)\right]' \equiv \lim_{s\to -1/2}\frac{\mathrm{d}}{\mathrm{d}s}\left[\left(s+\frac{1}{2}\right)^2 F(s)\right] \\ &= \lim_{s\to -1/2}\left[\frac{12\,s^3+38\,s^2+40\,s+15}{4(s^2+3\,s+2)}\right]' = 1, \end{aligned}$$

or as

$$\begin{aligned} c_4 &= \lim_{s\to -1/2}\left(s+\frac{1}{2}\right)\left(F(s) - \frac{c_1}{s+1} - \frac{c_2}{s+2} - \frac{c_3}{(s+1/2)^2}\right) \\ &= \lim_{s\to -1/2}\left(s+\frac{1}{2}\right)\left(\frac{12\,s^3+38\,s^2+40\,s+15}{4\,s^4+16\,s^3+21\,s^2+11\,s+2}\right. \\ &\qquad \left. - \frac{1}{s+1} - \frac{1}{s+2} - \frac{1}{(s+1/2)^2}\right) \\ &= 1. \end{aligned}$$

Generally for a group of partial fractions associated with the same repeated root $-\alpha$ (be $-\alpha$ a real root or a complex root)

$$A(s) = \frac{c_1}{(s+\alpha)^m} + \frac{c_2}{(s+\alpha)^{m-1}} + \cdots + \frac{c_m}{s+\alpha}.$$

Compute c_k $(1 \leq k \leq m)$ as

$$c_k = \lim_{s \to -\alpha} \frac{1}{(k-1)!}[(s+\alpha)^m A(s)]^{(k-1)} \equiv \lim_{s \to -\alpha} \frac{1}{(k-1)!} \frac{\mathrm{d}^{k-1}}{\mathrm{d}t^{k-1}}[(s+\alpha)^m A(s)],$$

or recursively as

$$\begin{aligned}
c_1 &= \lim_{s \to -\alpha} (s+\alpha)^m A(s), \\
A_2(s) &= A(s) - \frac{c_1}{(s+\alpha)^m}, \\
c_2 &= \lim_{s \to -\alpha} (s+\alpha)^{m-1} A_2(s), \\
A_3(s) &= A_2(s) - \frac{c_2}{(s+\alpha)^{m-1}}, \\
c_3 &= \lim_{s \to -\alpha} (s+\alpha)^{m-2} A_3(s), \\
&\vdots \\
A_m(s) &= A_{m-1}(s) - \frac{c_{m-1}}{(s+\alpha)^2}, \\
c_m &= \lim_{s \to -\alpha} (s+\alpha) A_m(s).
\end{aligned}$$

In summary, the s-domain function example described in (3.12) is decomposed as

$$F(s) = \frac{1}{s+1} + \frac{1}{s+2} + \frac{1}{(s+1/2)^2} + \frac{1}{s+1/2}.$$

Although only an example of partial fraction decomposition is demonstrated, above procedures of partial fraction decomposition, especially procedures for computing coefficients of partial fractions, can be easily generalized. It is worth noting that the general procedures of partial fraction decomposition can be automated in the spirit of symbolic operation.

3.3 Transfer Function

The Laplace transform is not only useful in the sense that it provides an effective and convenient methodology to solve differential equations of a control system, but also useful in the sense that it provides an effective and convenient methodology to abstract mathematical essence of the control system, especially mathematical essence related to important concerns for control purpose. In practical applications, we rarely take advantage of this first utility directly, whereas we normally take advantage of this second utility to model control systems.

Suppose that the control system's components are *linear time-invariant*, namely they can be modelled by linear time-invariant differential equations. Let a system component be modelled generically by a differential equation as

$$a_n \frac{d^n}{dt^n} y + a_{n-1} \frac{d^{n-1}}{dt^{n-1}} y + \cdots + a_0 y = b_m \frac{d^m}{dt^m} r + b_{m-1} \frac{d^{m-1}}{dt^{m-1}} r + \cdots + b_0 r, \tag{3.14}$$

where the differential equation coefficients $a_n, a_{n-1}, \ldots, a_0, b_m, b_{m-1}, \ldots, b_0$ are time-invariant namely constant in terms of time. The component model is illustrated in Fig. 3.1.

In practical applications, no system component is ideally linear time-invariant, yet many kinds of system components have proper linear time-invariant *approximations* for representative *operation ranges*. So the linear time-invariant assumption is rather fair for many practical applications and is widely adopted.

Given a linear time-invariant system component modelled by (3.14), perform the Laplace transform on both sides of (3.14) (without causing confusion, we abuse the same notation of a variable to denote both its time-domain function as well as its s-domain function). We have

$$\begin{aligned} &(a_n s^n + a_{n-1} s^{n-1} + \cdots + a_0) y(s) - A_0(s) \\ &= (b_m s^m + b_{m-1} s^{m-1} + \cdots + b_0) r(s) - B_0(s) \end{aligned}$$

and

$$y(s) = \frac{b_m s^m + b_{m-1} s^{m-1} + \cdots + b_0}{a_n s^n + a_{n-1} s^{n-1} + \cdots + a_0} r(s) + \frac{A_0(s) - B_0(s)}{a_n s^n + a_{n-1} s^{n-1} + \cdots + a_0}, \tag{3.15}$$

where $A_0(s)$ and $B_0(s)$ are constant polynomials determined by initial conditions of the output y and the input r respectively. The tail term

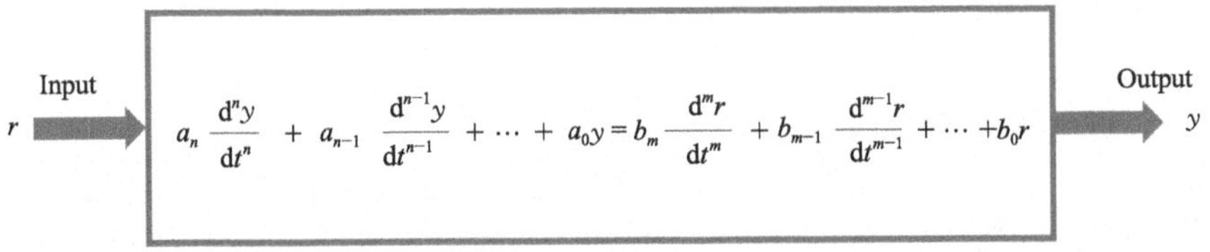

Fig. 3.1 Linear time-invariant system component

$$\frac{A_0(s) - B_0(s)}{a_n s^n + a_{n-1} s^{n-1} + \cdots + a_0}$$

represents the influence of input and output initial conditions on the output y.

For a stable control system,[1] any influence of control system initial conditions will fade away and usually fade away quickly. So for long-term operation, the tail term described in (3.15) can be discarded and hence (3.15) can be reduced to

$$y(s) = \frac{b_m s^m + b_{m-1} s^{m-1} + \cdots + b_0}{a_n s^n + a_{n-1} s^{n-1} + \cdots + a_0} r(s) \Longleftrightarrow$$
$$\frac{y(s)}{r(s)} = \frac{b_m s^m + b_{m-1} s^{m-1} + \cdots + b_0}{a_n s^n + a_{n-1} s^{n-1} + \cdots + a_0}, \tag{3.16}$$

where the constant fractional-polynomial s-domain function

$$G(s) = \frac{b_m s^m + b_{m-1} s^{m-1} + \cdots + b_0}{a_n s^n + a_{n-1} s^{n-1} + \cdots + a_0} \tag{3.17}$$

characterizes the inherent dynamics property that determines how the input r influences the output y. In other words, in terms of dynamics behaviour, a system component modelled by the differential equation described in (3.14) can be essentially characterized by the constant fractional-polynomial s-domain function $G(s)$. Such an s-domain function $G(s)$ described in (3.17) is called the **transfer function** of the system component.

3.4 Block Diagram Models

Thanks to the Laplace transform that substitutes easy *algebraic equations* for comparatively difficult *differential equations* in system modelling, we have an effective and convenient mathematical tool namely transfer functions to essentially characterize system components of a control system. For example, the rotating disk position proportional-derivative feedback control system illustrated in Fig. 2.20 can be concisely represented by a **transfer function block diagram** model illustrated in Fig. 3.2.

The mathematical tool of transfer functions can also facilitate analysis of the control system on the whole, which involves synthesis of a transfer function block diagram to form a holistic transfer function block equivalently.

[1] The concept of **control system stability** will be formally described in Chap. 4.

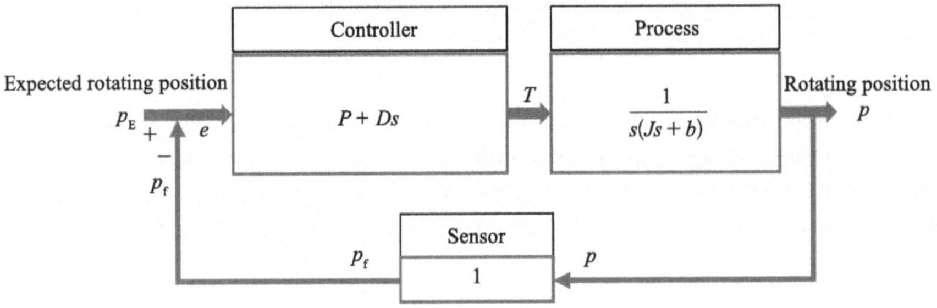

Fig. 3.2 Block diagram of rotating disk position proportional-derivative feedback control

3.4.1 Transfer Function Block Synthesis

All system components described in remaining sections of Chap. 3 are assumed linear time-invariant by default. For synthesis of two parallel transfer function blocks G_1 and G_2 illustrated in Fig. 3.3, the synthesized transfer function is $G_1(s) + G_2(s)$, which is obtained as

$$\frac{y(s)}{r(s)} = \frac{y_1(s) + y_2(s)}{r(s)} = \frac{y_1(s)}{r(s)} + \frac{y_2(s)}{r(s)} = G_1(s) + G_2(s).$$

For synthesis of two serial transfer function blocks G_1 and G_2 illustrated in Fig. 3.4, the synthesized transfer function is $G_1(s)G_2(s)$, which is obtained as

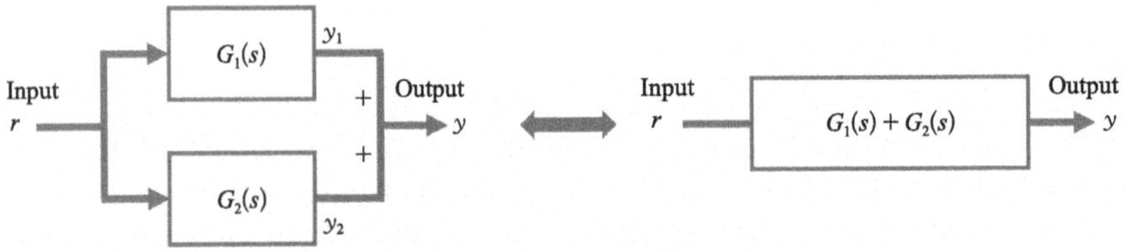

Fig. 3.3 Parallel transfer function block synthesis

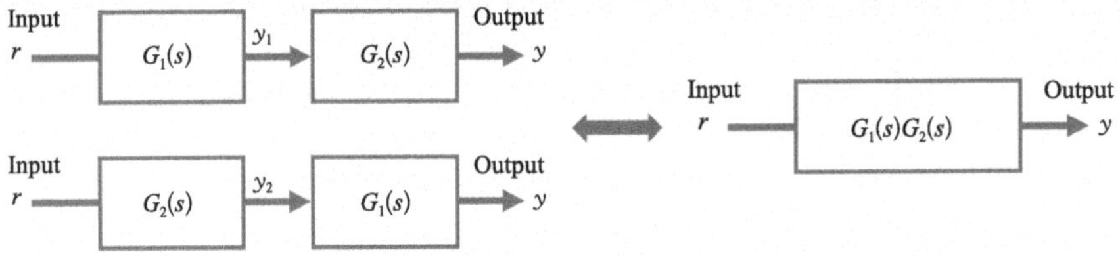

Fig. 3.4 Serial transfer function block synthesis

$$\frac{y(s)}{r(s)} = \frac{y_1(s)}{r(s)}\frac{y(s)}{y_1(s)} = G_1(s)G_2(s),$$

or

$$\frac{y(s)}{r(s)} = \frac{y_2(s)}{r(s)}\frac{y(s)}{y_2(s)} = G_2(s)G_1(s) = G_1(s)G_2(s),$$

which implies an interesting property of serial transfer function block synthesis, namely that the order of the transfer function blocks does not influence the synthesized transfer function block. In other words, if the transfer function blocks are exchanged in the serial block diagram, the same synthesized transfer function block will be obtained. This interesting property holds for theoretical analysis, though two system components normally cannot be exchanged physically in practical applications.

Above rules for both parallel transfer function block synthesis and serial transfer function block synthesis are rather intuitive. In contrast, the rule for synthesis of transfer function blocks in a closed-loop block diagram illustrated in Fig. 3.5 needs more derivation.

For the closed-loop *negative feedback* block diagram illustrated in the top half of Fig. 3.5, we can establish the following equations:

$$\begin{aligned} y_f(s) &= H(s)y(s), \\ e(s) &= r(s) - y_f(s), \\ y(s) &= G(s)e(s). \end{aligned}$$

Substitute the first equation into the second equation and then substitute the result into the third equation:

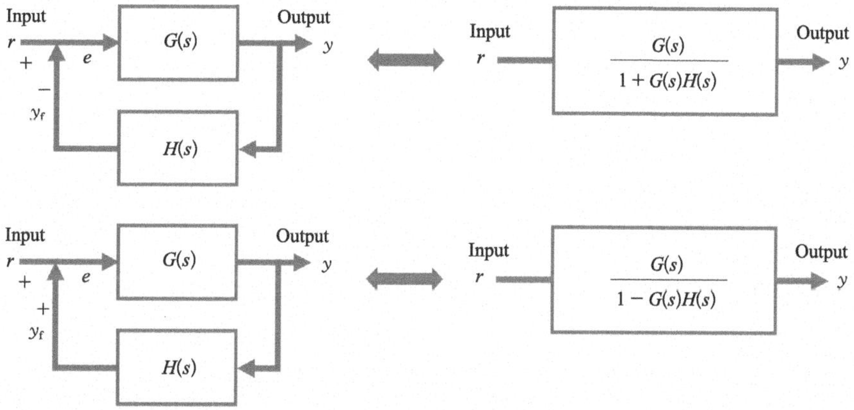

Fig. 3.5 Closed-loop transfer function block synthesis

$$e(s) = r(s) - H(s)y(s),$$
$$y(s) = G(s)e(s) = G(s)r(s) - G(s)H(s)y(s),$$

from which the synthesized transfer function can be derived as

$$\frac{y(s)}{r(s)} = \frac{G(s)}{1 + G(s)H(s)}. \tag{3.18}$$

For the closed-loop *positive feedback* block diagram illustrated in the low half of Fig. 3.5, we may regard it as the closed-loop negative feedback block diagram with feedback transfer function $-H(s)$. From (3.18), the synthesized transfer function for the closed-loop positive feedback block diagram can be obtained as

$$\frac{y(s)}{r(s)} = \frac{G(s)}{1 - G(s)H(s)}. \tag{3.19}$$

Rules for the three basic kinds of transfer function block synthesis namely the parallel transfer function block synthesis, the serial transfer function block synthesis, and the closed-loop transfer function block synthesis are summarized in Table 3.2.

The basic rules of transfer function block synthesis summarized in Table 3.2 enable us to synthesize transfer function block diagrams generally, in the spirit of **synthesis equivalence**. For example, consider the hybrid transfer function block diagram illustrated in Fig. 3.6. This transfer function block is hybrid in the sense that it does not belong to transfer function block diagrams of the three basic kinds (namely the parallel transfer function block diagram, the serial transfer function block diagram, and the closed-loop transfer function block diagram) but consists of them. The hybrid transfer function block diagram illustrated in Fig. 3.6 can be decomposed into transfer function block diagrams of the three basic kinds.

More specifically, the transfer function blocks G_1 and G_2 form a parallel transfer function block diagram locally, where the locally-synthesized transfer function is $G_{12}(s) = G_1(s) + G_2(s)$. The transfer function blocks G_3 and H form a closed-loop transfer function block diagram locally, where the locally-synthesized transfer function is

Table 3.2 Basic rules of transfer function block synthesis

Block diagram	Synthesis
G_1 and G_2 are parallel	$G_1(s) + G_2(s)$
G_1 and G_2 are serial	$G_1(s)G_2(s)$ or $G_2(s)G_1(s)$
G with H in feedback	$\frac{G(s)}{1+G(s)H(s)}$ (negative feedback)
	$\frac{G(s)}{1-G(s)H(s)}$ (positive feedback)

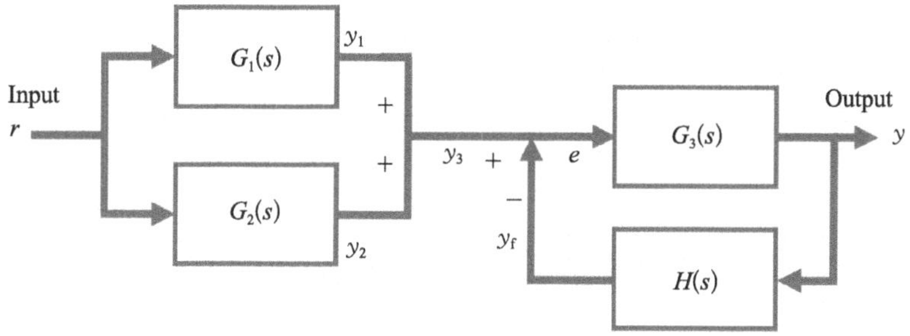

Fig. 3.6 Hybrid transfer function block diagram

$$G_{3H}(s) = \frac{G_3(s)}{1 + G_3(s)H(s)}.$$

The locally-synthesized transfer function blocks G_{12} and G_{3H} further form a serial transfer function block diagram. Decomposition of the hybrid transfer function block diagram is illustrated in Fig. 3.7.

Therefore, the synthesized transfer function for the hybrid transfer function block diagram is computed as

$$\frac{y(s)}{r(s)} = (G_1(s) + G_2(s))\frac{G_3(s)}{1 + G_3(s)H(s)}.$$

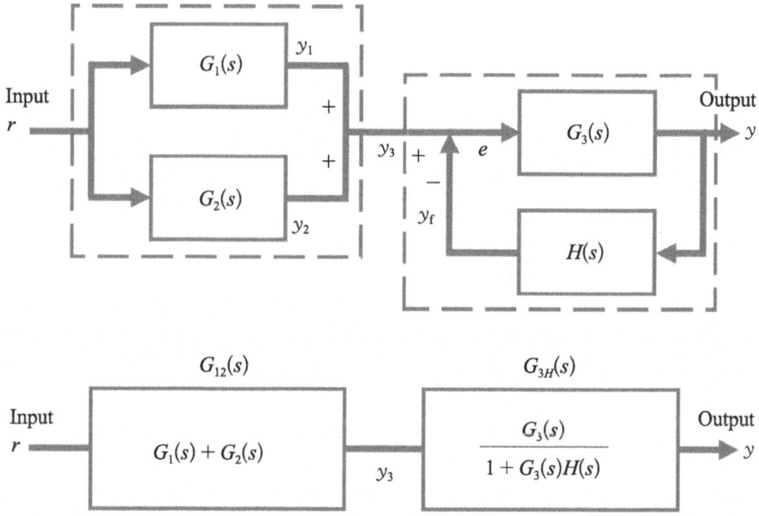

Fig. 3.7 Hybrid transfer function block diagram decomposition

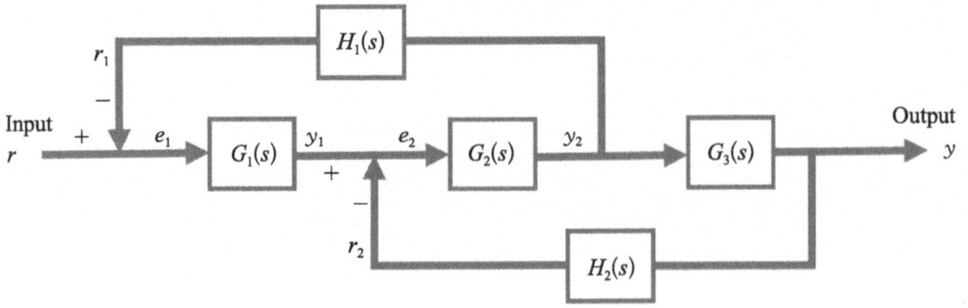

Fig. 3.8 Complex transfer function block diagram

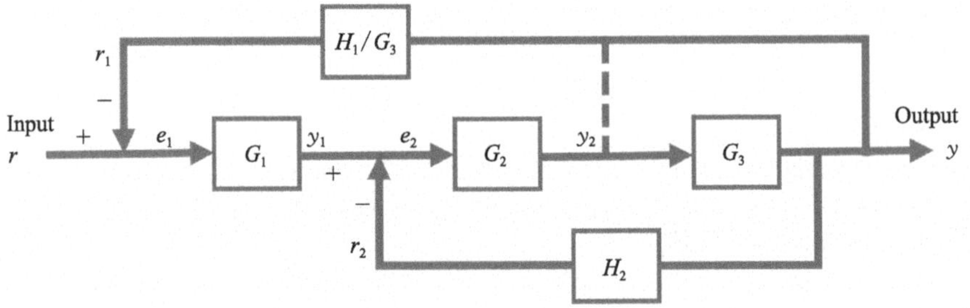

Fig. 3.9 Complex transfer function block diagram transform

We can see how the spirit of synthesis equivalence is reflected in synthesis of the hybrid transfer function block diagram: The local parallel transfer function block diagram G_1 and G_2 are replaced by the locally-synthesized transfer function block G_{12} in the spirit of synthesis equivalence. The local closed-loop transfer function block diagram G_3 and H are replaced by the locally-synthesized transfer function block G_{3H} also in the spirit of synthesis equivalence. Then the final synthesized transfer function can be easily computed, still in the spirit of synthesis equivalence.

Take a complex transfer function block diagram illustrated in Fig. 3.8 as another example. This complex transfer function block diagram cannot be simply decomposed into transfer function block diagrams of the three basic kinds, because the local closed-loop transfer function block diagram $\{G_1, G_2, H_1\}$ and the local closed-loop transfer function block diagram $\{G_2, G_3, H_2\}$ are entangled with each other.

Although the complex transfer function block diagram illustrated in Fig. 3.8 cannot be simply decomposed into transfer function block diagrams of the three basic kinds, we can transform it into another transfer function block diagram in the spirit of synthesis equivalence in order that the transformed transfer function block diagram can be simply decomposed into transfer function block diagrams of the three basic kinds or at least becomes easier for synthesis. We may transform the complex transfer function block diagram into another synthesis-equivalent transfer function block diagram illustrated in Fig. 3.9.

By so far, we have only been using the idea of **synthesis equivalence** implicitly without clarifying it explicitly, because examples encountered before the complex transfer function block diagram illustrated in Fig. 3.8 are simple and intuitive for understanding of synthesis equivalence. However, before we continue with the complex transfer function block diagram example, we had better clarify this idea explicitly.

Recall that a linear time-invariant system component's transfer function is *identified* by the ratio between the s-domain function of its output and that of its input, with influence of its initial conditions neglected. Readers may notice that we use the term "identified" instead of the term "defined". The reason is that the transfer function of the system component is an essential property inherent to the system component, regardless of existence of its input and its output. So we cannot use the term "defined". On the other hand, its transfer function can be revealed by examining the relationship between its output and its input, so we use the term "identified".

Given two system components, whenever they are given the same input, they will always have the same output. In other words, the two system components always have the same output-input relationship, and hence can be identified as mutually synthesis-equivalent. Based on such understanding, the idea of **synthesis equivalence** is explicitly clarified as follows.

Synthesis equivalence: *System components have synthesis equivalence, namely they are mutually synthesis-equivalent, if they always have the same output-input relationship.*

Given an original transfer function block diagram and a transformed version of it, if they have synthesis equivalence, then we can substitute the transformed transfer function block diagram for the original transfer function block diagram. To check whether they have synthesis equivalence, we can check whether they have the same output-input relationship.

Input to the two transfer function block diagrams is independent of them and can be arbitrary. The key point of synthesis equivalence checking is to check whether the transform from the original transfer function block diagram to the transformed diagram never changes output of the original diagram given arbitrary input. Note that the final output of a transfer function block diagram is potentially influenced only by output of its composing blocks. So if the transform never changes any output of the involved composing blocks, then the transform will never change the final diagram output and the two transfer function block diagrams before and after the transform will always have the same output-input relationship.

Based on above analysis, we can verify whether the transformed transfer function block diagram illustrated in Fig. 3.9 and that illustrated in Fig. 3.8 actually have synthesis equivalence. We can see that the transform illustrated in Fig. 3.9 only involves the original H_1 block as it only changes how the original H_1 block input is obtained. After the transform, the transfer function of the original H_1 block becomes H_1/G_3. Compute the output of the transformed H_1 block namely the H_1/G_3 block in the transformed diagram as

$$r_{1T}(s) = \frac{H_1(s)}{G_3(s)}y(s) = \frac{H_1(s)}{G_3(s)}(G_3(s)y_2(s)) = H_1(s)y_2(s) = r_{1O}(s),$$

where the subscripts T and O of r_1 correspond to the transformed H_1 block and the original H_1 block respectively. The derivation always holds regardless of diagram input r and hence implies the transform illustrated in Fig. 3.9 never changes the output of the involved composing block namely the H_1 block. So we can conclude that the two transfer function block diagrams illustrated in Figs. 3.8 and 3.9 do have synthesis equivalence.

Therefore, the original transfer function block diagram has a transformed version of synthesis equivalence, which can be simply decomposed into transfer function block diagrams of the three basic kinds, as illustrated in Fig. 3.10.

The transfer function blocks G_2 and G_3 form a serial transfer function block diagram which together with the transfer function block H_2 further form a closed-loop transfer function block diagram G_{23H} locally, where the locally-synthesized transfer function is

$$G_{23H}(s) = \frac{G_2(s)G_3(s)}{1 + G_2(s)G_3(s)H_2(s)}.$$

Then the locally-synthesized transfer function block G_{23H} and the transfer function block G_1 form a serial transfer function block diagram which together with the transfer function block H_1/G_3 further form a closed-loop transfer function block diagram locally. So the transfer function for the whole transfer function block diagram is obtained as

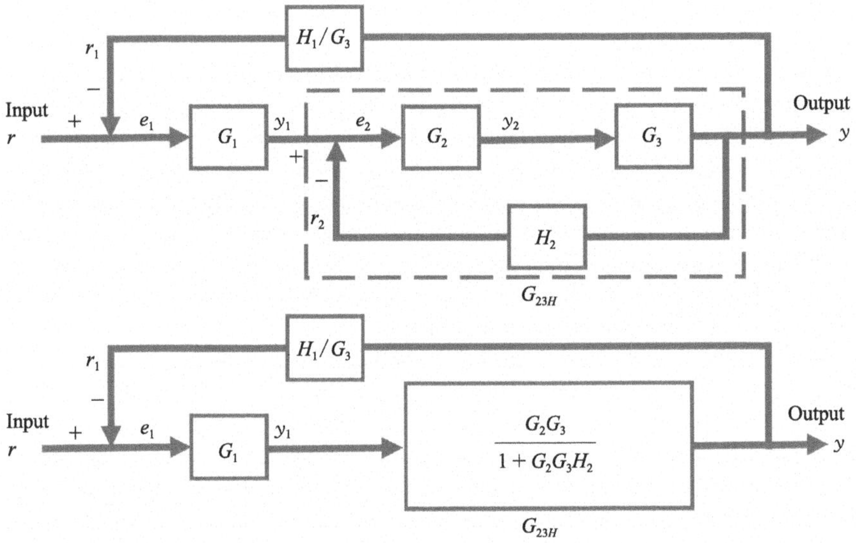

Fig. 3.10 Complex transfer function block diagram decomposition

$$\frac{y(s)}{r(s)} = \frac{G_1(s)G_{23H}(s)}{1 + G_1(s)G_{23H}(s)\frac{H_1(s)}{G_3(s)}} = \frac{G_1(s)G_2(s)G_3(s)}{1 + G_1(s)G_2(s)H_1(s) + G_2(s)G_3(s)H_2(s)}.$$

Synthesis of the complex transfer function block diagram illustrated in Fig. 3.8 is accomplished finally. It is worth noting that "complex" here only means that the transfer function block diagram illustrated in Fig. 3.8 is comparatively complex with respect to previous examples and cannot be simply decomposed into transfer function block diagrams of the three basic kinds. Transfer function block diagrams encountered in practical applications can be much more complex than that illustrated in Fig. 3.8. Therefore, we intend by no means to emphasize the so-called complexity of the transfer function block diagram illustrated in Fig. 3.8, but only intend to demonstrate by this suitable example how to flexibly apply the basic rules of transfer function block synthesis in the spirit of synthesis equivalence.

Above demonstrated method of transforming a complex transfer function block diagram gradually into easier and easier transfer function block diagrams for synthesis is called the **transfer function block diagram reduction** method.

3.4.2 Mason Signal-Flow Gain Formula

For synthesis of a transfer function block diagram, especially when the diagram has complicated and entangled closed-loops, we can also resort to the **Mason signal-flow gain formula** (Mason 1956) besides the transfer function block diagram reduction method. Before formal presentation of the Mason signal-flow gain formula, we need to introduce a number of basic concepts.

- **Signal-flow graph**: A transfer function block diagram can be treated as a signal-flow graph, where each block's input and output are treated as *nodes* and each block itself is treated as a *directed edge* weighted by its transfer function. For example, the transfer function block diagram illustrated in Fig. 3.8 can be treated as a signal-flow graph, with r, r_1, e_1, y_1, r_2, e_2, y_2, and y treated as graph nodes. The transfer functions G_1, G_2, G_3, H_1, and H_2 are treated as directed edges from the node e_1 to the node y_1, from the node e_2 to the node y_2, from the node y_2 to the node y, from the node y_2 to the node r_1, and from the node y to the node r_2 respectively. The directed edges are weighted by their corresponding transfer functions. The nodes r and r_1 point together to the node e_1 with the signal-flow relationship $e_1 = r - r_1$. Similarly, the nodes y_1 and r_2 point together to the node e_2 with the signal-flow relationship $e_2 = y_1 - r_2$.
- **Gain**: The gain of a path or a closed-loop is the product of all transfer functions on the path or the closed-loop (with the positive or negative sign taken into account). For example, each directed edge is a simple path whose gain is its corresponding transfer function or a signed unit. The gain of the path $e_1y_1e_2y_2$, namely the path from the node e_1 to the node y_2 through intermediate nodes y_1 and e_2, is G_1G_2. The gain of the path $re_1y_1e_2y_2y$ is $G_1G_2G_3$. The gain of the closed-loop $e_1y_1e_2y_2r_1e_1$

or simply $G_1G_2H_1$ is $-G_1G_2H_1$ (with the negative sign because r_1 serves as negative feedback to e_1). The gain of the closed-loop $G_2G_3H_2$ is $-G_2G_3H_2$.

- **Determinant**: The determinant of a signal-flow graph G is

$$\Delta_G(s) = 1 - \sum_{L_i \in L_G} L_i + \sum_{\substack{L_i, L_j \in L_G \\ \text{non-touching}}} L_i L_j - \sum_{\substack{L_i, L_j, L_k \in L_G \\ \text{non-touching}}} L_i L_j L_k$$
$$+ (-1)^m \sum_{\substack{L_{i_1}, L_{i_2}, \ldots, L_{i_m} \in L_G \\ \text{non-touching}}} L_{i_1} L_{i_2} \cdots L_{i_m} + \cdots, \tag{3.20}$$

where L_G denotes the set of all closed-loops in G as well as their signed closed-loop gains. The second sum involves all pairs of non-touching closed-loops in G, the third sum involves all triplets of non-touching closed-loops in G, and so on. Still taking the transfer function block diagram illustrated in Fig. 3.8 as example, there are totally two closed-loops $G_1G_2H_1$ and $G_2G_3H_2$ whose gains are $-G_1G_2H_1$ and $-G_2G_3H_2$ respectively. For the determinant of this signal-flow graph, the first sum is $-G_1G_2H_1 - G_2G_3H_2$, the second sum is zero because these two closed-loops are touching at nodes e_2 and y_2 and there is no pair of non-touching closed-loops, and sums after the second sum are all zero because there are no more than two closed-loops and of course no more than two non-touching closed-loops. Therefore, the determinant of the signal-flow graph is $1 + G_1G_2H_1 + G_2G_3H_2$.

After introduction of above basic concepts, the **Mason signal-flow gain formula** can be presented as

$$T^{r,y}(s) = \frac{\sum_{P \in P_G^{r,y}} P(s)\Delta_{G/P}(s)}{\Delta_G(s)}, \tag{3.21}$$

where $T^{r,y}$ denotes the transfer function from the diagram input r to the diagram output y and $P_G^{r,y}$ denotes all possible open-loop paths from r to y in the graph G. G/P denotes the imagined subgraph of G obtained by removing the closed-loops that touch the path P. Such removal does not mean that edges or nodes of the closed-loops touching the path P are actually removed from the graph G, but mean that these closed-loops are no longer treated as closed-loops in computation of the graph determinant via (3.20). This is why we call G/P an "imagined" subgraph of G.

It is worth noting that the Mason signal-flow gain formula described in (3.21) is generic for any pair of nodes in the signal-flow graph. In other words, we can replace the diagram input-output pair (r, y) by any two other nodes and the Mason signal-flow gain formula described in (3.21) still holds. A simple proof of the Mason signal-flow gain formula via *mathematical induction* is given below.

Proof It can be easily verified that the Mason signal-flow gain formula described in (3.21) holds for transfer function block diagrams without close-loop. We omit such trivial verification here and leave it to readers.

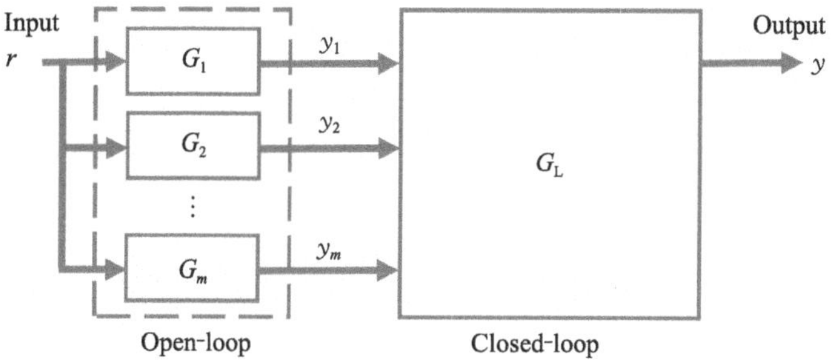

Fig. 3.11 Signal-flow graph decomposition

For a generic transfer function block diagram G, suppose that the Mason signal-flow gain formula described in (3.21) holds for all subgraphs of G. We intend to prove that the Mason signal-flow gain formula also holds for G. Suppose that the diagram input r passes a number of open-loop subgraphs $\{G_1, G_2, \ldots, G_m\}$ that connect into a closed-loop subgraph G_L to give the diagram output y, as illustrated in Fig. 3.11.

Since all closed-loops (if existing) are contained in the subgraph G_L, $\Delta_G(s) = \Delta_{G_\mathrm{L}}(s)$ and $\Delta_{G/P}(s) = \Delta_{G_\mathrm{L}/P}(s)$ for any path P. Apply the assumption of mathematical induction to the closed-loop subgraph G_L for pairs (y_k, y) $(k \in \{1, \ldots, m\})$ and obtain

$$T^{y_k,y}(s) = \frac{\sum_{P \in P_{G_\mathrm{L}}^{y_k,y}} P(s)\Delta_{G_\mathrm{L}/P}(s)}{\Delta_{G_\mathrm{L}}(s)} = \frac{\sum_{P \in P_G^{y_k,y}} P(s)\Delta_{G/P}(s)}{\Delta_G(s)}.$$

So

$$T^{r,y}(s) = \sum_{k=1}^{m} G_k(s)\frac{\sum_{P \in P_G^{y_k,y}} P(s)\Delta_{G/P}(s)}{\Delta_G(s)} = \frac{\sum_{P \in P_G^{r,y}} P(s)\Delta_{G/P}(s)}{\Delta_G(s)},$$

which implies that the Mason signal-flow gain formula described in (3.21) also holds for G.

Above derivation is valid except for the special case where the diagram input r enters a closed-loop subgraph without passing any open-loop subgraphs. In other words, except for the special case where the transfer function block diagram G forms a holistic closed-loop graph illustrated in Fig. 3.12.

Note that $e = r - T^{e,r_1}e$, so we have

$$\frac{e(s)}{r(s)} = \frac{1}{1 + T^{e,r_1}(s)}.$$

Apply the assumption of mathematical induction to the subgraph G_S for pairs (e, r_1) and (e, y) and obtain

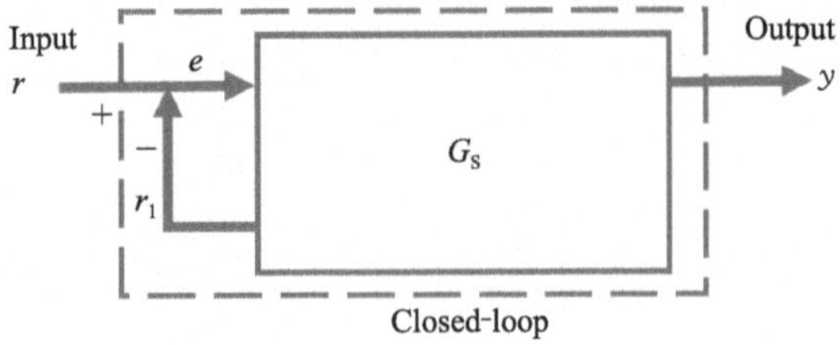

Fig. 3.12 Holistic closed-loop signal-flow graph

$$T^{e,r_1}(s) = \frac{\sum_{P \in P_{G_S}^{e,r_1}} P(s)\Delta_{G_S/P}(s)}{\Delta_{G_S}(s)},$$

$$T^{e,y}(s) = \frac{\sum_{P \in P_{G_S}^{e,y}} P(s)\Delta_{G_S/P}(s)}{\Delta_{G_S}(s)}.$$

So

$$T^{r,y}(s) = T^{e,y}(s)\frac{e(s)}{r(s)} = T^{e,y}(s)\frac{1}{1+T^{e,r_1}(s)}$$
$$= \frac{\sum_{P \in P_{G_S}^{e,y}} P(s)\Delta_{G_S/P}(s)}{\Delta_{G_S}(s) + \sum_{P \in P_{G_S}^{e,r_1}} P(s)\Delta_{G_S/P}(s)}. \tag{3.22}$$

Each path $P \in P_{G_S}^{e,y}$ corresponds to a unique path $(r, P) \in P_G^{r,y}$ and vice versa. Each pair of paths $P \in P_{G_S}^{e,y}$ and $(r, P) \in P_G^{r,y}$ have the same path gain $P(s)$. Besides, each closed-loop non-touching the path $P \in P_{G_S}^{e,y}$ in G_S is also a closed-loop non-touching the path $(r, P) \in P_G^{r,y}$ in G and vice versa, which implies $\Delta_{G_S/P}(s) = \Delta_{G/(r,P)}(s)$. So

$$\sum_{P \in P_{G_S}^{e,y}} P(s)\Delta_{G_S/P}(s) = \sum_{P \in P_G^{r,y}} P(s)\Delta_{G/P}(s). \tag{3.23}$$

Each path $P \in P_{G_S}^{e,r_1}$ corresponds to a unique closed-loop $L_{(e,P,r_1,e)} \in G$ with the associated closed-loop gain $L_{(e,P,r_1,e)} = -P(s)$ and vice versa. Since closed-loops in G_S are closed-loops in G without passing r_1, we have

$$\Delta_{G_S}(s) + \sum_{P \in P_{G_S}^{e,r_1}} P(s)\Delta_{G_S/P}(s) = \sum_{m=0}^{\infty}\left[(-1)^m \sum_{\substack{L_{i_1},\ldots,L_{i_m} \in L_{G_S} \\ \text{non-touching}}} L_{i_1}\cdots L_{i_m}\right]$$
$$+ \sum_{L_{(e,P,r_1,e)} \in G}\left[(-L_{(e,P,r_1,e)})\sum_{m=0}^{\infty}(-1)^m \sum_{\substack{L_{i_1},\ldots,L_{i_m} \in L_{G_S/P} \\ \text{non-touching}}} L_{i_1}\cdots L_{i_m}\right]$$

$$= \sum_{m=0}^{\infty} \left[(-1)^m \sum_{\substack{L_{i_1},\ldots,L_{i_m} \in L_{G_S} \\ \text{non-touching}}} L_{i_1} \cdots L_{i_m} \right]$$

(traverse non-touching closed-loops in G without passing r_1)

$$+ \sum_{m=0}^{\infty} \left[(-1)^{m+1} \sum_{\substack{L_{(e,P,r_1,e)} \in G \\ L_{i_1},\ldots,L_{i_m} \in L_{G_S/P} \\ \text{non-touching}}} L_{(e,P,r_1,e)} L_{i_1} \cdots L_{i_m} \right]$$

(traverse non-touching closed-loops in G with one passing r_1)

$$= \sum_{m=0}^{\infty} \left[(-1)^m \sum_{\substack{L_{i_1},\ldots,L_{i_m} \in L_G \\ \text{non-touching}}} L_{i_1} \cdots L_{i_m} \right] = \Delta_G(s). \tag{3.24}$$

(traverse non-touching closed-loops in G)

Note that each combination of non-touching closed-loops in G_S is also a combination of non-touching closed-loops in G without passing r_1 and vice versa. Also note that each combination of non-touching closed-loops formed by selecting a closed-loop passing r_1 and selecting the others from G_S is also a combination of non-touching closed-loops in G with one (and at most one) passing r_1 and vice versa. Then readers can easily understand derivation of (3.24). Substitute (3.23) and (3.24) into (3.22) and obtain

$$T^{r,y}(s) = \frac{\sum_{P \in P_G^{r,y}} P(s) \Delta_{G/P}(s)}{\Delta_G(s)}.$$

So the Mason signal-flow gain formula described in (3.21) also holds for the special case of G. Therefore, based on all above analysis, we can conclude via mathematical induction that the Mason signal-flow gain formula described in (3.21) holds generally for transfer function block diagrams. □

We now return to the transfer function block diagram illustrated in Fig. 3.8 and apply the Mason signal-flow gain formula described in (3.21) to synthesize it. There is only one open-loop path from the diagram input r to the diagram output y, namely the path $P_1 \equiv r e_1 y_1 e_2 y_2 y$ with the path gain $G_1 G_2 G_3$. There is no closed-loop non-touching P_1, so $\Delta_{G/P_1}(s) = 1$. We have already known that $\Delta_G(s) = 1 + G_1 G_2 H_1 + G_2 G_3 H_2$ (see previous introduction of signal-flow graph **determinant**). According to the Mason signal-flow gain formula described in (3.21), the synthesized transfer function is computed as

$$T^{r,y}(s) = \frac{G_1(s)G_2(s)G_3(s)}{1 + G_1(s)G_2(s)H_1(s) + G_2(s)G_3(s)H_2(s)},$$

which is consistent with the synthesized transfer function obtained by the transfer function block diagram reduction method presented in Sect. 3.4.1.

3.5 Note

In retrospect of Chap. 3, we have presented fundamentals of system modelling via Laplace transform in forms of transfer functions and transfer function block diagrams. Based on these fundamentals of control theory, we will present how to analyse and design control systems in Chaps. 4 and 5.

References

R. Feynman. *The Feynman lectures on physics (commemorative issue)*. Pearson Education, 2004.

S. Mason. Feedback theory-further properties of signal flow graphs. *Proceedings of the IRE*, 44(7):920–926, 1956.

Chapter 4
Control System Stability

For a control system, especially a closed-loop feedback control system, its stability is our first important concern for control purpose, though its stability is not the only important concern. Unstable control systems are obviously undesirable in practical applications and nobody likes unstable control systems. It is like in daily life, nobody is willing to drive a commercial car whose stability is not guaranteed. Although we have not clarified the term *stability* or *stable* by so far, the daily-life sense of *unstability* or *unstable* is already enough for readers to understand that unstable control systems should be avoided in practical applications.

A control system's stability may be regarded as the property that it should perform stably under influence of stable input.[1] However, this conception is still abstract for understanding because we have not clarified what "performing stably" means exactly. A more intuitive conception of control system stability may be that the control system should have no output or should at least have convergent output when it is given no input. In contrast, if a control system has oscillatory or even divergent output even when it is given no input, then it is an unstable control system.

For example, an inverted pendulum open-loop control system illustrated in Fig. 1.11 is unstable, because the inverted pendulum will have divergent output namely falling down when it is given no input—Readers may argue that there is a special initial condition namely the vertical position for the inverted pendulum to keep stationary. However, this special initial condition is too ideal, whereas *ad hoc* disturbance always exists in reality, which causes the inverted pendulum to deviate from the special initial condition of being absolutely vertical and finally fall down. Even supposing that no *ad hoc* disturbance exists, the inverted pendulum still cannot be located absolutely at the vertical position in reality according to the principle of *quantum physics*.

From the inverted pendulum example, we can see that a stable control system should have convergent output when it is given no input and with whatever initial conditions. In other words, for a stable control system, influence of whatever initial

[1] For a closed-loop feedback control system in practical applications, the input is usually certain expected process output.

H. Li, *Control Theory for Practical Applications*,
https://doi.org/10.1007/978-981-97-5008-5_4

conditions fades away. We say that a control system **converges** or is **convergent** if it has convergent output. Then **control system stability** refers to *the property of a control system that it converges or is convergent when it is given no input, regardless of its initial conditions.*

> **Control system stability**: *The property of a control system that it converges or is convergent when it is given no input, regardless of its initial conditions.*

4.1 Intuitive Conception of Control System Stability

To facilitate conception of control system stability, consider the concrete example of rotating disk position control presented in Sect. 2.5 in Chap. 2. The rotating disk position closed-loop feedback control system is illustrated in Fig. 4.1, where the friction coefficient is supposed to be zero. The friction coefficient being zero implies the ideal condition that the rotating disk is exempt from friction. The rotating disk position closed-loop feedback control system can be treated as a black box from customers' perspective. The input to the rotating disk position control black box is certain expected rotating position and the output is the rotating position.

Suppose that the proportional feedback control method described in (2.5) is adopted inside the rotating disk position control black box. We have already demonstrated in Sect. 2.5 in Chap. 2 that the rotating disk position control black box has severe oscillatory output when it is given the input $p_E = 1$. Such undesirable control performance is obvious to reflect that the rotating disk position control black box have problems. To further highlight that the problems are inherent to the rotating disk position control black box itself and have no relation at all with the input p_E, we set the input to zero (as if customers give no command to it) and observe its performance purely caused by its initial conditions.

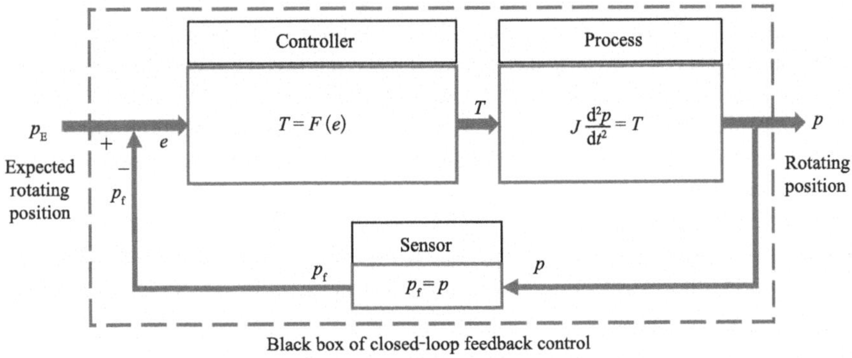

Fig. 4.1 Black box of rotating disk position closed-loop feedback control

Matlab code for simulation of rotating disk position proportional feedback control without external input p_E is given as follows. Suppose that the rotating disk initial condition is $p(0) = 0$ and $p^{(1)}(0) \equiv \frac{d}{dt}p(t)|_{t=0} = 4$.

RotatingDiskPositionPControlUnderInitialConditions.m

```
%% Rotating disk position control system parameters
J = 10; % Rotating disk load inertia
b = 0; % No friction
Tmax = 50; % Maximum torque

%% Simulation preliminary configuration
dt = 0.001; % Numerical computation step
tSpan = 0:dt:10; % Simulation time span
% Record rotating disk positions during simulation
pAll = zeros(size(tSpan)); k = 1;
pE = 0; % No expected rotating disk position
p = 0; v = 4; % Initial rotating disk position and speed

%% Simulation of rotating disk position control
for t = tSpan
    %% Control method
    e = pE-p; % Feedback error
    T = 1000*e; % Proportional feedback control
    T = max(min(T, Tmax), -Tmax); % Control law constraint

    %% Rotating disk dynamics
    dv = (T-b*v)/J;
    v = v + dv*dt;
    p = p + v*dt;
    pAll(k) = p; k = k+1;
end
plot(tSpan, pAll, 'LineWidth', 2); ylim([-2, 2]);
xlabel('Time'); ylabel('Rotating Disk Position'); grid on;
```

The simulation result is demonstrated in Fig. 4.2, where the horizontal axis represents the simulation time span and the vertical axis represents the rotating position during simulation. The rotating disk position control system has severe oscillatory rotating position output as well, even when no external input p_E is given to it. This tricky phenomenon reflects that the rotating disk position control system must itself have inherent problems if the proportional feedback control method described in (2.5) is adopted. In other words, the rotating disk position proportional feedback control system itself is unstable, regardless of its external input.

On the other hand, suppose that the proportional-derivative feedback control method described in (2.6) is adopted inside the rotating disk position control black box. Matlab code for simulation of rotating disk position proportional-derivative feedback control without external input p_E is given as follows. Suppose that the rotating disk initial condition is the same to previous configuration.

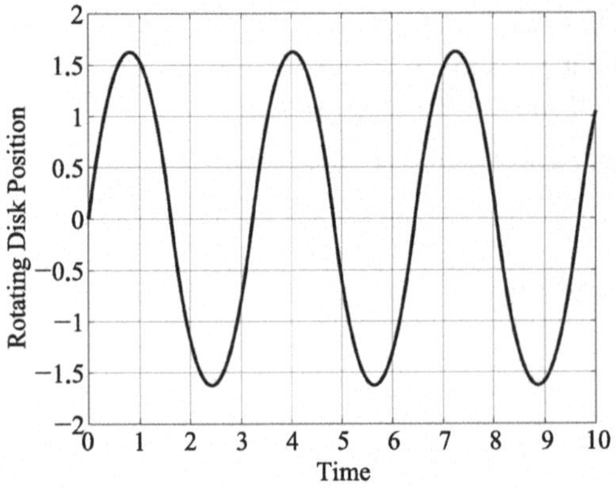

Fig. 4.2 Rotating disk position proportional feedback control (influence of the initial condition $p(0^-) = 0$ and $p^{(1)}(0^-) = 4$)

RotatingDiskPositionPDControlUnderInitialConditions.m

```
%% Rotating disk position control system parameters
J = 10; % Rotating disk load inertia
b = 0; % No friction
Tmax = 50; % Maximum torque

%% Simulation preliminary configuration
dt = 0.001; % Numerical computation step
tSpan = 0:dt:10; % Simulation time span
% Record rotating disk positions during simulation
pAll = zeros(size(tSpan)); k = 1;
pE = 0; % No expected rotating disk position
p = 0; v = 4; % Initial rotating disk position and speed

%% Simulation of rotating disk position control
for t = tSpan
    %% Control method
    e = pE-p; % Feedback error
    de = -v; % Feedback error derivative
    T = 1000*e + 200*de; % Proportional-derivative feedback
    control
    T = max(min(T, Tmax), -Tmax); % Control law constraint

    %% Rotating disk dynamics
    dv = (T-b*v)/J;
    v = v + dv*dt;
    p = p + v*dt;
    pAll(k) = p; k = k+1;
end
plot(tSpan, pAll, 'LineWidth', 2); ylim([-2, 2]);
xlabel('Time'); ylabel('Rotating Disk Position'); grid on;
```

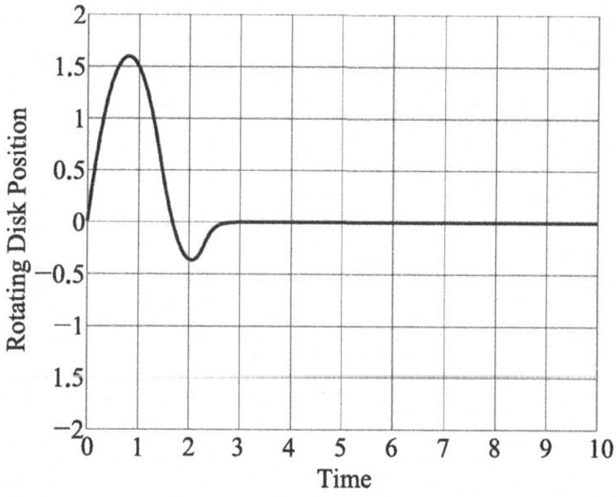

Fig. 4.3 Rotating disk position proportional-derivative feedback control (influence of the initial condition $p(0^{-}) = 0$ and $p^{(1)}(0^{-}) = 4$)

The simulation result is demonstrated in Fig. 4.3. The rotating disk position control system has convergent rotating position output. Readers can try more rounds of simulation with different rotating disk initial conditions and will find that the rotating disk position control system always has convergent output. In other words, the rotating disk position proportional-derivative feedback control system itself is convergent, regardless of its initial conditions. So it is stable.

Above simulation experiments with the two rotating disk position control system examples can facilitate intuitive conception of control system stability. Simulation experiments can also reveal control system unstability such as unstability of the rotating disk position proportional feedback control system. However, simulation experiments and even field experiments *cannot prove* control system stability, just like a billion of orthogonal triangles cannot prove the Gou-Gu theorem or Pythagoras theorem that an orthogonal triangle's hypotenuse square equals the sum of its two side squares. To study whether and **why** a control system is essentially stable or unstable, we resort to theoretical analysis according to control theory.

4.2 Theoretical Analysis of Control System Stability

4.2.1 Theoretical Analysis of Closed-Loop Feedback Control Examples

For theoretical analysis of the rotating disk position proportional feedback control system and the rotating disk position proportional-derivative feedback control system, we resort to the methodology of system modelling presented in Chap. 3. The rotating disk position proportional feedback control system (where the friction coefficient is ideally supposed to be zero) can be modelled by a transfer function block diagram illustrated in Fig. 4.4.

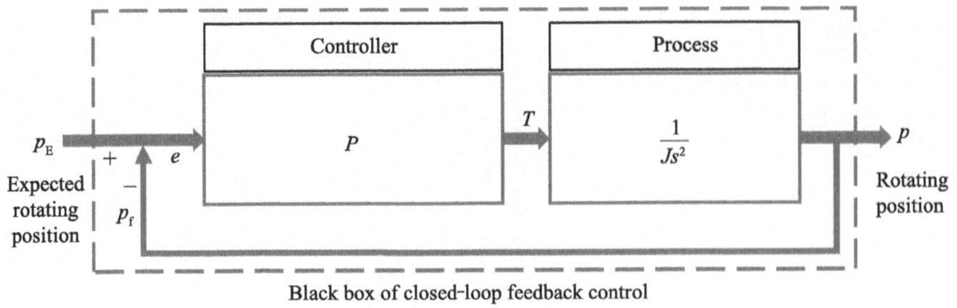

Fig. 4.4 Transfer function block diagram of rotating disk position proportional feedback control

By rules of transfer function block synthesis introduced in Sect. 3.4 in Chap. 3, we can compute the holistic transfer function of the rotating disk position proportional feedback control system as

$$\frac{p(s)}{p_E(s)} = \frac{P\frac{1}{Js^2}}{1 + P\frac{1}{Js^2}} = \frac{P}{Js^2 + P}, \tag{4.1}$$

from which we can re-establish the holistic dynamics relationship between the control system input p_E and the control system output p by a differential equation as

$$J\frac{d^2p}{dt^2} + Pp = Pp_E. \tag{4.2}$$

Perform the Laplace transform on both sides of (4.2) with initial conditions taken into account and obtain

$$J(s^2p(s) - sp(0^-) - p^{(1)}(0^-)) + Pp(s) = Pp_E(s).$$

So

$$p(s) = \frac{P}{Js^2 + P}p_E(s) + \frac{J(sp(0^-) + p^{(1)}(0^-))}{Js^2 + P}.$$

If there is no control system input, namely $p_E = 0$, then we have

$$p(s) = \frac{sp(0^-) + p^{(1)}(0^-)}{s^2 + P/J}, \tag{4.3}$$

which reflects dynamics of the control system purely caused by its initial conditions.

Adopt the same control system parameters and initial conditions as in simulation presented in Sect. 4.1, namely $J = 10$, $P = 1000$, $p(0^-) = 0$, and $p^{(1)}(0^-) = 4$. Substitute them into (4.3) and obtain

$$p(s) = \frac{4}{s^2 + 100},$$
$$p(t) = 0.4\sin(10t), \quad \text{(check Table 3.1)} \tag{4.4}$$

which implies theoretically the rotating position output p is indeed oscillatory, consistent with the simulation result demonstrated in Fig. 4.2. We can put forward theoretical analysis by considering generic control system configurations and derive (4.5) from (4.3):

$$p(t) = p(0^-)\cos\left(\sqrt{\frac{P}{J}}t\right) + \frac{p^{(1)}(0^-)}{\sqrt{P/J}}\sin\left(\sqrt{\frac{P}{J}}t\right), \tag{4.5}$$
$$\text{(check Table 3.1)}$$

which implies theoretically the rotating disk position proportional feedback control system *definitely* has oscillatory output when its initial condition is non-zero, with the oscillatory amplitude determined by the initial condition. Above theoretical analysis *proves* that the rotating disk position proportional feedback control system is unstable, regardless of concrete values of the rotating disk load inertia J and the proportional coefficient P.

In practical applications, such theoretical analysis can help us avoid blind trials with the rotating disk position proportional feedback control system which has inherent problems. Generally, theoretical analysis based on control theory can help us avoid blind trials with control systems that have inherent problems. This is right an important aspect of power of control theory.

> **Control theory power I**: *Theoretical analysis based on control theory can help us avoid blind trials with control systems that have inherent problems.*

Readers may be confused by a small detail: theoretical analysis described in (4.4) coveys that the oscillatory amplitude of the control system output is 0.4, whereas the simulation result demonstrated in Fig. 4.2 conveys that the oscillatory amplitude of the control system output is a bit more than 1.5. Why does such considerable discrepancy exist?

The reason is as follows: In simulation, we set a maximum allowable controller output torque $T_{\max}$ according to field practices (note that no controller output torque can be unlimited). So the proportional controller is truncated by the maximum controller output and is not ideally linear. However, according to system modelling via Laplace transform, each system component including the controller is assumed ideally linear, which implies the controller output torque T can be larger than $T_{\max}$ (or smaller than $-T_{\max}$) when the feedback error e is with large enough absolute value. In other words, the controller in simulation is not as ideal as that in system modelling via Laplace transform, so the control system performance in simulation is even worse than that derived by theoretical analysis.

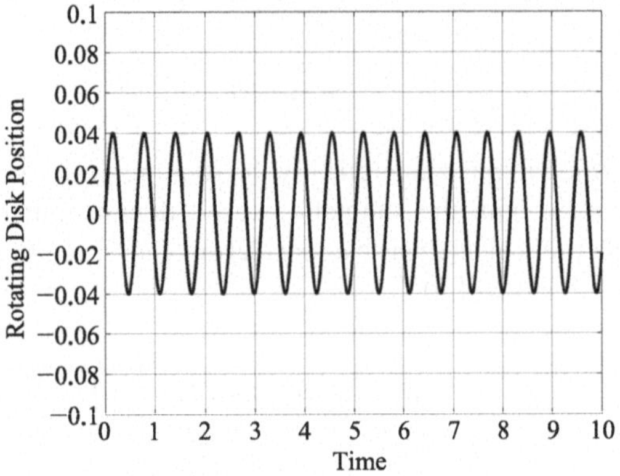

Fig. 4.5 Rotating disk position proportional feedback control (influence of the initial condition $p(0^-) = 0$ and $p^{(1)}(0^-) = 0.4$)

If the absolute value of $p^{(1)}(0^-)$ is reduced, the oscillatory amplitude of the control system output is reduced accordingly and the controller output torque also becomes unlikely to exceed the maximum controller output torque limit. So the controller becomes closer to an ideal linear system component and simulation results would be closer to results derived by theoretical analysis. For example, we keep all control system configurations except the initial condition $p^{(1)}(0^-)$ which is reduced from 4 to 0.4. On one hand, we obtain $p(t) = 0.04\sin(10t)$ theoretically according to (4.5). On the other hand, we obtain the simulation result demonstrated in Fig. 4.5, which is consistent with the theoretical result—Not only the oscillatory amplitude but also the oscillatory frequency are consistent with the theoretical result.

The fact that the controller is constrained by the maximum controller output torque limit and hence is not ideally linear does not imply that the methodology of system modelling via Laplace transform is invalid. In fact, system modelling via Laplace transform is completely valid for the most crucial operation range namely the operation range about the stable operation point or equilibrium point (In the example of rotating disk position control, the stable operation point is $p = p_E$ with $e = 0$). Besides, even for other operation ranges where system component linearity is violated, system modelling via Laplace transform may still help us infer certain control performance tendency.

On the other hand, we have to admit that we should by no means exaggerate validity of system modelling via Laplace transform, especially when the control system is severely nonlinear itself or in operation conditions where its linearity is severely violated. Nonlinear control system examples will be encountered in this book.

Above discussions stemming from discrepancy between the theoretical result described in (4.4) and the simulation result demonstrated in Fig. 4.2 may make readers feel confused and at a loss about validity of system modelling via Laplace transform. It seems that system modelling via Laplace transform is both valid and invalid in practical applications and there is no conclusion about its validity. In fact, not only system modelling via Laplace transform, but also any aspect of control theory

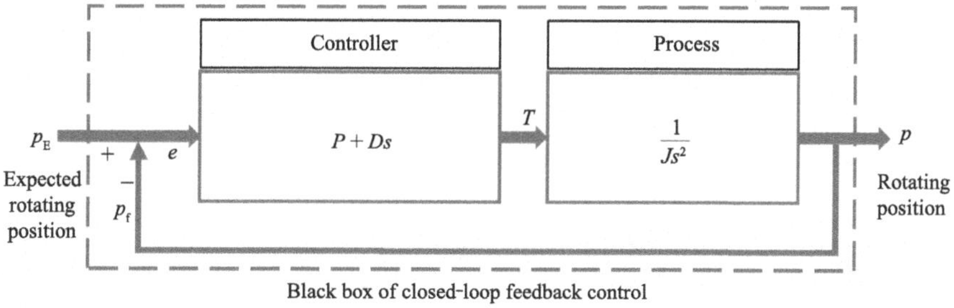

Fig. 4.6 Transfer function block diagram of rotating disk position proportional-derivative feedback control

and even any aspect of engineering-oriented mathematics-style knowledge *necessitates dialectic conception and understanding of the relationship between theory and practical applications and necessitates flexible treatment of theory and practical applications holistically*. This is an essential ability and quality of high-level engineers.

We return from complementary discussions to theoretical analysis of the second control system example presented in Sect. 4.1. The rotating disk position proportional-derivative feedback control system (where the friction coefficient is ideally supposed to be zero) can be modelled by a transfer function block diagram illustrated in Fig. 4.6.

Similar to theoretical analysis of the rotating disk position proportional feedback control system, we can compute the holistic transfer function of the rotating disk position proportional-derivative feedback control system as

$$\frac{p(s)}{p_{\mathrm{E}}(s)} = \frac{(P+Ds)\frac{1}{Js^2}}{1+(P+Ds)\frac{1}{Js^2}} = \frac{Ds+P}{Js^2+Ds+P}, \tag{4.6}$$

from which we can re-establish the holistic dynamics relationship between the control system input p_{E} and the control system output p by a differential equation as

$$J\frac{\mathrm{d}^2 p}{\mathrm{d}t^2} + D\frac{\mathrm{d}p}{\mathrm{d}t} + Pp = D\frac{\mathrm{d}p_{\mathrm{E}}}{\mathrm{d}t} + Pp_{\mathrm{E}}. \tag{4.7}$$

Perform the Laplace transform on both sides of (4.7) with initial conditions taken into account and obtain

$$\begin{aligned} &J(s^2p(s) - sp(0^-) - p^{(1)}(0^-)) + D(sp(s) - p(0^-)) + Pp(s) \\ &= D(sp_{\mathrm{E}}(s) - p_{\mathrm{E}}(0^-)) + Pp_{\mathrm{E}}(s). \end{aligned}$$

So

$$p(s) = \frac{Ds + P}{Js^2 + Ds + P} p_{\mathrm{E}}(s) + \frac{(Js + D)p(0^-) + Jp^{(1)}(0^-) - Dp_{\mathrm{E}}(0^-)}{Js^2 + Ds + P}.$$

If there is no control system input, namely $p_{\mathrm{E}} = 0$, then we have

$$p(s) = \frac{(Js + D)p(0^-) + Jp^{(1)}(0^-)}{Js^2 + Ds + P}, \tag{4.8}$$

which reflects dynamics of the control system purely caused by its initial conditions.

Adopt the same control system parameters and initial conditions as in simulation presented in Sect. 4.1, namely $J = 10$, $P = 1\,000$, $D = 200$, $p(0^-) = 0$, and $p^{(1)}(0^-) = 4$. Substitute them into (4.8) and obtain

$$p(s) = \frac{4}{s^2 + 20\,s + 100} = \frac{4}{(s + 10)^2},$$
$$p(t) = 4\mathrm{e}^{-10t}t, \quad \text{(check Table 3.1)} \tag{4.9}$$

which implies theoretically that the rotating position output p is indeed convergent (note that $\mathrm{e}^{-10t}t \to 0$ as $t \to \infty$), consistent with the simulation result demonstrated in Fig. 4.3. There is small discrepancy between the theoretical result described in (4.9) and the simulation result demonstrated in Fig. 4.3—We believe readers now can understand why such discrepancy exists—If we keep all control system configurations except the initial condition $p^{(1)}(0^-)$ which is reduced from 4 to 0.4, then on one hand, we obtain $p(t) = 0.4\mathrm{e}^{-10t}t$ theoretically. On the other hand, we obtain the simulation result demonstrated in Fig. 4.7, which is consistent with the theoretical result—We also believe readers now can understand why the discrepancy between the theoretical result and the simulation result diminishes once the initial condition $p^{(1)}(0^-)$ is reduced from 4 to 0.4.

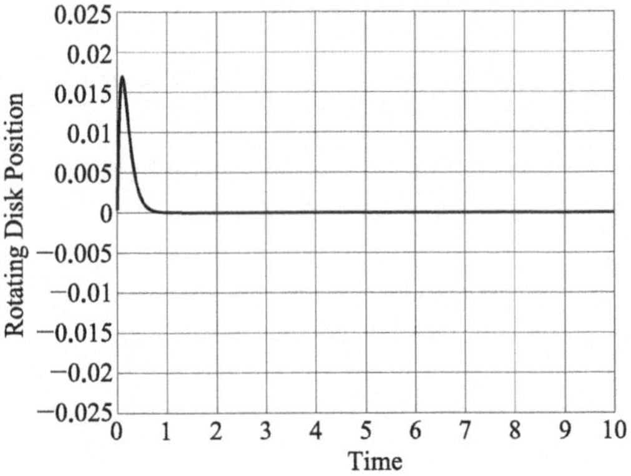

Fig. 4.7 Rotating disk position proportional-derivative feedback control (influence of the initial condition $p(0^-) = 0$ and $p^{(1)}(0^-) = 0.4$)

For a generic initial condition $p(0^-)$ and $p^{(1)}(0^-)$, from (4.8) we have (still with control system parameters $J = 10$, $P = 1\,000$, and $D = 200$)

$$\begin{aligned} p(t) &= L^{-1}\left\{\frac{(s+20)p(0^-)+p^{(1)}(0^-)}{s^2+20s+100}\right\} \\ &= L^{-1}\left\{\frac{p(0^-)}{s+10}\right\}+L^{-1}\left\{\frac{10p(0^-)+p^{(1)}(0^-)}{(s+10)^2}\right\} \\ &= p(0^-)e^{-10t}+(10p(0^-)+p^{(1)}(0^-))e^{-10t}t, \end{aligned} \tag{4.10}$$

which implies theoretically that the rotating disk position proportional-derivative feedback control system converges when it is with no matter what initial condition. In other words, above theoretical analysis *proves* that the rotating disk position proportional-derivative feedback control system is essentially stable, regardless of its initial condition. *To prove* is an important *guarantee* that cannot be achieved purely by experiments, be it simulation or field experiments, no matter how many rounds of experiments under various conditions can be done. In practical applications, such theoretical analysis can help us know whether the rotating disk position proportional-derivative feedback control system is guaranteed to be stable and *why* it is essentially stable. Generally, theoretical analysis based on control theory can help us know whether and why a control system essentially possesses certain property. This is right another important aspect of power of control theory.

Control theory power II: *Theoretical analysis based on control theory can help us know whether and why a control system essentially possesses certain property.*

4.2.2 Control System Characteristic Polynomial and Poles for Stability Analysis

We can extend theoretical analysis of concrete control systems presented previously to theoretical analysis of a generic control system that follows the *linear time-invariant* assumption explained in Sect. 3.3 in Chap. 3. Suppose that the control system's holistic transfer function after transfer function block synthesis is denoted generically as

$$\frac{y(s)}{r(s)} = \frac{b_m s^m + b_{m-1}s^{m-1} + \cdots + b_0}{a_n s^n + a_{n-1}s^{n-1} + \cdots + a_0}, \tag{4.11}$$

from which we can re-establish the holistic dynamics relationship between the control system input r and the control system output y by a generic differential equation as

$$a_n \frac{\mathrm{d}^n}{\mathrm{d}t^n} y + a_{n-1} \frac{\mathrm{d}^{n-1}}{\mathrm{d}t^{n-1}} y + \cdots + a_0 y = b_m \frac{\mathrm{d}^m}{\mathrm{d}t^m} r + b_{m-1} \frac{\mathrm{d}^{m-1}}{\mathrm{d}t^{m-1}} r + \cdots + b_0 r. \tag{4.12}$$

Perform the Laplace transform on both sides of (4.12) with initial conditions taken into account and obtain

$$\begin{aligned}&(a_n s^n + a_{n-1} s^{n-1} + \cdots + a_0) y(s) - A_0(s) \\ &= (b_m s^m + b_{m-1} s^{m-1} + \cdots + b_0) r(s) - B_0(s).\end{aligned}$$

So

$$y(s) = \frac{b_m s^m + b_{m-1} s^{m-1} + \cdots + b_0}{a_n s^n + a_{n-1} s^{n-1} + \cdots + a_0} r(s) + \frac{A_0(s) - B_0(s)}{a_n s^n + a_{n-1} s^{n-1} + \cdots + a_0}, \tag{4.13}$$

where $A_0(s)$ and $B_0(s)$ are constant polynomials determined by initial conditions of the output y and the input r respectively. If there is no control system input, namely $r = 0$, then we have

$$y_i(s) = \frac{A_0(s)}{a_n s^n + a_{n-1} s^{n-1} + \cdots + a_0}, \tag{4.14}$$

which reflects dynamics of the control system purely caused by its initial conditions.

The control system output purely caused by initial conditions is called the **initial response** of the control system, and y_i described in (4.14) is the s-domain function of the control system initial response, from which the original time-domain function of the control system initial response can be recovered.

The numerator $A_0(s)$ of the s-domain function $y_i(s)$ described in (4.14) is *ad hoc* determined by concrete initial conditions of the control system, whereas the denominator of $y_i(s)$ is uniquely determined once the control system is given. The denominator of $y_i(s)$ is the same to the denominator of the control system's holistic transfer function described in (4.11), namely

$$a_n s^n + a_{n-1} s^{n-1} + \cdots + a_0, \tag{4.15}$$

which is called the **characteristic polynomial** of the control system. The following equation

$$a_n s^n + a_{n-1} s^{n-1} + \cdots + a_0 = 0 \tag{4.16}$$

is called the **characteristic equation** of the control system. The roots of the characteristic polynomial described in (4.15) or the characteristic equation described in (4.16) are called **poles** of the control system.

Suppose that the control system's characteristic polynomial described in (4.15) is factored as

$$a_n s^n + a_{n-1} s^{n-1} + \cdots + a_0 = a_n (s + \alpha_1)^{m_1} (s + \alpha_2)^{m_2} \cdots (s + \alpha_N)^{m_N},$$

where $-\alpha_1, -\alpha_2, \ldots, -\alpha_N$ denote control system poles and $m_1, m_2, \ldots, m_N$ denote their numbers respectively.

In the spirit of partial fraction decomposition presented in Sect. 3.2.3 in Chap. 3, the s-domain function $y_i(s)$ described in (4.14) can be decomposed into the sum of partial fractions as

$$y_i(s) = \frac{A_0(s)}{a_n (s + \alpha_1)^{m_1} (s + \alpha_2)^{m_2} \cdots (s + \alpha_N)^{m_N}} = \sum_{j=1}^{N} \sum_{k=1}^{m_j} \frac{c_{j,k}}{(s + \alpha_j)^k}, \tag{4.17}$$

where the generically-denoted coefficients $c_{j,k}$ ($j \in \{1, \ldots, N\}$, $k \in \{1, \ldots, m_j\}$) can be *ad hoc* determined once concrete initial conditions of the control system are given.

From (4.17), the original time-domain function of $y_i(s)$ can be recovered as

$$y_i = \sum_{j=1}^{N} \sum_{k=1}^{m_j} c_{j,k} \mathrm{e}^{-\alpha_j t} \frac{t^{k-1}}{(k-1)!} \quad \text{(check Table 3.1)}, \tag{4.18}$$

where each term $\mathrm{e}^{-\alpha_j t} t^{k-1}$ converges to zero if and only if the real part of each $-\alpha_j$ is negative. If the real parts of all the poles are negative, namely $\mathrm{Re}(-\alpha_1), \mathrm{Re}(-\alpha_2), \ldots, \mathrm{Re}(-\alpha_N)$ are negative, then no matter for what initial conditions of the control system, y_i will always converge to zero. In other words, the control system always converges if the real parts of all its poles are negative, regardless of its initial conditions. On the other hand, if the real part of at least one pole is non-negative, then the control system definitely has oscillatory or divergent output caused by certain initial conditions. Therefore, we can conclude that *a linear time-invariant control system is stable if and only if the real parts of all its poles are negative.*

Control system stability criterion: *A linear time-invariant control system is stable if and only if the real parts of all its poles are negative.*

4.3 Routh-Hurwitz Criterion

4.3.1 *Parametrized Characteristic Polynomial*

If a linear time-invariant control system's characteristic polynomial is determined, then we may resort to analytical methods or numerical methods to compute its poles and hence know whether it is stable or not. However, in practical applications, we often encounter *parametrized characteristic polynomials* in control system design. Take the rotating disk position proportional-derivative feedback control system presented in Sect. 4.2 as an example. Let the proportional-derivative controller be modelled generically by $P + Ds$ with the proportional coefficient P and the derivative coefficient D. The characteristic polynomial is given as

$$C_{PD}^{RDP}(s) = Js^2 + Ds + P, \tag{4.19}$$

which is a parametrized characteristic polynomial. More specifically, $C_{PD}^{RDP}(s)$ is with parametrized coefficients namely P and D for the constant term and the first-order term respectively.

For control system design, we need to determine the range of controller parameters P and D for the control system to be stable or determine whether such range of P and D ever exists. In other words, we need to determine the range of P and D such that all poles of the control system, namely all roots of the characteristic polynomial described in (4.19), are with negative real part or whether such requirement can ever be satisfied. The roots of the characteristic polynomial described in (4.19) are expressed by its parametrized coefficients as

$$s_{1,2} = \frac{-D \pm \sqrt{D^2 - 4JP}}{2J}.$$

It is easy to verify that $P > 0$ and $D > 0$ form the necessary and sufficient condition for the rotating disk position proportional-derivative feedback control system to be stable (note that the rotating disk load J is always positive in real applications).

However, many control system characteristic polynomials encountered in practical applications are not as simple as the second-order polynomial described in (4.19). For example, consider a third-order parametrized characteristic polynomial as

$$C_{3K}(s) = s^3 + 3s^2 + 3Ks + 1, \tag{4.20}$$

where the coefficient for the first-order term is a parametrized coefficient $3K$. The characteristic polynomial $C_{3K}(s)$ is the characteristic polynomial of a control system with the controller parameter K. A polynomial with real-value coefficients that has roots all with negative real part is obviously a polynomial with positive real-value coefficients, so $K > 0$ is a necessary condition for $C_{3K}(s)$ to have roots all with negative real part.

For control system stability, we need to determine the range of K within $(0, \infty)$ such that the characteristic polynomial $C_{3K}(s)$ has roots all with negative real part. Similar to how we handle the second-order parametrized characteristic polynomial described in (4.19), we solve the third-order characteristic equation $C_{3K}(s) = 0$ and express its roots by parametrized coefficients in terms of K. To solve a third-order polynomial equation analytically, we need more skills than solving a second-order polynomial equation. Denote $s + 1 = u + v$, substitute it into (4.20) and obtain

$$
\begin{aligned}
&(s+1)^3 + (3K-3)(s+1) + 3 - 3K = 0 \\
\Longleftrightarrow\ &(u+v)^3 + (3K-3)(u+v) + 3 - 3K = 0 \\
\Longleftrightarrow\ &u^3 + v^3 + 3 - 3K + (3uv + 3K - 3)(u+v) = 0.
\end{aligned}
$$

Let $u^3 + v^3 + 3 - 3K = 0$ and $3uv + 3K - 3 = 0$ (so $uv = 1 - K$ and $u^3v^3 = (1-K)^3$), from which we know that u^3 and v^3 are the two roots of the second-order polynomial $x^2 + (3-3K)x + (1-K)^3$ and hence are solved as

$$
\begin{aligned}
u^3 &= \frac{3(K-1) + \sqrt{5+4K}|K-1|}{2}, \\
v^3 &= \frac{3(K-1) - \sqrt{5+4K}|K-1|}{2}.
\end{aligned}
$$

So

$$
\begin{aligned}
u &= \sqrt[3]{\frac{3(K-1) + \sqrt{5+4K}|K-1|}{2}}, \\
v &= \sqrt[3]{\frac{3(K-1) - \sqrt{5+4K}|K-1|}{2}},
\end{aligned}
$$

where both u and v are real values as $K > 0$.

Since $uv = 1 - K$ must be satisfied, the three roots of $C_{3K}(s)$ described in (4.20) are respectively

$$
\begin{aligned}
s_1 &= -1 + u + v \\
&= -1 + \sqrt[3]{\frac{3(K-1) + \sqrt{5+4K}|K-1|}{2}} + \sqrt[3]{\frac{3(K-1) - \sqrt{5+4K}|K-1|}{2}}, \\
s_2 &= -1 + u\omega + v\omega^2 \\
&= -1 + \sqrt[3]{\frac{3(K-1) + \sqrt{5+4K}|K-1|}{2}}\omega + \sqrt[3]{\frac{3(K-1) - \sqrt{5+4K}|K-1|}{2}}\omega^2, \\
s_3 &= -1 + u\omega^2 + v\omega
\end{aligned}
\tag{4.21}
$$

$$= -1 + \sqrt[3]{\frac{3(K-1) + \sqrt{5+4K}|K-1|}{2}}\omega^2 + \sqrt[3]{\frac{3(K-1) - \sqrt{5+4K}|K-1|}{2}}\omega,$$

where $\omega = e^{2\pi j/3} = \frac{-1+\sqrt{3}j}{2}$. Since both u and v are real values, the real parts of the three roots s_1, s_2, s_3 are respectively

$$\begin{aligned}
\mathrm{Re}(s_1) &= -1 + u + v \\
&= -1 + \sqrt[3]{\frac{3(K-1) + \sqrt{5+4K}|K-1|}{2}} + \sqrt[3]{\frac{3(K-1) - \sqrt{5+4K}|K-1|}{2}}, \\
\mathrm{Re}(s_{2,3}) &= -1 - \frac{1}{2}(u + v) \qquad (4.22)\\
&= -1 - \frac{1}{2}\left(\sqrt[3]{\frac{3(K-1) + \sqrt{5+4K}|K-1|}{2}} + \sqrt[3]{\frac{3(K-1) - \sqrt{5+4K}|K-1|}{2}}\right).
\end{aligned}$$

If $K \geqslant 1$, we have

$$\begin{aligned}
u &= \sqrt[3]{\frac{3+\sqrt{5+4K}}{2}}\sqrt[3]{K-1} \equiv \alpha\sqrt[3]{K-1}, \\
v &= -\sqrt[3]{\frac{\sqrt{5+4K}-3}{2}}\sqrt[3]{K-1} \equiv \beta\sqrt[3]{K-1}, \\
u + v &= (\alpha - \beta)\sqrt[3]{K-1} > 0, \\
&\left(\alpha \equiv \sqrt[3]{\frac{3+\sqrt{5+4K}}{2}} > \beta \equiv \sqrt[3]{\frac{\sqrt{5+4K}-3}{2}}\right)
\end{aligned}$$

and

$$\begin{aligned}
\mathrm{Re}(s_{2,3}) &= -1 - \frac{1}{2}(u + v) < 0, \\
\mathrm{Re}(s_1) &= -1 + (\alpha - \beta)\sqrt[3]{K-1} \\
&= -1 + \frac{3\sqrt[3]{K-1}}{\alpha^2 + \alpha\beta + \beta^2} \quad (\alpha^3 - \beta^3 = 3) \\
&< -1 + \frac{3\sqrt[3]{K-1}}{3\alpha\beta} \quad (\alpha^2 + \beta^2 > 2\alpha\beta > 0) \\
&= -1 + \frac{\sqrt[3]{K-1}}{\sqrt[3]{K-1}} = 0.
\end{aligned}$$

Therefore, the characteristic polynomial $C_{3K}(s)$ has roots all with negative real part if $K \geqslant 1$.

If $0 < K < 1$, we have

$$u = -\sqrt[3]{\frac{3-\sqrt{5+4K}}{2}}\sqrt[3]{1-K} \equiv -\gamma\sqrt[3]{1-K} < 0,$$

$$v = -\sqrt[3]{\frac{3+\sqrt{5+4K}}{2}}\sqrt[3]{1-K} \equiv -\alpha\sqrt[3]{1-K} < 0,$$

and

$$\mathrm{Re}(s_1) = -1 + u + v < 0,$$

$$\mathrm{Re}(s_{2,3}) = -1 - \frac{1}{2}(u+v) = -1 + \frac{1}{2}(\gamma+\alpha)\sqrt[3]{1-K}.$$

Note that

$$\frac{\mathrm{d}}{\mathrm{d}K}(\gamma+\alpha) = -\frac{1}{3\sqrt{5+4K}}\left(\frac{1}{\gamma^2} - \frac{1}{\alpha^2}\right) < 0,$$

$$\frac{\mathrm{d}}{\mathrm{d}K}\sqrt[3]{1-K} = -\frac{1}{3(\sqrt[3]{1-K})^2} < 0,$$

so we have

$$\frac{\mathrm{d}}{\mathrm{d}K}\mathrm{Re}(s_{2,3}) = \frac{1}{2}\left[\sqrt[3]{1-K}\frac{\mathrm{d}}{\mathrm{d}K}(\gamma+\alpha) + (\gamma+\alpha)\frac{\mathrm{d}}{\mathrm{d}K}\sqrt[3]{1-K}\right] < 0.$$

Also note that $\mathrm{Re}(s_{2,3})|_{K=1/9} = 0$ (the solution $K = 1/9$ can be gradually located by binary search Bonnans et al. 2006), so we have $\mathrm{Re}(s_{2,3}) \geqslant 0$ for $0 < K \leqslant 1/9$ whereas $\mathrm{Re}(s_{2,3}) < 0$ for $1/9 < K < 1$.

Finally, for the third-order parametrized characteristic polynomial $C_{3K}(s)$ described in (4.20), we have succeeded in determining the range of its parameter K for it to have roots all with negative real part, namely $K > 1/9$. Above method of determining the parameter range can be summarized as follows: first solve the parametrized characteristic polynomial equation analytically and then check the analytical roots directly to determine the range of parameters for the parametrized characteristic polynomial to have roots all with negative real part or determine whether such parameter range ever exists. We may call such method *analytical root direct checking method*.

Although the analytical root direct checking method succeeds in determining the range of K for $C_{3K}(s)$ to have roots all with negative real part, readers may begin to feel difficulty in applying the method. For fourth-order parametrized characteristic polynomials, it is even more difficult to apply the method. For a fifth-order or above fifth-order parametrized characteristic polynomial, the analytical root direct checking method becomes *infeasible*, because we cannot obtain generic analytical solutions of its roots in terms of its parametrized coefficients. Readers need knowledge of modern algebra to understand this point. A representative book worth recommendation is Lang's *Algebra* (Lang 2002).

4.3.2 Locating Poles Qualitatively According to Characteristic Polynomial Coefficients

For a generic parametrized characteristic polynomial, the analytical root direct checking method is infeasible, the reason for which has just been mentioned above. However, we may reflect on a basic question: *for a generic polynomial, is it necessary to solve its roots first before determining whether its roots are all with negative real part*?

Fortunately, the answer to above basic question is *no*. In fact, we can locate a generic polynomial's roots qualitatively according to its coefficients. More specifically, we can locate according to its coefficients how many roots are in the half plane on the right side of the imaginary axis, on the imaginary axis, or in the half plane on the left side of the imaginary axis.[2] So for a linear time-invariant control system, we can locate its poles qualitatively according to its characteristic polynomial (or parametrized characteristic polynomial) coefficients.

A representative method of locating a generic polynomial's roots qualitatively according to the polynomial coefficients is the **Routh-Hurwitz criterion** method which was developed independently by E. Routh and A. Hurwitz in the late nineteenth century (Dorf and Bishop 2008). The Routh-Hurwitz criterion method is based on checking an array called **Routh array** that is completed iteratively from the ordered coefficients of the characteristic polynomial. Consider a generic characteristic polynomial (with $a_n > 0$ or even $a_n = 1$):

$$a_n s^n + a_{n-1} s^{n-1} + \cdots + a_0. \tag{4.23}$$

Order its coefficients into the first two rows of the following Routh array:

[2] Studies on locating polynomial roots qualitatively can date back to Hermite's works in 1850s, in his paper *Sur le nombre des racines d'une équation algébrique comprise entre des limites données*, originally published on *Journal de Crelle* and also collected in *Oeuvres de Charles Hermite* (Hermite 2009).

$$\begin{array}{c|llll} s^n & a_n & a_{n-2} & a_{n-4} & \cdots \\ s^{n-1} & a_{n-1} & a_{n-3} & a_{n-5} & \cdots \\ s^{n-2} & b_{n-2} & b_{n-4} & b_{n-6} & \cdots \\ s^{n-3} & c_{n-3} & c_{n-5} & c_{n-7} & \cdots \\ \vdots & \vdots & \vdots & \vdots & \\ s^0 & z_0 & & & \end{array}$$

where rows after the second row, namely rows initiated by $s^{n-2}, s^{n-3}, \ldots, s^0$, are computed iteratively as

$$b_{n-2} = -\frac{1}{a_{n-1}} \begin{vmatrix} a_n & a_{n-2} \\ a_{n-1} & a_{n-3} \end{vmatrix} = \frac{a_{n-1}a_{n-2} - a_n a_{n-3}}{a_{n-1}},$$

$$b_{n-4} = -\frac{1}{a_{n-1}} \begin{vmatrix} a_n & a_{n-4} \\ a_{n-1} & a_{n-5} \end{vmatrix}, \; b_{n-6} = -\frac{1}{a_{n-1}} \begin{vmatrix} a_n & a_{n-6} \\ a_{n-1} & a_{n-7} \end{vmatrix}, \; \ldots$$

$$c_{n-3} = -\frac{1}{b_{n-2}} \begin{vmatrix} a_{n-1} & a_{n-3} \\ b_{n-2} & b_{n-4} \end{vmatrix}, \; c_{n-5} = -\frac{1}{b_{n-2}} \begin{vmatrix} a_{n-1} & a_{n-5} \\ b_{n-2} & b_{n-6} \end{vmatrix}, \; \ldots$$

Then we can conclude that *the number of characteristic polynomial roots with positive real part is equal to the number of sign changes in the first column of the Routh array*. Purely from mathematical perspective, the term "characteristic" can be omitted in above description. However, we keep the mathematically unnecessary term "characteristic" to emphasize the practical sense of such polynomials in the context of control systems.

Routh-Hurwitz criterion: *The number of characteristic polynomial roots with positive real part is equal to the number of sign changes in the first column of the Routh array.*

For example, the characteristic polynomial

$$(s + 1)(s^2 + 2s + 2) = s^3 + 3s^2 + 4s + 2 \tag{4.24}$$

has three roots -1 and $-1 \pm j$, all with negative real part. So the number of its roots with positive real part is zero. The Routh array associated with it is

$$\begin{array}{c|cc} s^3 & 1 & 4 \\ s^2 & 3 & 2 \\ s^1 & 10/3 & \\ s^0 & 2 & \end{array}$$

The number of sign changes in the first column of the Routh array is zero and is equal to the number of characteristic polynomial roots with positive real part.

For another example, the characteristic polynomial

$$(s+1)(s^2-2s+2) = s^3 - s^2 + 2 \tag{4.25}$$

has a negative real root -1 and two complex roots $1 \pm \mathrm{j}$ with positive real part. So the number of its roots with positive real part is two. The Routh array associated with it is

$$\begin{array}{c|cc} s^3 & 1 & 0 \\ s^2 & -1 & 2 \\ s^1 & 2 & \\ s^0 & 2 & \end{array}$$

The number of sign changes in the first column of the Routh array is two and is equal to the number of characteristic polynomial roots with positive real part.

A simplified version of proof of the Routh-Hurwitz criterion, which catches the essential spirit of the proof given in Anagnost and Desoer (1991), is given below.

Proof Define $p_n(s), p_{n-1}(s), p_{n-2}(s), \cdots, p_0(s)$ as

$$\begin{aligned} p_n(s) &\equiv a_n s^n + a_{n-1}s^{n-1} + a_{n-2}s^{n-2} + a_{n-3}s^{n-3} + \cdots, \\ p_{n-1}(s) &\equiv a_{n-1}s^{n-1} + b_{n-2}s^{n-2} + a_{n-3}s^{n-3} + b_{n-4}s^{n-4} + \cdots, \\ p_{n-2}(s) &\equiv b_{n-2}s^{n-2} + c_{n-3}s^{n-3} + b_{n-4}s^{n-4} + c_{n-5}s^{n-5} + \cdots, \\ &\vdots \\ p_0(s) &\equiv z_0. \end{aligned}$$

Suppose that $a_n, a_{n-1}, b_{n-2}, c_{n-3}, \ldots, z_0$ namely the elements in the first column of the Routh array are non-zero. Even if zero exists in the first column of the Routh array, we can always find a way to vary original polynomial coefficients by arbitrarily infinitesimal amount to avoid existence of zero in the first column of the Routh array—Imagine the original polynomial coefficients $(a_n, a_{n-1}, \ldots, a_0)$ as a free point in an arbitrarily infinitesimal space of $n+1$ dimensions and imagine the condition of an element in the first column of the Routh array being zero as a constraint of *hyper plane* or *hyper curved surface* which is definitely below $n+1$ dimensions in the $n+1$ dimensional space. A finite number of hyper planes or hyper curved surfaces below $n+1$ dimensions in together are still below $n+1$ dimensions. Therefore, in the arbitrarily infinitesimal space of $n+1$ dimensions we can definitely find a point of $(a_n, a_{n-1}, \ldots, a_0)$ that avoids all the constraints and hence avoid existence of zero in the first column of the Routh array—Since polynomials are arbitrary-order-continuous functions, varying original polynomial coefficients by arbitrarily infinitesimal amount will cause variation of root locations by arbitrarily infinitesimal amount and hence will not change distribution of original polynomial roots that are *not located on the imaginary axis*. In other words, such variation will neither change the location of a root from the left-imaginary-axis side to the right-imaginary-axis side nor vice versa. Therefore, supposing that the elements in the first column of the Routh array are non-zero will not influence crucial judgements such as whether the original polynomial belongs to the *stability case*.

$p_n(s)$ is the original polynomial with order n, and $p_n(s), p_{n-1}(s), p_{n-2}(s), \ldots$ have decreasing order. Each polynomial since the second row in the sequence is with one less order than its precedent polynomial and has the iterative relationship with its precedent polynomial as

$$\begin{aligned}
p_{n-1}(s) &= a_{n-1}s^{n-1} + a_{n-3}s^{n-3} + \cdots + \left(a_{n-2} - \frac{a_n}{a_{n-1}}a_{n-3}\right)s^{n-2} \\
&\quad + \left(a_{n-4} - \frac{a_n}{a_{n-1}}a_{n-5}\right)s^{n-4} + \cdots \\
&= (a_{n-1}s^{n-1} + a_{n-3}s^{n-3} + \cdots) \\
&\quad + (a_n s^n + a_{n-2}s^{n-2} + a_{n-4}s^{n-4} + \cdots) \\
&\quad - \frac{a_n}{a_{n-1}}s(a_{n-1}s^{n-1} + a_{n-3}s^{n-3} + a_{n-5}s^{n-5} + \cdots) \\
&= p_n(s) - \frac{a_n}{a_{n-1}}s(a_{n-1}s^{n-1} + a_{n-3}s^{n-3} + a_{n-5}s^{n-5} + \cdots).
\end{aligned}$$

Similarly, we have

$$\begin{aligned}
p_{n-2}(s) &= p_{n-1}(s) - \frac{a_{n-1}}{b_{n-2}}s(b_{n-2}s^{n-2} + b_{n-4}s^{n-4} + b_{n-6}s^{n-6} + \cdots), \\
p_{n-3}(s) &= p_{n-2}(s) - \frac{b_{n-2}}{c_{n-3}}s(c_{n-3}s^{n-3} + c_{n-5}s^{n-5} + c_{n-7}s^{n-7} + \cdots), \\
&\vdots
\end{aligned}$$

or (distinguishing the case where k is odd and the case where k is even)

$$\begin{aligned}
p_{k-1}(s) &= p_k(s) - \frac{p_k^{\text{prime}}}{p_{k-1}^{\text{prime}}}sp_k^{\text{even}}(s), \quad \text{(if } k \text{ is odd)} \\
p_{k-1}(s) &= p_k(s) - \frac{p_k^{\text{prime}}}{p_{k-1}^{\text{prime}}}sp_k^{\text{odd}}(s), \quad \text{(if } k \text{ is even)}
\end{aligned}$$

where $p^{\text{odd}}(s)$ and $p^{\text{even}}(s)$ with superscripts *odd* and *even* denote the sub-polynomials of $p(s)$ with *odd* orders and *even* orders respectively. p^{prime} denotes the prime coefficient of $p(s)$, so $p_n^{\text{prime}} = a_n$, $p_{n-1}^{\text{prime}} = a_{n-1}$, $p_{n-2}^{\text{prime}} = b_{n-2}$, $p_{n-3}^{\text{prime}} = c_{n-3}, \ldots, p_0^{\text{prime}} = z_0$.

For a pure imaginary number $s = wj$, we have

$$
\begin{aligned}
s = w\mathrm{j} \text{ is a root of } p_k(s) &\iff p_k(s) = 0 \iff p_k^{\mathrm{odd}}(s) = p_k^{\mathrm{even}}(s) = 0 \\
&\implies p_{k-1}(s) = 0 \iff p_{k-1}^{\mathrm{odd}}(s) = p_{k-1}^{\mathrm{even}}(s) = 0 \\
&\implies p_{k-2}(s) = 0 \iff p_{k-2}^{\mathrm{odd}}(s) = p_{k-2}^{\mathrm{even}}(s) = 0 \\
&\quad \vdots \\
&\implies p_0(s) = 0 \iff s = w\mathrm{j} \text{ is a root of } p_0(s).
\end{aligned}
$$

Since $p_0(s) \equiv z_0$ which belongs to the first column of the Routh array is assumed non-zero and hence has no pure imaginary root, all $p_n(s), p_{n-1}(s), \ldots, p_0(s)$ have no pure imaginary root. Consider the curve C_I that starts from the origin $s = 0$ and goes along the imaginary axis to the positive infinity, namely $s = w\mathrm{j}$ with w varying from 0 to $+\infty$. Then $p_k(s) \neq 0$ for any $s \in C_I$ (where k is $n, n-1, \ldots$).

For the case where k is odd, we have

$$
p_{k-1}^{\mathrm{even}}(s) = p_k^{\mathrm{even}}(s),
$$

$$
p_{k-1}^{\mathrm{odd}}(s) = p_k^{\mathrm{odd}}(s) - \frac{p_k^{\mathrm{prime}}}{p_{k-1}^{\mathrm{prime}}} s p_k^{\mathrm{even}}(s),
$$

from which we have

$$
p_{k-1}^{\mathrm{even}}(0) = p_k^{\mathrm{even}}(0), \quad p_{k-1}^{\mathrm{odd}}(0) = p_k^{\mathrm{odd}}(0) \quad \text{hence} \quad p_{k-1}(0) = p_k(0)
$$

and

$$
p_k^{\mathrm{even}}(s) = p_{k-1}^{\mathrm{even}}(s) = 0 \implies p_{k-1}^{\mathrm{odd}}(s) = p_k^{\mathrm{odd}}(s).
$$

This last statement implies that $p_k(s)$ and $p_{k-1}(s)$ always cross the same points on the imaginary axis in the same order as s traverses the whole curve C_I. Let $s = w_1\mathrm{j}, w_2\mathrm{j}, \ldots, w_m\mathrm{j}$ $(0 < w_1 < w_2 < \cdots < w_m)$ on the curve C_I correspond to all such cross points on the imaginary axis.

Note that neither $p_k(s)$ nor $p_{k-1}(s)$ crosses the imaginary axis as s varies in the intervals $(0\mathrm{j}, w_1\mathrm{j}), (w_1\mathrm{j}, w_2\mathrm{j}), \cdots, (w_m\mathrm{j}, +\infty\mathrm{j})$ and consider the *phase angles* of $p_k(s)$ and $p_{k-1}(s)$ as s traverses the whole curve C_I. Then we have[3]

$$
\begin{aligned}
&\angle p_k(+\infty\mathrm{j}) - \angle p_{k-1}(+\infty\mathrm{j}) = \angle p_k(s)|_{w_m\mathrm{j}}^{+\infty\mathrm{j}} + \angle p_k(s)|_{w_{m-1}\mathrm{j}}^{w_m\mathrm{j}} + \cdots + \\
&\angle p_k(s)|_{0\mathrm{j}}^{w_1\mathrm{j}} + \angle p_k(0) - \angle p_{k-1}(s)|_{w_m\mathrm{j}}^{+\infty\mathrm{j}} - \angle p_{k-1}(s)|_{w_{m-1}\mathrm{j}}^{w_m\mathrm{j}} - \cdots - \\
&\angle p_{k-1}(s)|_{0\mathrm{j}}^{w_1\mathrm{j}} - \angle p_{k-1}(0) \\
&= (\angle p_k(s)|_{w_m\mathrm{j}}^{+\infty\mathrm{j}} - \angle p_{k-1}(s)|_{w_m\mathrm{j}}^{+\infty\mathrm{j}}) \in (-\pi, \pi).
\end{aligned}
$$

[3] Here, the *phase angle* takes into account clockwise and counter-clockwise tours around the origin. Each clockwise tour corresponds to minus 2π whereas each counter-clockwise tour corresponds to plus 2π.

For the case where k is *even*, we have

$$p_{k-1}^{\text{odd}}(s) = p_k^{\text{odd}}(s),$$

$$p_{k-1}^{\text{even}}(s) = p_k^{\text{even}}(s) - \frac{p_k^{\text{prime}}}{p_{k-1}^{\text{prime}}} s p_k^{\text{odd}}(s),$$

from which we have

$$p_{k-1}^{\text{odd}}(0) = p_k^{\text{odd}}(0), \quad p_{k-1}^{\text{even}}(0) = p_k^{\text{even}}(0) \text{ hence } p_{k-1}(0) = p_k(0)$$

and

$$p_k^{\text{odd}}(s) = p_{k-1}^{\text{odd}}(s) = 0 \implies p_{k-1}^{\text{even}}(s) = p_k^{\text{even}}(s).$$

This last statement implies that $p_k(s)$ and $p_{k-1}(s)$ always cross the same points on the real axis in the same order as s traverses the whole curve C_I. Let $s = w_1\text{j}, w_2\text{j}, \ldots, w_m\text{j}$ $(0 < w_1 < w_2 < \cdots < w_m)$ on the curve C_I correspond to all such cross points on the real axis.

Note that neither $p_k(s)$ nor $p_{k-1}(s)$ crosses the real axis as s varies in the intervals $(0\text{j}, w_1\text{j}), (w_1\text{j}, w_2\text{j}), \ldots, (w_m\text{j}, +\infty\text{j})$ and consider the *phase angles* of $p_k(s)$ and $p_{k-1}(s)$ as s traverses the whole curve C_I. Then we have

$$\begin{aligned}
&\angle p_k(+\infty\text{j}) - \angle p_{k-1}(+\infty\text{j}) = \angle p_k(s)|_{w_m\text{j}}^{+\infty\text{j}} + \angle p_k(s)|_{w_{m-1}\text{j}}^{w_m\text{j}} + \cdots + \\
&\quad \angle p_k(s)|_{0\text{j}}^{w_1\text{j}} + \angle p_k(0) - \angle p_{k-1}(s)|_{w_m\text{j}}^{+\infty\text{j}} - \angle p_{k-1}(s)|_{w_{m-1}\text{j}}^{w_m\text{j}} - \cdots - \\
&\quad \angle p_{k-1}(s)|_{0\text{j}}^{w_1\text{j}} - \angle p_{k-1}(0) \\
&= (\angle p_k(s)|_{w_m\text{j}}^{+\infty\text{j}} - \angle p_{k-1}(s)|_{w_m\text{j}}^{+\infty\text{j}}) \in (-\pi, \pi).
\end{aligned}$$

In summary, $p_{k-1}(0) = p_k(0)$ and $\angle p_k(+\infty\text{j}) - \angle p_{k-1}(+\infty\text{j}) \in (-\pi, \pi)$. On the other hand, we have

$$\begin{aligned}
&\angle p_k(+\infty\text{j}) - \angle p_{k-1}(+\infty\text{j}) \mod 2\pi = \angle \frac{p_k(+\infty\text{j})}{p_{k-1}(+\infty\text{j})} \mod 2\pi \\
&= \angle \frac{p_k^{\text{prime}} s^k}{p_{k-1}^{\text{prime}} s^{k-1}}|_{s=+\infty\text{j}} \mod 2\pi = \angle \frac{p_k^{\text{prime}}}{p_{k-1}^{\text{prime}}} + \angle(+\infty\text{j}) \mod 2\pi \\
&= \text{sign}\left(\frac{p_k^{\text{prime}}}{p_{k-1}^{\text{prime}}}\right) \frac{\pi}{2} \mod 2\pi
\end{aligned}$$

and hence

$$\angle p_k(+\infty\text{j}) - \angle p_{k-1}(+\infty\text{j}) = \text{sign}\left(\frac{p_k^{\text{prime}}}{p_{k-1}^{\text{prime}}}\right) \frac{\pi}{2}.$$

Let $L_{p(s)}$ and $R_{p(s)}$ denote the number of $p(s)$'s roots with negative real part and that with positive real part respectively. Since $p_k(s)$ and $p_{k-1}(s)$ have no pure imaginary root, $L_{p_k(s)} + R_{p_k(s)} = k$ and $L_{p_{k-1}(s)} + R_{p_{k-1}(s)} = k - 1$. Note that

$$\angle p_k(s)|_{0j}^{+\infty j} = \frac{\pi}{2}(L_{p_k(s)} - R_{p_k(s)}),$$
$$\angle p_{k-1}(s)|_{0j}^{+\infty j} = \frac{\pi}{2}(L_{p_{k-1}(s)} - R_{p_{k-1}(s)}),$$

so we have

$$\begin{aligned} \operatorname{sign}\left(\frac{p_k^{\text{prime}}}{p_{k-1}^{\text{prime}}}\right)\frac{\pi}{2} &= \angle p_k(+\infty j) - \angle p_{k-1}(+\infty j) = \angle p_k(s)|_{0j}^{+\infty j} - \angle p_{k-1}(s)|_{0j}^{+\infty j} \\ &= \frac{\pi}{2}(L_{p_k(s)} - L_{p_{k-1}(s)}) - \frac{\pi}{2}(R_{p_k(s)} - R_{p_{k-1}(s)}) \end{aligned}$$

and

$$\frac{p_k^{\text{prime}}}{p_{k-1}^{\text{prime}}} > 0 \iff L_{p_{k-1}(s)} = L_{p_k(s)} - 1 \text{ and } R_{p_{k-1}(s)} = R_{p_k(s)},$$
$$\frac{p_k^{\text{prime}}}{p_{k-1}^{\text{prime}}} < 0 \iff L_{p_{k-1}(s)} = L_{p_k(s)} \text{ and } R_{p_{k-1}(s)} = R_{p_k(s)} - 1.$$

In other words, $p_{k-1}(s)$ has one less root with negative real part than $p_k(s)$ if and only if there is no sign change between the prime coefficients of $p_k(s)$ and $p_{k-1}(s)$, whereas $p_{k-1}(s)$ has one less root with positive real part than $p_k(s)$ if and only if there is sign change between the prime coefficients of $p_k(s)$ and $p_{k-1}(s)$. Based on above analysis and the obvious condition that $L_{p_0(s)} = R_{p_0(s)} = 0$, we can finally conclude the Routh-Hurwitz criterion. □

According to the proof, a control system's characteristic polynomial has roots all with negative real part (namely the control system is *stable*) if and only if neither zero nor sign change exists in the first column of its associated Routh array, and has roots all with positive real part (namely the control system is *extremely unstable*) if and only if there is sign change between any two consecutive elements in the first column of its associated Routh array.[4]

[4] For a control system, whenever it is stable or extremely unstable, zero can definitely be avoided in the first column of its associated Routh array. Zero exists in the first column of the Routh array only if its characteristic polynomial has pure imaginary roots (namely it is *marginal unstable*) or its characteristic polynomial has both roots with positive real part and roots with negative real part. Above remark may be incomprehensible to readers, yet we have no intention to explain it but leave it to readers for reflection. After all, the remark about the stability case (which is our most important concern) is unnecessary in above proof, because as already mentioned, even if zero is assumed to exist in the first column of the Routh array, we can always find a way to vary the characteristic polynomial coefficients by arbitrarily infinitesimal amount to avoid existence of zero in the first

Routh-Hurwitz stability criterion: *A linear time-invariant control system is stable if and only if neither zero nor sign change exists in the first column of the Routh array associated with its characteristic polynomial.*

Now we review the characteristic polynomial $C_{3K}(s)$ described in (4.20) in Sect. 4.3.1 and take advantage of the Routh-Hurwitz stability criterion to determine the range of K for $C_{3K}(s)$ to have roots all with negative real part. The characteristic polynomial $C_{3K}(s)$ is

$$C_{3K}(s) = s^3 + 3\,s^2 + 3\,Ks + 1,$$

and its associated Routh array is

$$\begin{array}{c|cc} s^3 & 1 & 3\,K \\ s^2 & 3 & 1 \\ s^1 & 3\,K - 1/3 & \\ s^0 & 1 & \end{array}$$

According to the Routh-Hurwitz stability criterion, the necessary and sufficient condition for $C_{3K}(s)$ to have roots all with negative real part is $3K - 1/3 > 0$ namely $K > 1/9$. From this reviewed example, we can easily see enormous convenience brought by the Routh-Hurwitz stability criterion based method, in contrast with the method presented in Sect. 4.3.1.

4.3.3 Handling Zeros in the First Column of the Routh Array

We have explained that even if zero exists in the first column of the Routh array, we can always find a way to vary original characteristic polynomial coefficients by arbitrarily infinitesimal amount to avoid existence of zero in the first column of the Routh array, which does not influence crucial judgements. Readers may still find such idea somewhat abstract and may wonder a concrete way to vary original characteristic polynomial coefficients.

A concrete way to vary original characteristic polynomial coefficients is simply based on shifting the Laplace variable s by an arbitrarily infinitesimal offset ϵ. For example, consider the characteristic polynomial $C_{3K}(s)$ described in (4.20) with K set to $1/9$, as

$$C_{1/3}(s) = s^3 + 3\,s^2 + s/3 + 1.$$

---

column of the Routh array, which does not influence crucial judgements such as whether the control system is stable.

Shift the Laplace variable s by the offset ϵ and obtain

$$C_{1/3}(s+\epsilon) = s^3 + (3+3\epsilon)s^2 + (1/3+6\epsilon+3\epsilon^2)s + 1 + \epsilon/3 + 3\epsilon^2 + \epsilon^3.$$

The Routh array associated with $C_{1/3}(s+\epsilon)$ is

$$\begin{array}{c|cc} s^3 & 1 & 1/3+6\epsilon+3\epsilon^2 \\ s^2 & 3+3\epsilon & 1+\epsilon/3+3\epsilon^2+\epsilon^3 \\ s^1 & 8\epsilon(7+9\epsilon+3\epsilon^2)/(9+9\epsilon) & \\ s^0 & 1+\epsilon/3+3\epsilon^2+\epsilon^3 & \end{array}$$

As $\epsilon \to 0$, the Routh array associated with $C_{1/3}(s+\epsilon)$ becomes

$$\begin{array}{c|cc} s^3 & 1 & 1/3 \\ s^2 & 3 & 1 \\ s^1 & (56/9)\epsilon & \\ s^0 & 1 & \end{array}$$

If $\epsilon \to 0^+$, there is no sign change in the first column of the Routh array, so $C_{1/3}(s+\epsilon)$ with $\epsilon \to 0^+$ has roots all with negative real part. If $\epsilon \to 0^-$, there are two sign changes in the first column of the Routh array, so $C_{1/3}(s+\epsilon)$ with $\epsilon \to 0^-$ has two roots with positive real part. In other words, if we shift the characteristic polynomial roots to the left by an arbitrarily infinitesimal offset, then the roots will be all with negative real part. If we shift the characteristic polynomial roots to the right by an arbitrarily infinitesimal offset, then the characteristic polynomial will have two roots with positive real part. From these two aspects of analysis, we can infer that the characteristic polynomial $C_{1/3}(s)$ has one root with negative real part and two pure imaginary roots.

For another example, the characteristic polynomial

$$(s+1)^2(s-1) = s^3 + s^2 - s - 1 \tag{4.26}$$

has two repeated negative real roots -1 and one positive real root $+1$. So the number of its roots with positive real part is one. Shift the Laplace variable s by the offset ϵ and obtain

$$s^3 + (1+3\epsilon)s^2 + (-1+2\epsilon+3\epsilon^2)s - 1 - \epsilon + \epsilon^2 + \epsilon^3.$$

The Routh array associated with the characteristic polynomial is

$$\begin{array}{c|cc} s^3 & 1 & -1+2\epsilon+3\epsilon^2 \\ s^2 & 1+3\epsilon & -1-\epsilon+\epsilon^2+\epsilon^3 \\ s^1 & 8\epsilon^2(1+\epsilon)/(1+3\epsilon) & \\ s^0 & -1-\epsilon+\epsilon^2+\epsilon^3 & \end{array}$$

As $\epsilon \to 0$, the Routh array becomes

$$\begin{array}{c|cc} s^3 & 1 & -1 \\ s^2 & 1 & -1 \\ s^1 & 8\epsilon^2 & \\ s^0 & -1 & \end{array}$$

No matter for $\epsilon \to 0^+$ or $\epsilon \to 0^-$, we always have $8\epsilon^2 > 0$ and there is exactly one sign change in the first column of the Routh array. Therefore, the characteristic polynomial described in (4.26) has one root with positive real part and two roots with negative real part.

We believe that above two examples are enough for readers to catch the spirit of shifting the Laplace variable s by an arbitrarily infinitesimal offset ϵ to handle zeros (if existing) in the first column of the Routh array. It is worth noting that no matter for $\epsilon \to 0^+$ or $\epsilon \to 0^-$, we can always find an arbitrarily infinitesimal offset ϵ such that no zero exists in the first column of the Routh array—In fact, each element in the first column of the Routh array is a non-zero fractional polynomial in terms of ϵ.[5] So the set of ϵ that incurs existence of zero in the first column of the Routh array is a *countable* set, whereas any arbitrarily infinitesimal open set about zero is an *uncountable* set and must contain certain ϵ that does not incur any zero in the first column of the Routh array.

4.4 Note

In retrospect of Chap. 4, we have clarified the concept of control system stability which is the first important concern for control purpose, and have presented control theory on how to analyse control system stability. Other important concerns include control system accuracy and control system responsiveness, control theory on which will be presented in Chap. 5.

References

J. Anagnost, C. Desoer. An elementary proof of the routh-hurwitz stability criterion. *Circuits Systems Signal Process*, 10(1):101–114, 1991.

J. Bonnans, J. Gilbert, C. Lemarechal, et al. *Numerical optimization: theoretical and practical aspects*. Springer Science & Business Media, 2006.

[5] For a generic characteristic polynomial $C(s)$, there is a large enough positive ϵ such that $C(s+\epsilon)$ has roots all with negative real part. In this case, according to the Routh-Hurwitz stability criterion, no zero exists in the first column of the Routh array associated with $C(s+\epsilon)$. Besides, each element in the first column of the Routh array is apparently a fractional polynomial in terms of ϵ. So each element in the first column of the Routh array is a non-zero fractional polynomial in terms of ϵ.

R. Dorf, R. Bishop. *Modern control systems*. Pearson Prentice Hall, 2008.
C. Hermite. *Oeuvres de Charles Hermite*. Cambridge university press, 2009.
S. Lang. *Algebra*. Springer, 2002.

Chapter 5
Control System Performance and Design

For design of a control system, its stability is the first important concern. Besides, its accuracy and responsiveness are also important concerns.

5.1 Control System Accuracy

5.1.1 Theoretical Analysis of Closed-Loop Feedback Control Examples

To facilitate conception of control system accuracy, we consider the concrete example of rotating disk speed control presented in Sect. 2.3 in Chap. 2. The rotating disk speed closed-loop feedback control system is illustrated in Fig. 5.1, where torque disturbance is omitted. The rotating disk speed closed-loop feedback control system can be treated as a black box from customers' perspective. The input to the rotating disk speed control black box is certain expected rotating speed and the output is the rotating speed.

Suppose that the proportional feedback control method described in (2.4) is adopted inside the rotating disk speed control black box. Matlab code for simulation of rotating disk speed proportional feedback control is given as follows.

RotatingDiskSpeedPControlAccuracy.m

```
%% Rotating disk speed control system parameters
J = 10; % Rotating disk load inertia
b = 5; % Friction coefficient
Tmax = 50; % Maximum torque

%% Simulation preliminary configuration
dt = 0.001; % Numerical computation step
tSpan = 0:dt:10; % Simulation time span
% Record rotating disk speeds during simulation
```


H. Li, *Control Theory for Practical Applications*,
https://doi.org/10.1007/978-981-97-5008-5_5

```
vAll = zeros(size(tSpan)); k = 1;
v = 0; % Initial rotating disk speed
vE = 1; % Expected rotating disk speed

%% Simulation of rotating disk speed control
for t = tSpan
    %% Control method
    e = vE-v; % Feedback error
    T = 100*e; % Proportional feedback control
    T = max(min(T, Tmax), -Tmax); % Control law constraint

    %% Rotating disk dynamics
    dv = (T-b*v)/J;
    v = v + dv*dt;
    vAll(k) = v; k = k+1;
end
plot(tSpan, vAll, 'LineWidth', 2); ylim([0, 2]);
xlabel('Time'); ylabel('Rotating Disk Speed'); grid on;
fprintf('Steady Rotating Disk Speed Error: %f\n', vE-v);
```

Simulation experiments with various proportional coefficients are done. For each proportional coefficient, its corresponding steady rotating speed error namely the error between the expected rotating speed and the final stable rotating speed is recorded. Steady rotating speed errors corresponding to various proportional coefficients are listed in Table 5.1, which demonstrates a phenomenon that the rotating disk speed proportional feedback control system always has a steady rotating speed error and another phenomenon that the steady rotating speed error is almost inversely proportional to the proportional coefficient.

For theoretical analysis of the rotating disk speed proportional feedback control system, we resort to the methodology of system modelling presented in Chap. 3. The rotating disk speed proportional feedback control system can be modelled by a transfer function block diagram illustrated in Fig. 5.2.

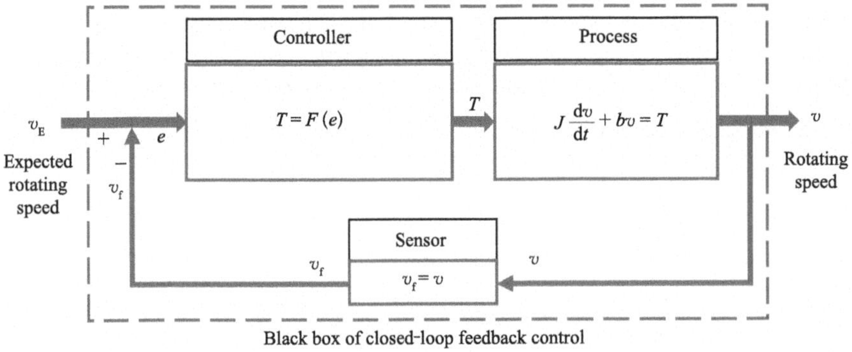

Fig. 5.1 Black box of rotating disk speed closed-loop feedback control

Table 5.1 Rotating disk speed proportional feedback control error

Proportional coefficient	Steady rotating speed error
10000	0.000500
1000	0.004975
500	0.009901
200	0.024390
100	0.047619

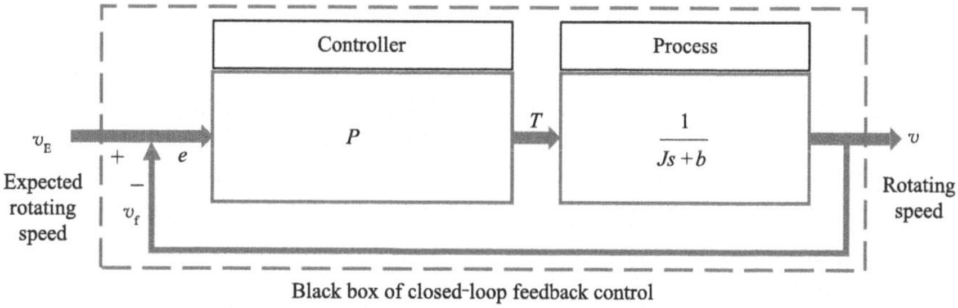

Fig. 5.2 Transfer function block diagram of rotating disk speed proportional feedback control

By rules of transfer function block synthesis introduced in Sect. 3.4 in Chap. 3, we can compute the holistic transfer function of the rotating disk speed proportional feedback control system as

$$\frac{v(s)}{v_E(s)} = \frac{P\frac{1}{Js+b}}{1 + P\frac{1}{Js+b}} = \frac{P}{Js + P + b}. \tag{5.1}$$

The rotating disk speed proportional feedback control system has a unique pole $s = -(P + b)/J$ which is negative (note that the proportional coefficient P is by default positive), so it is a stable control system. If its initial condition is neglected, the rotating speed output can be revealed via (5.1) as

$$v(s) = \frac{P}{Js + P + b} v_E(s) = \frac{v_E P}{s(Js + P + b)} \quad \left(\text{note } L\{v_E(t) \equiv v_E\} = \frac{v_E}{s}\right)$$
$$= \frac{v_E P}{P + b}\left(\frac{1}{s} - \frac{1}{s + \frac{P+b}{J}}\right).$$

So we have

$$v(t) = \frac{v_E P}{P + b} - \frac{v_E P}{P + b} e^{-\frac{P+b}{J}t} \quad \text{(check Table 3.1)}$$

and

$$v_S \equiv v(\infty) = \frac{P}{P+b} v_E, \tag{5.2}$$

$$e_S = v_E - v_S = \frac{b}{P+b} v_E \approx \frac{b}{P} v_E \quad (\text{if } P \gg b), \tag{5.3}$$

which implies theoretically that the rotating disk speed proportional feedback control system indeed has a steady rotating speed error e_S that is almost inversely proportional to the proportional coefficient P.

Theoretically speaking, if we set $P \to \infty$, then $e_S \to 0$. In other words, we can almost eliminate the steady rotating speed error by setting a very large proportional coefficient P. However, in practical applications, if we set a too large proportional coefficient P, the proportional feedback controller tends to function like the intuitive feedback controller presented in Chap. 2 that switches frequently and abruptly between two extreme controller output torques $T_{\max}$ and $-T_{\max}$ and may be too severe and harmful to the control system. Therefore, we had better not set a too large proportional coefficient P.

Increasing the proportional coefficient is not the only way to decrease the steady rotating speed error. In fact, to eliminate the steady rotating speed error, we can replace the proportional feedback control method by another feedback control method namely the proportional-integral feedback control method as

$$T = \begin{cases} T_{\max} & Pe + I\int_0^t e\mathrm{d}t > T_{\max} \\ Pe + I\int_0^t e\mathrm{d}t & -T_{\max} \leqslant Pe + I\int_0^t e\mathrm{d}t \leqslant T_{\max} \\ -T_{\max} & Pe + I\int_0^t e\mathrm{d}t < -T_{\max} \end{cases} \tag{5.4}$$

Compared with proportional feedback control, proportional-integral feedback control involves one more term namely the integral term $I\int_0^t e\mathrm{d}t$ where I is the integral coefficient. Matlab code for simulation of rotating disk speed proportional-integral feedback control is given as follows.

RotatingDiskSpeedPIControlAccuracy.m

```
%% Rotating disk speed control system parameters
J = 10; % Rotating disk load inertia
b = 5; % Friction coefficient
Tmax = 50; % Maximum torque

%% Simulation preliminary configuration
dt = 0.001; % Numerical computation step
tSpan = 0:dt:10; % Simulation time span
% Record rotating disk speeds during simulation
vAll = zeros(size(tSpan)); k = 1;
v = 0; % Initial rotating disk speed
vE = 1; % Expected rotating disk speed
eSum = 0;
```

```
%% Simulation of rotating disk speed control
for t = tSpan
    %% Control method
    e = vE-v; % Feedback error
    eSum = eSum + e*dt; % Rotating disk speed error integral
    T = 100*e + 200*eSum; % Proportional-integral feedback
    control
    T = max(min(T, Tmax), -Tmax); % Control law constraint

    %% Rotating disk dynamics
    dv = (T-b*v)/J;
    v = v + dv*dt;
    vAll(k) = v; k = k+1;
end
plot(tSpan, vAll, 'LineWidth', 2); ylim([0, 2]);
xlabel('Time'); ylabel('Rotating Disk Speed'); grid on;
fprintf('Steady Rotating Disk Speed Error: %f\n', vE-v);
```

We suggest that readers try above Matlab simulation code to verify the ability of proportional-integral feedback control in eliminating steady rotating speed error. Like in Sect. 2.3 in Chap. 2, we can add various aspects of uncertainty into the rotating disk speed control system in simulation. Matlab code for simulation of rotating disk speed proportional-integral feedback control with various aspects of uncertainty incorporated is given as follows.

RotatingDiskSpeedPIControlAccuracy2.m

```
%% Rotating disk speed control system parameters
J = random('Uniform', 5, 20); % Unknown rotating disk load
    inertia
b = random('Uniform', 2, 10); % Unknown Friction coefficient
Td = random('Uniform', -5, 5); % Unknown torque disturbance
Tmax = 50; % Maximum torque

%% Simulation preliminary configuration
dt = 0.001; % Numerical computation step
tSpan = 0:dt:10; % Simulation time span
% Record rotating disk speeds during simulation
vAll = zeros(size(tSpan)); k = 1;
v = 0; % Initial rotating disk speed
vE = 1; % Expected rotating disk speed
eSum = 0;

%% Simulation of rotating disk speed control
for t = tSpan
    %% Control method
    e = vE-v; % Feedback error
    eSum = eSum + e*dt; % Rotating disk speed error integral
    T = 100*e + 200*eSum; % Proportional-integral feedback
    control
    T = max(min(T, Tmax), -Tmax); % Control law constraint

    %% Rotating disk dynamics suffering from uncertainty
    dv = (T-Td-b*v)/J;
```

```
    v = v + dv*dt;
    vAll(k) = v; k = k+1;
end
plot(tSpan, vAll, 'LineWidth', 2); ylim([0, 2]);
xlabel('Time'); ylabel('Rotating Disk Speed'); grid on;
fprintf('Steady Rotating Disk Speed Error: %f\n', vE-v);
```

Readers can verify via above simulation code the ability of rotating disk speed proportional-integral feedback control in eliminating steady rotating speed error, even when the control system is with various aspects of uncertainty.

For theoretical analysis, the rotating disk speed proportional-integral feedback control system can be modelled by a transfer function block diagram illustrated in Fig. 5.3.

By rules of transfer function block synthesis introduced in Sect. 3.4 in Chap. 3, we can compute the holistic transfer function of the rotating disk speed proportional-integral feedback control system as

$$\frac{v(s)}{v_E(s)} = \frac{(P + \frac{I}{s})\frac{1}{Js+b}}{1 + (P + \frac{I}{s})\frac{1}{Js+b}} = \frac{Ps + I}{Js^2 + (P + b)s + I}. \tag{5.5}$$

The rotating disk speed proportional-integral feedback control system has two poles

$$\alpha_{1,2} = \frac{-(P + b) \pm \sqrt{(P + I)^2 - 4JI}}{2J}$$

which both have negative real part (note that the proportional coefficient P and the integral coefficient I are by default positive), so it is a stable control system. Its stability can also be verified according to the Routh-Hurwitz criterion: Establish its associated Routh array as

$$\begin{array}{c|cc} s^2 & J & I \\ s^1 & P + b & \\ s^0 & I & \end{array}$$

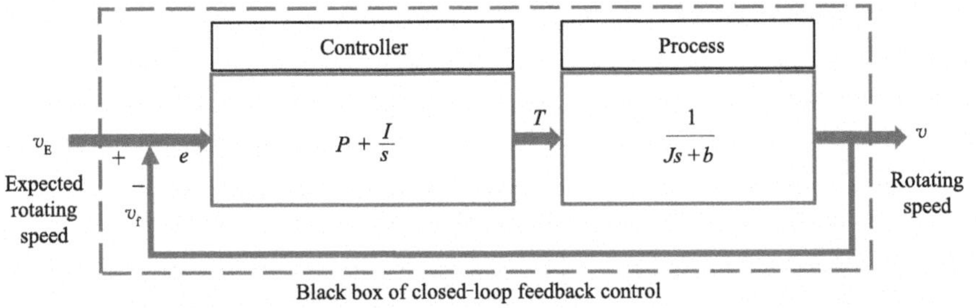

Fig. 5.3 Transfer function block diagram of rotating disk speed proportional-integral feedback control

There is neither zero nor sign change in the first column of the Routh array, so the control system stability is guaranteed.

If its initial condition is neglected, the rotating speed output can be revealed via (5.5) as (when $\alpha_1 \neq \alpha_2$)

$$v(s) = \frac{Ps + I}{Js^2 + (P + b)s + I} v_E(s) = \frac{v_E(Ps + I)}{s(Js^2 + (P + b)s + I)}$$

$$(\text{note } L\{v_E(t) \equiv v_E\} = \frac{v_E}{s})$$

$$= \frac{v_E}{s} - \frac{v_E \frac{P\alpha_1 + I}{(P+b)\alpha_1 + 2I}}{s - \alpha_1} - \frac{v_E \frac{P\alpha_2 + I}{(P+b)\alpha_2 + 2I}}{s - \alpha_2}.$$

So we have

$$v(t) = v_E - \frac{v_E(P\alpha_1 + I)}{(P + b)\alpha_1 + 2I} e^{\alpha_1 t} - \frac{v_E(P\alpha_2 + I)}{(P + b)\alpha_2 + 2I} e^{\alpha_2 t}$$

and

$$v_S \equiv v(\infty) = v_E \quad (\text{note } \mathrm{Re}(\alpha_{1,2}) < 0), \tag{5.6}$$

$$e_S = v_E - v_S = 0. \tag{5.7}$$

If $\alpha_1 = \alpha_2$, the same conclusion can also be inferred (details of derivation are left to readers for practice). Equations (5.6) and (5.7) imply theoretically that the rotating disk speed proportional-integral feedback control system indeed has zero steady rotating speed error.

We can take various aspects of uncertainty into account in above theoretical analysis. The rotating disk speed proportional-integral feedback control system with uncertainty can be modelled by a transfer function block diagram illustrated in Fig. 5.4.

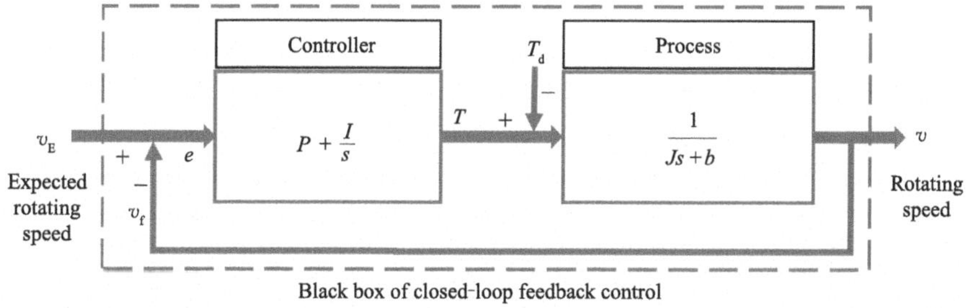

Fig. 5.4 Transfer function block diagram of rotating disk speed proportional-integral feedback control with uncertainty

If its initial condition is neglected, the rotating speed output can be derived as

$$v(s) = \frac{Ps + I}{Js^2 + (P + b)s + I} v_E(s) + \frac{s}{Js^2 + (P + b)s + I} T_d(s).$$

Let $v_E(t) \equiv v_E$ and $T_d(t) \equiv T_d$ be constant and hence obtain (when $\alpha_1 \neq \alpha_2$)

$$\begin{aligned} v(s) &= \frac{v_E(Ps + I)}{s(Js^2 + (P + b)s + I)} + \frac{T_d}{Js^2 + (P + b)s + I} \\ &= \frac{v_E}{s} - \frac{v_E \frac{P\alpha_1 + I}{(P+b)\alpha_1 + 2I}}{s - \alpha_1} - \frac{v_E \frac{P\alpha_2 + I}{(P+b)\alpha_2 + 2I}}{s - \alpha_2} + \frac{\frac{T_d}{J(\alpha_1 - \alpha_2)}}{s - \alpha_1} + \frac{\frac{T_d}{J(\alpha_2 - \alpha_1)}}{s - \alpha_2}. \end{aligned}$$

So we have

$$\begin{aligned} v(t) = v_E &- \frac{v_E(P\alpha_1 + I)}{(P + b)\alpha_1 + 2I} e^{\alpha_1 t} - \frac{v_E(P\alpha_2 + I)}{(P + b)\alpha_2 + 2I} e^{\alpha_2 t} \\ &+ \frac{T_d}{J(\alpha_1 - \alpha_2)} e^{\alpha_1 t} + \frac{T_d}{J(\alpha_2 - \alpha_1)} e^{\alpha_2 t} \end{aligned}$$

and

$$v_S \equiv v(\infty) = v_E \quad (\text{note } \mathrm{Re}(\alpha_{1,2}) < 0), \tag{5.8}$$

$$e_S = v_E - v_S = 0. \tag{5.9}$$

If $\alpha_1 = \alpha_2$, the same conclusion can also be inferred (details of derivation are left to readers for practice). Equations (5.8) and (5.9) imply theoretically that the rotating disk speed proportional-integral feedback control system indeed has zero steady rotating speed error, regardless of concrete rotating disk load inertia J, concrete friction coefficient b, and concrete torque disturbance T_d. In other words, the rotating disk speed proportional-integral feedback control system even with these aspects of uncertainty can guarantee its control accuracy without any steady rotating speed error.

5.1.2 *Final Value Theorem*

To obtain the output **steady state** of a control system, we may transform the output s-domain function into its original time-domain function and then compute its limit at time infinity namely $t \to \infty$, as presented in Sect. 5.1.1. In fact, we have a much more convenient method to obtain the output steady state of a control system. The method is based on the **final value theorem** which states that the limit of a time-domain function at time infinity is equivalent to the limit of its associated s-domain function at the s origin.

Final value theorem: $\lim_{t\to\infty} y(t) = \lim_{s\to 0^+} sY(s)$ where $Y(s) = L\{y(t)\}$.

Proof In practical applications, we are concerned about output steady states of stable control systems only, because it is meaningless to compute output steady states of unstable control systems. A stable control system naturally has bounded output, so we have $|y(t)| < M$ where M is certain bound limit. In following derivation, suppose s is by default in the s-domain of absolute convergence for all relevant integrals and $s \in \mathbb{R}^+$. We have

$$s\int_{0^-}^{\infty} e^{-st}dt = 1,$$
$$s\int_{a}^{\infty} e^{-st}dt \leqslant s\int_{0^-}^{\infty} e^{-st}dt = 1,$$

and

$$\begin{aligned}
|sY(s) - y(\infty)| &= \left| s\int_{0^-}^{\infty} y(t)e^{-st}dt - y(\infty)s\int_{0^-}^{\infty} e^{-st}dt \right| \\
&= \left| s\int_{0^-}^{\infty} (y(t) - y(\infty))e^{-st}dt \right| \\
&\leqslant \left| s\int_{0^-}^{a} (y(t) - y(\infty))e^{-st}dt \right| + \left| s\int_{a}^{\infty} (y(t) - y(\infty))e^{-st}dt \right| \\
&\quad \left(\text{note } s \in R^+ \text{ and } |s|\int_{a}^{\infty} |e^{-st}|dt = s\int_{a}^{\infty} e^{-st}dt \leqslant 1\right) \\
&\leqslant |s|(2Ma) + \max\{|y(t) - y(\infty)|, t \geqslant a\}.
\end{aligned}$$

For any arbitrarily infinitesimal ϵ, we can always find a such that $|y(t) - y(\infty)| < \frac{\epsilon}{2}$ for $t \geqslant a$. Then for any $s \in \left(0, \frac{\epsilon}{4Ma}\right)$ we have $|sY(s) - y(\infty)| < \epsilon$. By definition of limit, we can conclude the final value theorem. □

We can apply the final value theorem to facilitate theoretical analysis of the rotating disk speed proportional feedback control system and the rotating disk speed proportional-integral feedback control system. Compute the output steady state of the rotating disk speed proportional feedback control system as

$$v_S \equiv \lim_{t\to\infty} v(t) = \lim_{s\to 0^+} sv(s) = \lim_{s\to 0^+} \frac{v_E P}{Js + P + b} = \frac{P}{P + b} v_E,$$

which is consistent with (5.3). Compute the output steady state of the rotating disk speed proportional-integral feedback control system as

$$v_S = \lim_{s\to 0^+} s v(s) = \lim_{s\to 0^+} s\left[\frac{v_E(Ps+I)}{s(Js^2+(P+b)s+I)} + \frac{T_d}{Js^2+(P+b)s+I}\right] = v_E,$$

which is consistent with (5.8). Readers can see the convenience of using the final value theorem to obtain output stead states of control systems.

5.2 Control System Responsiveness

Control system responsiveness is another important concern for control purpose. A stable control system's responsiveness refers to the speed of its convergence to the steady state.[1] The quicker the control system converges to its steady state, the higher the control system responsiveness is. The slower the control system converges to its steady state, the lower the control system responsiveness is. The concept of control system responsiveness is already encountered in Sect. 2.4 in Chap. 2, where the concrete closed-loop feedback control system is demonstrated to outperform the concrete open-loop control system in terms of their responsiveness.

5.2.1 *Control System Characteristic Polynomial and Poles for Responsiveness Analysis*

Given a generic stable control system that follows the linear time-invariant assumption explained in Sect. 3.3 in Chap. 3, suppose that the control system's holistic transfer function after transfer function block synthesis is denoted generically as

$$\frac{y(s)}{r(s)} = \frac{b_m s^m + b_{m-1}s^{m-1} + \cdots + b_0}{a_n s^n + a_{n-1}s^{n-1} + \cdots + a_0} \equiv \frac{B(s)}{A(s)}. \tag{5.10}$$

The denominator of the holistic transfer function namely

$$A(s) \equiv a_n s^n + a_{n-1}s^{n-1} + \cdots + a_0 \tag{5.11}$$

is the characteristic polynomial of the control system and let it be factored as

$$a_n s^n + a_{n-1}s^{n-1} + \cdots + a_0 = a_n(s+\alpha_1)^{m_1}(s+\alpha_2)^{m_2}\cdots(s+\alpha_N)^{m_N},$$

where $-\alpha_1, -\alpha_2, \ldots, -\alpha_N$ denote control system poles namely roots of the characteristic polynomial described in (5.11) and $m_1, m_2, \ldots, m_N$ denote their numbers respectively. Since the control system is stable, the poles $-\alpha_1, -\alpha_2, \ldots, -\alpha_N$ are

[1] The concept of **control system responsiveness** only applies to stable control systems, whereas it is meaningless to talk about responsiveness of unstable control systems.

all with negative real part, namely $\mathrm{Re}(-\alpha_1) < 0, \mathrm{Re}(-\alpha_2) < 0, \ldots, \mathrm{Re}(-\alpha_N) < 0$. Suppose that these poles are ordered according to their real parts in descending order as $0 > \mathrm{Re}(-\alpha_1) \geqslant \mathrm{Re}(-\alpha_2) \geqslant \cdots \geqslant \mathrm{Re}(-\alpha_N)$.

Let $r(t) \equiv r_0$ be constant and neglect the control system's initial conditions. Then, the control system output is derived as

$$y(s) = \frac{B(s)}{A(s)} \frac{r_0}{s} = \frac{r_0(b_m s^m + b_{m-1}s^{m-1} + \cdots + b_0)}{a_n s(s+\alpha_1)^{m_1}(s+\alpha_2)^{m_2} \cdots (s+\alpha_N)^{m_N}}. \tag{5.12}$$

In the spirit of partial fraction decomposition presented in Sect. 3.2.3 in Chap. 3, the s-domain function $y(s)$ described in (5.12) can be decomposed into the sum of partial fractions as

$$y(s) = r_0 \left(\frac{c_0}{s} + \sum_{j=1}^{N} \sum_{k=1}^{m_j} \frac{c_{j,k}}{(s+\alpha_j)^k} \right), \tag{5.13}$$

where the generically-denoted coefficients c_0 and $c_{j,k}$ ($j \in \{1, \ldots, N\}, k \in \{1, \cdots, m_j\}$) are determined according to the control system's holistic transfer function. Following the method presented in Sect. 3.2.3 in Chap. 3, we can compute these coefficients as

$$c_0 = \lim_{s \to 0} s \frac{B(s)}{sA(s)} = \lim_{s \to 0} \frac{B(s)}{A(s)} = \frac{b_0}{a_0} \tag{5.14}$$

and

$$c_{j,k} = \lim_{s \to -\alpha_j} \frac{1}{(m_j - k)!} \frac{\mathrm{d}^{m_j - k}}{\mathrm{d}t^{m_j - k}} \left[(s+\alpha_j)^{m_j} \frac{B(s)}{sA(s)} \right]. \tag{5.15}$$

From (5.13), the original time-domain function of $y(s)$ is recovered as

$$y = r_0 \left(c_0 + \sum_{j=1}^{N} \sum_{k=1}^{m_j} c_{j,k} \mathrm{e}^{-\alpha_j t} \frac{t^{k-1}}{(k-1)!} \right). \quad \text{(check Table 3.1)} \tag{5.16}$$

As $0 > \mathrm{Re}(-\alpha_1) \geqslant \mathrm{Re}(-\alpha_2) \geqslant \cdots \geqslant \mathrm{Re}(-\alpha_N)$, all the terms $\mathrm{e}^{-\alpha_j t} t^{k-1}$ ($j \in \{1, \cdots, N\}, k \in \{1, \ldots, m_j\}$) converge to zero at time infinity $t \to \infty$. So together with (5.12) and (5.14) we have

$$y_S \equiv y(\infty) = r_0 c_0 = r_0 \lim_{s \to 0} s \frac{B(s)}{sA(s)} = \lim_{s \to 0} s \frac{r_0 B(s)}{sA(s)} = \lim_{s \to 0} s y(s),$$

which is consistent with the final value theorem.

From (5.16) we can also see that the control system's speed of convergence to the output steady state y_S depends on the speed of the exponential terms $\mathrm{e}^{-\alpha_1 t}, \mathrm{e}^{-\alpha_2 t}, \cdots,$

$e^{-\alpha_N t}$ decaying to zero, which further depends on the control system poles $-\alpha_1, -\alpha_2, \cdots, -\alpha_N$, more specifically, on their real parts $\mathrm{Re}(-\alpha_1), \mathrm{Re}(-\alpha_2), \cdots, \mathrm{Re}(-\alpha_N)$. For each exponential term $e^{-\alpha_j t}$ ($j \in \{1, \dots, N\}$), the smaller $\mathrm{Re}(-\alpha_j)$ is, namely the further the control system pole $-\alpha_j$ is to the left of the imaginary axis, the faster it decays to zero.

Control system poles comparatively close to the left of the imaginary axis, especially the *rightmost pole* $-\alpha_1$, are called **dominant control system poles**, because they dominate the **transient response** of the control system. In other words, terms associated with control system poles other than dominant control system poles normally decay much faster to zero than terms associated with dominant control system poles. Therefore, a control system's responsiveness normally depends on its dominant poles: *The further the dominant control system poles are to the left of the imaginary axis, the higher the control system responsiveness is namely the more responsive the control system is. The closer the dominant control system poles are to the left of the imaginary axis, the lower the control system responsiveness is namely the less responsive the control system is.*

Control system responsiveness qualitative criterion: *For a stable linear time-invariant control system, its responsiveness depends on its dominant poles: The further the dominant poles are to the left of the imaginary axis, the more responsive it is. The closer the dominant poles are to the left of the imaginary axis, the less responsive it is.*

5.2.2 Control System Response Time

Qualitative theoretical analysis of a control system's responsiveness has been presented in Sect. 5.2.1. Besides qualitative theoretical analysis, we also need quantitative theoretical analysis of a control system's responsiveness, which involves the concept of **control system response time**. For a control system, ideally speaking, its response time refers to the time that it needs to converge to its steady state. However, as implied in (5.16), the control system output y will converge asymptotically to the output steady state $y_S = r_0 c_0$ at time infinity $t \to \infty$ namely an imaginary time point, but will not converge exactly to y_S at a practical time point.

In practical applications, a control system's response time normally refers to the time that it needs to converge within certain relative error tolerance range about its steady state, namely

$$T_R = \min\left\{T : \left|\frac{y(t)}{y_S} - 1\right| \leqslant r_{tol} \text{ for } t \geqslant T\right\}, \tag{5.17}$$

where the relative error tolerance range r_{tol} can be set flexibly according to concrete applications. For examples, $r_{\mathrm{tol}} = 0.05$ and $r_{\mathrm{tol}} = 0.01$ are typical choices for practical applications.

It is worth noting that when we talk about the response time of a control system, we assume by default that the control system takes constant input $r(t) \equiv r_0$ namely *step input*. In other words, its response time actually refers to its *step response time*. Although the control system may encounter input of various types, its responsiveness which is one of its inherent characteristics can be well reflected by its step response. Therefore, we just adopt the term *response time* without explicitly mentioning "step".

Follow (5.16) and compute the relative error of $y(t)$ as

$$\left| \frac{y(t)}{y_{\mathrm{S}}} - 1 \right| = \left| \frac{1}{c_0} \sum_{j=1}^{N} \sum_{k=1}^{m_j} c_{j,k} \mathrm{e}^{-\alpha_j t} \frac{t^{k-1}}{(k-1)!} \right|. \tag{5.18}$$

So the control system response time is

$$T_{\mathrm{R}} = \min \left\{ T : \left| \frac{1}{c_0} \sum_{j=1}^{N} \sum_{k=1}^{m_j} c_{j,k} \mathrm{e}^{-\alpha_j t} \frac{t^{k-1}}{(k-1)!} \right| \leqslant r_{\mathrm{tol}} \text{ for } t \geqslant T \right\}. \tag{5.19}$$

From (5.18) and (5.19) we can see that the control system response time is irrelevant to the input amplitude r_0. So we can identify a control system's response time by checking its arbitrary step response via experiments.

According to the *control system responsiveness qualitative criterion* presented in Sect. 5.2.1, we infer qualitatively that a stable linear time-invariant control system's response time depends on its dominant poles: The further the dominant poles are to the left of the imaginary axis, the less the response time is. The closer the dominant poles are to the left of the imaginary axis, the more the response time is.

It seems desirable to enhance control system responsiveness as much as possible. For this purpose, we may design a control system in a way to keep its dominant poles as far away as possible from the left of the imaginary axis. However, it is not so in practical applications. Reasons are two-folds: First, keeping a control system's dominant poles (and hence all its poles) too far away from the left of the imaginary axis usually incurs large operational cost. Second, a more responsive control system tends to be more susceptible to control system noises that are inevitable—This second point will be clarified in Sect. 5.3—In practical applications, we do not aim at enhancing control system responsiveness as much as possible. In fact, we enhance control system responsiveness until the control system response time satisfies certain requirement. Then it is unnecessary to further enhance control system responsiveness.

5.3 Frequency Response Analysis

We can analyse a control system's responsiveness from its **frequency response** namely its steady-state sinusoidal response to sinusoidal input.

5.3.1 Frequency Transfer Function and Bode Plots

Given a stable linear time-invariant control system with its holistic transfer function denoted as

$$T(s) \equiv \frac{y(s)}{r(s)} = \frac{b_m s^m + b_{m-1}s^{m-1} + \cdots + b_0}{a_n s^n + a_{n-1}s^{n-1} + \cdots + a_0}, \tag{5.20}$$

let its characteristic polynomial be factored as

$$a_n s^n + a_{n-1}s^{n-1} + \cdots + a_0 = a_n(s+\alpha_1)^{m_1}(s+\alpha_2)^{m_2}\cdots(s+\alpha_N)^{m_N},$$

where $-\alpha_1, -\alpha_2, \cdots, -\alpha_N$ denote control system poles and $m_1, m_2, \ldots, m_N$ denote their numbers respectively. Since the control system is stable, the poles $-\alpha_1$, $-\alpha_2, \cdots, -\alpha_N$ are all with negative real part, namely $\mathrm{Re}(-\alpha_1) < 0$, $\mathrm{Re}(-\alpha_2) < 0, \ldots, \mathrm{Re}(-\alpha_N) < 0$.

Let $r(t) = A\sin(\omega t)$ be sinusoidal input and neglect the control system's initial conditions. Then the control system output is derived as

$$y(s) = T(s)\frac{A\omega}{s^2+\omega^2}. \tag{5.21}$$

In the spirit of partial fraction decomposition presented in Sect. 3.2.3 in Chap. 3, the s-domain function $y(s)$ described in (5.21) can be decomposed into the sum of partial fractions as

$$y(s) = A\left(\frac{c_1}{s+\omega \mathrm{j}} + \frac{c_2}{s-\omega \mathrm{j}} + \sum_{j=1}^{N}\sum_{k=1}^{m_j}\frac{c_{j,k}}{(s+\alpha_j)^k}\right), \tag{5.22}$$

where the generically-denoted coefficients c_1, c_2, and $c_{j,k}$ ($j \in \{1, \ldots, N\}, k \in \{1, \ldots, m_j\}$) are determined according to the control system's holistic transfer function. Following the method presented in Sect. 3.2.3 in Chap. 3, we can compute these coefficients as

$$c_1 = \lim_{s\to -\omega j}(s+\omega j)T(s)\frac{\omega}{s^2+\omega^2} = \frac{\omega T(-\omega j)}{-2\omega j} = \frac{j}{2}[T_E(\omega j) - T_O(\omega j)], \quad (5.23)$$

$$c_2 = \lim_{s\to \omega j}(s-\omega j)T(s)\frac{\omega}{s^2+\omega^2} = \frac{\omega T(\omega j)}{2\omega j} = -\frac{j}{2}[T_E(\omega j) + T_O(\omega j)], \quad (5.24)$$

$$c_{j,k} = \lim_{s\to -\alpha_j}\frac{1}{(m_j-k)!}\frac{d^{m_j-k}}{dt^{m_j-k}}\left[(s+\alpha_j)^{m_j}T(s)\frac{\omega}{s^2+\omega^2}\right], \quad (5.25)$$

where $T_O(s)$ and $T_E(s)$ with subscripts O and E denote the sub-polynomials of $T(s)$ with odd orders and even orders respectively ($T(s)$ may be viewed as infinite power series). Substitute (5.23) and (5.24) into (5.22) and obtain

$$\begin{aligned} y(s) &= A\frac{T_E(\omega j)\omega - T_O(\omega j)js}{s^2+\omega^2} + A\sum_{j=1}^{N}\sum_{k=1}^{m_j}\frac{c_{j,k}}{(s+\alpha_j)^k} \\ &= A\frac{T_R(\omega j)\omega + T_I(\omega j)s}{s^2+\omega^2} + A\sum_{j=1}^{N}\sum_{k=1}^{m_j}\frac{c_{j,k}}{(s+\alpha_j)^k}, \end{aligned} \quad (5.26)$$

where $T_R(\omega j)$ and $T_I(\omega j)$ denote the *real part* and the *imaginary part* of $T(\omega j)$ respectively, namely

$$T(\omega j) = T_R(\omega j) + T_I(\omega j)j.$$

Since $T_E(\omega j)$ contributes to the real part of $T(\omega j)$ whereas $T_O(\omega j)$ contributes to the imaginary part of $T(\omega j)$, we have

$$\begin{aligned} T_E(\omega j) &= T_R(\omega j), \\ T_O(\omega j) &= T_I(\omega j)j \iff T_I(\omega j) = -T_O(\omega j)j, \end{aligned}$$

which are used in derivation of (5.26). From (5.26), the original time-domain function of $y(s)$ is recovered as

$$y = A(T_R(\omega j)\sin(\omega t) + T_I(\omega j)\cos(\omega t)) + A\sum_{j=1}^{N}\sum_{k=1}^{m_j}c_{j,k}e^{-\alpha_j t}\frac{t^{k-1}}{(k-1)!}. \quad (5.27)$$

As the poles $-\alpha_1, -\alpha_2, \ldots, -\alpha_N$ are all with negative real part, all the terms $e^{-\alpha_j t}t^{k-1}$ ($j \in \{1, \ldots, N\}, k \in \{1, \ldots, m_j\}$) converge to zero at time infinity $t \to \infty$ and we have

$$\begin{aligned} \lim_{t\to\infty} y = y_S &\equiv A(T_R(\omega j)\sin(\omega t) + T_I(\omega j)\cos(\omega t)) \\ &= A|T(\omega j)|\sin(\omega t + \angle T(\omega j)), \end{aligned} \quad (5.28)$$

where

$$|T(\omega j)| = \sqrt{T_R(\omega j)^2 + T_I(\omega j)^2}$$

is the **magnitude** of $T(\omega j)$ and $\angle T(\omega j)$ that satisfies

$$\sin(\angle T(\omega j)) = T_I(\omega j)/|T(\omega j)|, \quad \cos(\angle T(\omega j)) = T_R(\omega j)/|T(\omega j)|$$

is the **phase** or **phase angle** of $T(\omega j)$.

Equation (5.28) conveys that the control system transforms sinusoidal input asymptotically into steady-state sinusoidal output via the **frequency transfer function** $T(\omega j)$. More specifically, the effect of the frequency transfer function $T(\omega j)$ consists in two aspects: First, it scales the amplitude of the sinusoidal input by a **gain** that is the same to its magnitude. Second, it shifts the phase of the sinusoidal input by an **offset** that is the same to its phase. It is worth noting that the frequency transfer function $T(\omega j)$ is obtained simply by substituting $s = \omega j$ into the transfer function $T(s)$.

The frequency transfer function $T(\omega j)$ varies as the frequency ω varies. To visualize how the magnitude and phase of the frequency transfer function $T(\omega j)$ varies according to the frequency ω, we normally resort to two **Bode plots** (Oppenheim et al. 1997), one of which is the **magnitude Bode plot** and the other is the **phase Bode plot**. In the magnitude Bode plot, the *logarithmic gain* namely the *logarithmic magnitude* of the frequency transfer function $T(\omega j)$ such as

$$\text{Logarithmic gain} = 20 \log_{10} |T(\omega j)| \tag{5.29}$$

is expressed with respect to the *logarithmic frequency*. It is worth noting that units of logarithmic gains are a kind of *decibels* (*dB*). In the phase Bode plot, the phase or phase angle of the frequency transfer function $T(\omega j)$ is expressed with respect to the logarithmic frequency (why logarithmic plots are used is left to readers for reflection). An example of Bode plots for the frequency transfer function $T_1(\omega j) = 1/\left(1 + \frac{\omega}{10} j\right)$ is illustrated in Fig. 5.5.

The frequency transfer function in above example corresponds to the transfer function $T_1(s) = 1/\left(1 + \frac{s}{10}\right)$ which has a pole $-\alpha = -10$. Qualitatively speaking, as demonstrated in Fig. 5.5, it seems that $\omega = |-\alpha| = 10$ is a turning frequency point, because it is at this frequency point that the magnitude of the frequency transfer function starts to decrease noticeably and it is also at this frequency point that the phase of the frequency transfer function has the locally sharpest change. Quantitatively speaking, the magnitude Bode plot curve has two *asymptotic lines* as $\omega \to 0$ and $\omega \to \infty$ respectively and these two asymptotic lines intersect exactly at the **turning frequency point**. In fact, for a control system, magnitudes of its zeros and poles do correspond to turning frequency points in Bode plots of its frequency transfer function. This facilitates frequency response based control system identification, clarification of which is postponed to Sect. 5.3.3.

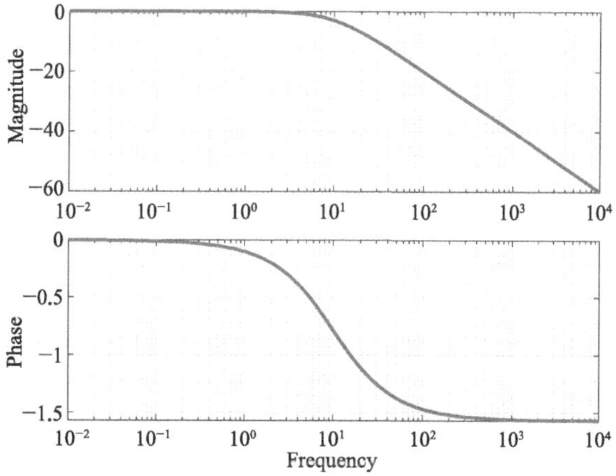

Fig. 5.5 Bode plots for $T_1(\omega \mathrm{j}) = 1/\left(1 + \frac{\omega}{10}\mathrm{j}\right)$

5.3.2 *Bandwidth and Responsiveness*

As demonstrated in Fig. 5.5, the frequency transfer function $T_1(\omega \mathrm{j}) = 1/\left(1 + \frac{\omega}{10}\mathrm{j}\right)$ serves as a **low-pass filter** (Oppenheim et al. 1997) which allows low-frequency signals to pass while suppressing high-frequency signals. More specifically, for sinusoidal input with frequency smaller or even much smaller than the turning frequency point $\omega = |-\alpha| = 10$, the corresponding gain has a square close to 1 or even almost the same to 1, which implies that most part or even almost all part of the sinusoidal input energy is passed to output. In contrast, for sinusoidal input with frequency higher or even much higher than the turning frequency point, the corresponding gain has a square close to 0 or even almost the same to 0, which implies that most part or even almost all part of the sinusoidal input energy is suppressed without being passed to output.

For the frequency transfer function $T_1(\omega \mathrm{j}) = 1/\left(1 + \frac{\omega}{10}\mathrm{j}\right)$, its **pass-band** or **band** can be regarded as the frequency range between 0 and the turning frequency point $\omega = |-\alpha| = 10$, and hence its **bandwidth** is $|-\alpha|$, the same to the magnitude of the pole of its associated transfer function $T_1(s) = 1/\left(1 + \frac{s}{10}\right)$.

Consider a generic first-order control system modelled by $T_{1\mathrm{g}}(s) = 1/\left(1 + \frac{s}{\alpha_1}\right)$ whose step response has no steady-state error. Its pole is $-\alpha_1$. The further the pole $-\alpha_1$ is to the left of the imaginary axis namely the larger the pole magnitude α_1 is, the more responsive it is.

Its frequency transfer function is $T_{1\mathrm{g}}(\omega \mathrm{j}) = 1/\left(1 + \frac{\omega}{\alpha_1}\mathrm{j}\right)$ that has Bode plots similar to those demonstrated in Fig. 5.5 and has a turning frequency point at $\omega = \alpha_1$. The generic frequency transfer function $T_{1\mathrm{g}}(\omega \mathrm{j}) = 1/\left(1 + \frac{\omega}{\alpha_1}\mathrm{j}\right)$ is a low-pass filter with bandwidth α_1. Therefore, for the control system modelled by $T_{1\mathrm{g}}(s) = 1/\left(1 + \frac{s}{\alpha_1}\right)$, we can establish a relationship between its *responsiveness* and its *bandwidth* namely the bandwidth of its associated frequency transfer function: *The more responsive it* *is, which implies the further its pole $-\alpha_1$ is to the left of the imaginary axis namely the larger the pole magnitude α_1 is, the larger its bandwidth is.*

By analogue to above generic first-order control system, qualitatively speaking, a properly-designed control system in practical applications normally functions like a low-pass filter (the reason will be explained soon in this section) and has the following mutual relationship between its responsiveness and its bandwidth: *The more responsive it is, the larger its bandwidth is. The less responsive it is, the smaller its bandwidth is.*

> **Control system responsiveness-bandwidth qualitative criterion**: *For a stable linear time-invariant control system, the more responsive it is, the larger its bandwidth is. The less responsive it is, the smaller its bandwidth is.*

Above qualitative criterion may also be understood as follows: For a control system, the more responsive it is, the better its output tends to follow high-frequency input, which implies that the larger is the range of frequencies that it allows to pass. The less responsive it is, the worse its output tends to follow high-frequency input, which implies that the smaller is the range of frequencies that it allows to pass. To further understand the point, readers may compare two extreme cases: $T_0(\omega j) = \delta(\omega)$ (impulse function) with zero bandwidth, which allows no signal change and hence is not responsive at all, in contrast with $T_\infty(\omega j) = 1$ with infinite bandwidth, which passes input to output steady state immediately without any delay and hence is most responsive.

In practical applications, a control system's output change usually comes from two sources: operational changes, and noisy changes. Operational changes are intended and should be realized by the control system, whereas noisy changes are undesirable and should be suppressed. For this purpose from signal processing perspective, the control system had better maintain its frequency transfer function gain (typically being *one*) for operational changes while suppressing its frequency transfer function gain to zero for noisy changes. If operational changes were similar to noisy changes, then these two aspects of expectation would contradict each other.

Fortunately, in practical applications, operational changes are normally of low frequency, whereas noisy changes are normally of high frequency. Therefore, we normally *design the intended closed-loop feedback control system to function like a low-pass filter with bandwidth boundary well separating low frequencies associated with operational changes and high frequencies associated with noisy changes* so that operational changes can be properly realized whereas noisy changes would be suppressed.

In other words, the control system bandwidth *should be large enough* for the pass-band to cover low frequencies associated with operational changes on one hand, and *should not be too large* for the pass-band to also cover high frequencies associated with noisy changes on the other hand. According to the *control system responsiveness-bandwidth qualitative criterion*, the control system *should be responsive enough* to follow operational changes on one hand, and *should not be too responsive* to be susceptible to noisy changes on the other hand. **Moderate responsiveness** namely being neither under-responsive nor over-responsive is an important concern for control purpose.

> **Moderate responsiveness concern**: *A control system should be neither under-responsive nor over-responsive for practical applications.*

5.3.3 Bode Plots for Control System Identification

As briefly mentioned in Sect. 5.3.1, for a control system, magnitudes of its zeros and poles correspond to turning frequency points in Bode plots of its frequency transfer function. We clarify this point with further examples. For this purpose, we use following Matlab code to conveniently generate Bode plots for frequency transfer functions.

BodePlots.m

```
% Transfer function: T(s) = (b0 + b1*s + b2*s^2 + ...)/(a0 + a1*
    s + a2*s^2 + ...)
B = [1, 1/10]; % Numerator definition [b0, b1, b2, ...]
A = [1, 101/100, 1/100]; % Denominator definition [a0, a1, a2,
    ...]

wSpan = 10.^(-2:0.1:4); idx = 0; % Frequency plotting span
% Record magnitude and phase
Mag = zeros(size(wSpan)); Phs = zeros(size(wSpan));
for w = wSpan
    idx = idx+1;
    % T(w*j) = (Br + Bi*j)/(Ar + Ai*j)
    % B real part
    Br = sum(B(1:2:end).*(-w^2).^((0:2:(length(B)-1))/2));
    % B imaginary part
    Bi = sum(B(2:2:end).*w.*(-w^2).^((0:2:(length(B)-2))/2));
    % A real part
    Ar = sum(A(1:2:end).*(-w^2).^((0:2:(length(A)-1))/2));
    % A imaginary part
    Ai = sum(A(2:2:end).*w.*(-w^2).^((0:2:(length(A)-2))/2));

    Mag(idx) = sqrt(Br^2+Bi^2)/sqrt(Ar^2+Ai^2); % Magnitude gain
    Phs(idx) = atan2(Bi, Br) - atan2(Ai, Ar); % Phase offset
end

Mag = 20*log10(Mag); % Logarithmic gain
% Magnitude Bode plot
subplot(2,1,1); semilogx(wSpan, Mag, 'LineWidth', 2);
ylabel('Magnitude'); grid on;
% Phase Bode plot
subplot(2,1,2); semilogx(wSpan, Phs, 'LineWidth', 2);
```

```
xlabel('Frequency'); ylabel('Phase'); grid on;

% Matlab built-in
% T = tf(B(end:-1:1), A(end:-1:1)); figure(2), bode(T);
```

It is worth noting that Matlab has built-in functions for Bode plots generation, as indicated at the last line of the Matlab code. If Matlab is unavailable, readers can adapt the Matlab code *BodePlots.m* to implementation by other programming languages. The Matlab code *BodePlots.m* generates Bode plots for the frequency transfer function

$$T_c(\omega j) = \frac{1 + \frac{\omega}{10}j}{(1 + \omega j)\left(1 + \frac{\omega}{100}j\right)}, \tag{5.30}$$

whose associated transfer function is

$$T_c(s) = \frac{1 + \frac{s}{10}}{1 + \frac{101\,s}{100} + \frac{s^2}{100}} = \frac{1 + \frac{s}{10}}{(1 + s)\left(1 + \frac{s}{100}\right)}. \tag{5.31}$$

If readers want to generate Bode plots for another frequency transfer function and its associated transfer function, readers can simply replace specification of arrays B and A in the Matlab code by another specification, where B and A specify numerator coefficients and denominator coefficients of the transfer function respectively. For example, to generate Bode plots for the frequency transfer function $T_1(\omega j) = 1/\left(1 + \frac{\omega}{10}j\right)$ and its associated transfer function $T_1(s) = 1/\left(1 + \frac{s}{10}\right)$, readers can replace specification of arrays B and A to $B = [1]$ and $A = [1, 1/10]$. This is how Bode plots illustrated in Fig. 5.5 are generated.

More examples of Bode plots for various frequency transfer functions are given below. Bode plots for the frequency transfer function $T_2(\omega j) = 1 + \frac{\omega}{10}j$ and its associated transfer function $T_2(s) = 1 + \frac{s}{10}$ are illustrated in Fig. 5.6. Bode plots for the frequency transfer function $T_3(\omega j) = 1/\left(1 + \frac{\omega}{10}\,j - \frac{\omega^2}{100}\right)$ and its associated transfer function $T_3(s) = 1/\left(1 + \frac{s}{10} + \frac{s^2}{100}\right)$ are illustrated in Fig. 5.7. Bode plots for the frequency transfer function $T_4(\omega j) = 1 + \frac{\omega}{10}\,j - \frac{\omega^2}{100}$ and its associated transfer function $T_4(s) = 1 + \frac{s}{10} + \frac{s^2}{100}$ are illustrated in Fig. 5.8.

The transfer function $T_2(s)$ has a zero $\beta = -10$ without any pole—**Zeros** refer to numerator roots of a control system's transfer function, in contrast with **poles** which refer to denominator roots of its transfer function—As demonstrated in Fig. 5.6, the magnitude of the zero β corresponds to a turning frequency point of Bode plots for $T_2(\omega j)$. Different from the *pole turning frequency point* in Fig. 5.5 which turns the magnitude down, this *zero turning frequency point* in Fig. 5.6 turns the magnitude up.

The same tendency of turning frequency points can also be observed in Fig. 5.7 where the transfer function $T_3(s)$ has complex poles and in Fig. 5.8 where the transfer function $T_4(s)$ has complex zeros. It is worth noting that each pair of conjugate

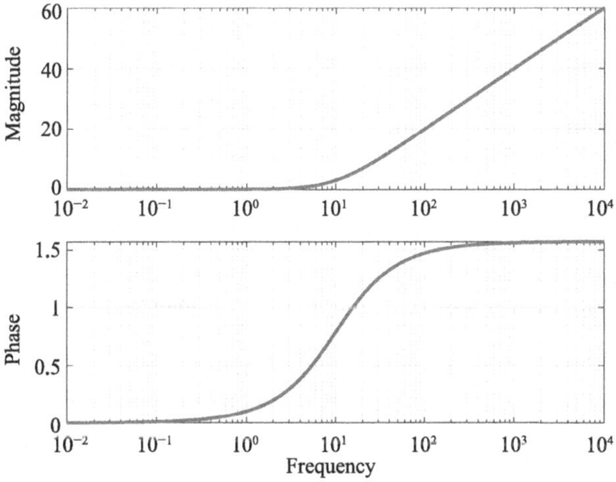

Fig. 5.6 Bode plots for $T_2(\omega j) = 1 + \frac{\omega}{10}j$

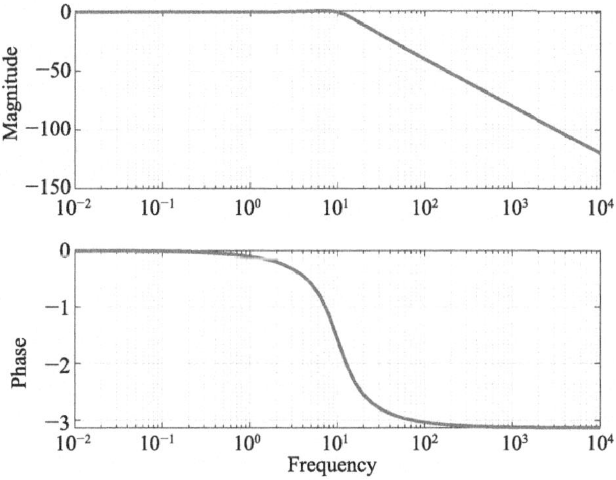

Fig. 5.7 Bode plots for $T_3(\omega j) = 1/\left(1 + \frac{\omega}{10}j - \frac{\omega^2}{100}\right)$

complex poles or conjugate complex zeros have the same magnitude and hence correspond to the same turning frequency point.

Generally speaking, for a linear time-invariant control system, a pole turning frequency point tends to turn the frequency transfer function gain down, whereas a zero turning frequency point tends to turn the frequency transfer function gain up. Besides, the turning effects of pole turning frequency points and zero turning frequency points can be *superposed*. For example, compared with the turning frequency point associated with a single pole, the turning frequency point associated with a double pole or a pair of conjugate complex poles tends to turn the frequency transfer function gain

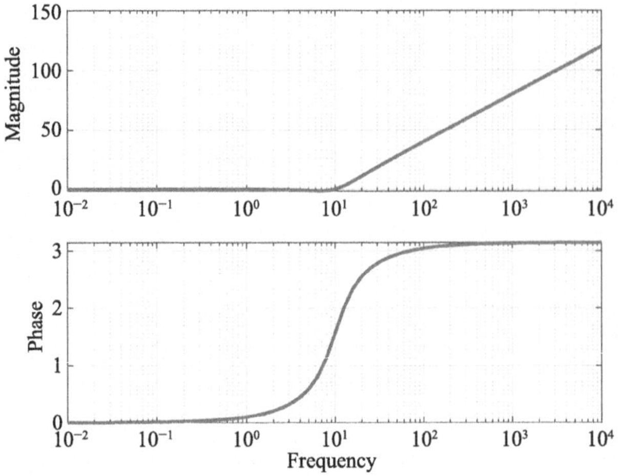

Fig. 5.8 Bode plots for $T_4(\omega \mathrm{j}) = 1 + \frac{\omega}{10}\mathrm{j} - \frac{\omega^2}{100}$

down "twice" sharply. Similarly, compared with the turning frequency point associated with a single zero, the turning frequency point associated with a double zero or a pair of conjugate complex zeros tends to turn the frequency transfer function gain up "twice" sharply. Readers can observe such kinds of effect superposition by comparing Bode plots in Figs. 5.5 and 5.7 and by comparing Bode plots in Figs. 5.6 and 5.8.

> **Turning frequency point criterion**: *For a linear time-invariant control system, a pole turning frequency point tends to turn the frequency transfer function gain down, whereas a zero turning frequency point tends to turn the frequency transfer function gain up. The turning effects of pole turning frequency points and zero turning frequency points can be superposed.*

The turning frequency point criterion can facilitate frequency response based control system identification. More specifically, we stimulate the control system by testing sinusoidal input with various frequencies and amplitudes, and record corresponding sinusoidal output steady states. By such experiments for the control system, we can determine its frequency transfer function gains and phase offsets associated with various frequencies. Then we can further depict Bode plots for its frequency transfer function, from which we can inversely infer zeros and poles of the control system by observing turning frequency points in the Bode plots.

For example, consider Bode plots illustrated in Fig. 5.9, in which we can observe three turning frequency points about the frequencies $\omega_1 = 1$, $\omega_2 = 10$, and $\omega_3 = 100$ respectively.

The first turning frequency point namely the one about the frequency ω_1 turns the magnitude down at a slope of 20 decibels per frequency decade, which implies

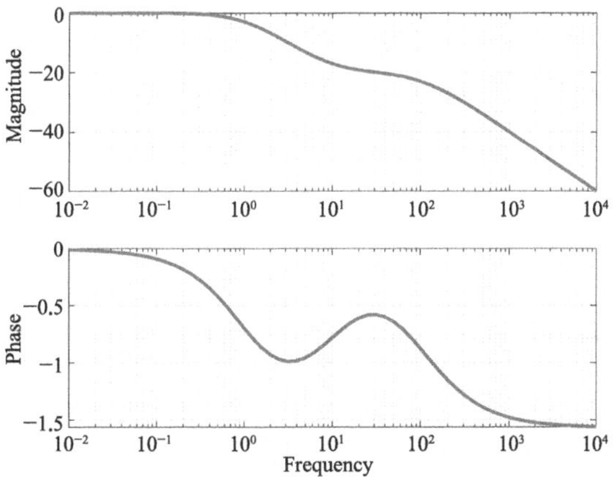

Fig. 5.9 Frequency response based control system identification

that the first turning frequency point corresponds to a single control system pole. The second turning frequency point namely the one about the frequency ω_2 stops the magnitude falling tendency and turns the magnitude flat, which implies that the second turning frequency point corresponds to a single control system zero that tends to turn the magnitude up and hence counteracts the falling tendency brought by the first turning frequency point. The third turning frequency point namely the one about the frequency ω_3 turns the magnitude down again at a slope of 20 decibels per frequency decade, which implies that the third turning frequency point corresponds to another single control system pole. Besides, the frequency transfer function gain at the frequency $\omega = 0$ is 1. Therefore, we can infer that the transfer function associated with Bode plots illustrated in Fig. 5.9 is about

$$\frac{1+\frac{s}{10}}{(1+s)\left(1+\frac{s}{100}\right)},$$

which is the transfer function $T_c(s)$ described in (5.31). In fact, the transfer function ground-truth associated with Bode plots illustrated in Fig. 5.9 is indeed the transfer function $T_c(s)$ described in (5.31).

The frequency response based control system identification technique demonstrated in above example of Fig. 5.9 may somewhat seem ideal. For more complicated control systems especially for those involving conjugate complex poles and zeros, we need more delicate and systematic control system identification techniques. However, we have no intention to deviate far on this topic because it is beyond the focus of this book.

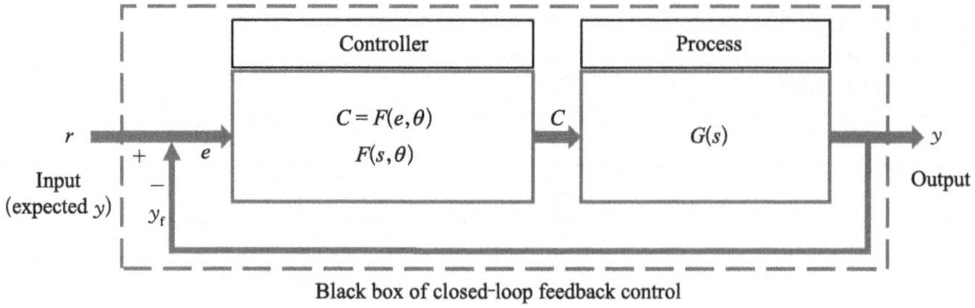

Fig. 5.10 Generic controller design for closed-loop feedback control

5.4 Control System Design

After clarification of important concerns namely **stability**, **accuracy**, and **responsiveness** for control purpose, we now consider how to design a control system (normally a closed-loop feedback control system) so that its **stability**, its **accuracy**, and its **moderate responsiveness** can be achieved.

5.4.1 Controller Parameter Design

Control system design normally refers to design of the controller in a closed-loop feedback control system, namely in mathematical sense to design the controller mapping $C = F(e)$ between the input e and the output C, as illustrated in Fig. 5.10. In practical applications, we do not design a controller blindly but normally design it according to certain control methodology. Controllers namely control methods that share the same control methodology can be regarded as a **family of controllers** which usually share a common control architecture and differ from each other only by a set of tunable **controller parameters**. The expression $C = F(e, \boldsymbol{\theta})$ in Fig. 5.10 denotes a family of controllers generically, where $\boldsymbol{\theta}$ denotes the set of controller parameters.

For example, the family of **proportional-integral-derivative (PID)** controllers is illustrated in Fig. 5.11. This is the most famous and popular family of controllers (Samad 2017) that share the same control methodology of generating the control law of C namely the controller output by combining linearly a proportional term, an integral term, and a derivative term of the feedback error e namely the controller input as

$$C = Pe + I \int e\mathrm{d}t + D\frac{\mathrm{d}}{\mathrm{d}t}e, \tag{5.32}$$

$$C(s) = \left(P + \frac{I}{s} + Ds\right) e(s),$$

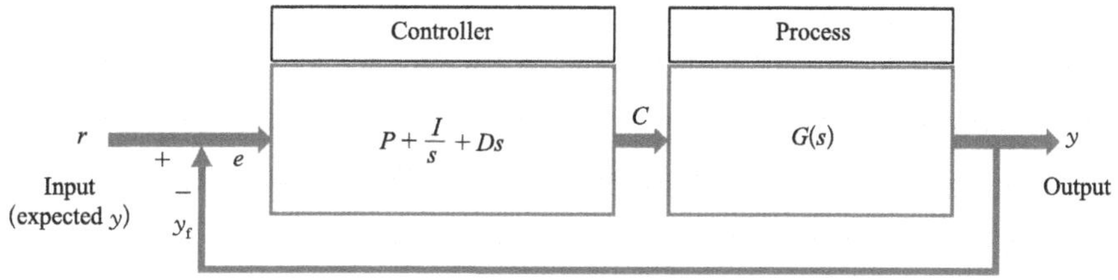

Fig. 5.11 Proportional-integral-derivative (PID) controller design

where the set of controller parameters for the PID control method[2] is the set of proportional, integral, and derivative coefficients (P, I, D).

It is worth noting that the PID control method can also be called the **proportional-integral-differential** method, where D means *differential*. In practical applications where digital processors are commonly adopted nowadays, the control period or numeric computation period dt can be naturally incorporated into the D-coefficient and the I-coefficient in implementation, and hence the PID control method is transformed into

$$C = Pe + \bar{I}\sum e + \bar{D}\mathrm{d}e, \tag{5.33}$$

where the *derivative* term becomes the *differential* term and the *integral* term becomes the *sum* term. The PID coefficient $\bar{I}$ can be imagined as the original I-coefficient in (5.32) multiplied by $\mathrm{d}t$, namely $\bar{I} \equiv I\mathrm{d}t$. A sum can be treated as a special kind of integral, so *PID* still keeps its acronym letter I representing *integral*, without replacement of the acronym letter I by the acronym letter S that represents *sum*. In (5.33), the PID coefficient $\bar{D}$ can be imagined as the original D-coefficient in (5.32) divided by $\mathrm{d}t$, namely $\bar{D} \equiv D/\mathrm{d}t$. Since there is no essential difference between the PID control method described in (5.32) and that described in (5.33), there is no essential difference between interpretation of *PID* as *proportional-integral-derivative* and interpretation of *PID* as *proportional-integral-differential*.

Once the control method such as the PID control method is determined, controller design actually becomes **controller parameter design**. More specifically, given a determined family of controllers, design the set of controller parameters to obtain certain concrete controller that enables the control system to achieve stability, accuracy, and moderate responsiveness.

For the generic closed-loop feedback control system with parametrized controller $F(s, \boldsymbol{\theta})$ illustrated in Fig. 5.10, it is comparatively easy to analyse its accuracy. Let the control system take step input $r(t) = r_0$. Then the output steady state is derived according to the *final value theorem* as

[2] For expression convenience yet without causing confusion, we sometimes abuse the term **control method** to mean both the general control methodology shared by a family of controllers and certain concrete controller with specific controller parameters in this family.

$$y_S = \lim_{s \to 0^+} sy(s) = r_0 \lim_{s \to 0^+} \frac{F(s, \theta)G(s)}{1 + F(s, \theta)G(s)}.$$

So the feedback error steady state is

$$e_S = r_0 \lim_{s \to 0^+} \frac{1}{1 + F(s, \theta)G(s)} \equiv r_0 G_e(\theta).$$

It is usually not difficult to determine proper ranges for controller parameters θ such that $G_e(\theta) \to 0$ or $G_e(\theta)$ is within certain relative error tolerance range r_{tol} namely $|G_e(\theta)| \leqslant r_{tol}$, where the relative error tolerance range r_{tol} can be set flexibly according to concrete applications.

Both a control system's stability and responsiveness depend on locations of its poles especially its *dominant poles*. It is comparatively difficult to analyse how controller parameters θ influence locations of control system poles and determine proper ranges for controller parameters θ such that control system poles especially dominant control system poles are located desirably in the s-domain.

The characteristic equation of the control system can be described equivalently as

$$C_F(s, \theta) \equiv 1 + F(s, \theta)G(s) = 0, \tag{5.34}$$

where $C_F(s, \theta)$ is a characteristic polynomial with parametrized coefficients in terms of controller parameters θ. As already clarified in Sect. 4.3 in Chap. 4, it is normally difficult and even impossible to directly express control system poles analytically in terms of these parametrized coefficients. To analyse how controller parameters θ influence locations of control system poles, we may resort to *indirect* methods. Among these indirect methods, a representative method is the root locus method, and another representative method is the Routh-Hurwitz criterion based axis shift method.

5.4.2 Root Locus Method

The spirit of the root locus method (Evans 1954) is to analyse how controller parameters θ influence locations of control system poles by exploiting zeros and poles of an open-loop transfer function. Although the control system may have multiple controller parameters to design, *each round of applying the root locus method focuses on only one controller parameter, namely focuses on how an individual controller parameter influences locations of control system poles.* So for presentation convenience, we suppose that the controller has only one parameter to design. In other words, we consider such **controller-unique-parameter design** problem.

Designing the proportional coefficient P for a proportional feedback control system illustrated in Fig. 5.12 is a typical example of controller-unique-parameter design.

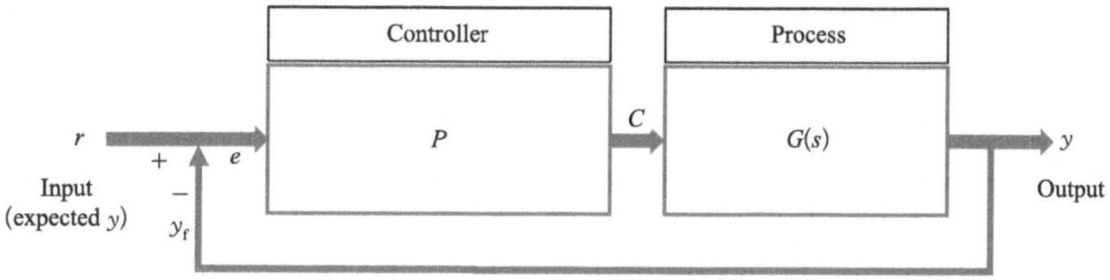

Fig. 5.12 Proportional controller design

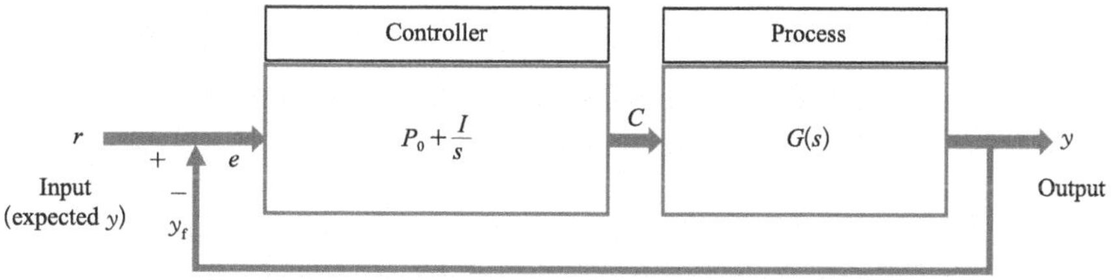

Fig. 5.13 Proportional-integral controller design with proportional coefficient fixed

For another example, consider the proportional-integral feedback control system illustrated in Fig. 5.13. Suppose that the proportional coefficient P is fixed to certain proper value P_0 whereas the integral coefficient I is left for design. Designing the integral coefficient I for the proportional-integral controller with the proportional coefficient P fixed as $P \equiv P_0$ also belongs to controller-unique-parameter design.

For controller-unique-parameter design, the parametrized characteristic equation of the control system can always be transformed into the form described in (5.35) with K being different from the original controller parameter only by a constant scaling factor:

$$1 + KF(s) = 0, \tag{5.35}$$

where

$$F(s) = \frac{\prod_{i=1}^{m}(s + z_i)}{\prod_{j=1}^{n}(s + p_j)}.$$

It is worth noting that the poles $-p_j$ and zeros $-z_i$ are respectively poles and zeros of the open-loop transfer function $F(s)$, but not those of the closed-loop feedback control system's holistic transfer function.

We focus on determining the locus of roots of the parametrized characteristic equation described in (5.35) when K varies as $0 \leqslant K < \infty$, whereas the case when K varies as $-\infty < K \leqslant 0$ can be analysed similarly. Rewrite (5.35) as

$$\prod_{j=1}^{n}(s+p_j)+K\prod_{i=1}^{m}(s+z_i)=0. \tag{5.36}$$

Equations (5.35) and (5.36) give two different forms to describe the characteristic equation.

When $K=0$, (5.36) is reduced to

$$\prod_{j=1}^{n}(s+p_j)=0,$$

which implies that control system poles associated with $K=0$ coincide with the poles of the open-loop transfer function $F(s)$.

Conversely, when $K\to\infty$, (5.36) is reduced to

$$\frac{1}{K}\prod_{j=1}^{n}(s+p_j)+\prod_{i=1}^{m}(s+z_i)\to\prod_{i=1}^{m}(s+z_i)=0,$$

which implies that control system poles associated with $K\to\infty$ coincide with the zeros of the open-loop transfer function $F(s)$. Therefore, *the locus of the roots of the characteristic equation described in* (5.35) *begins at the poles of the open-loop transfer function* $F(s)$ *and ends at the zeros of* $F(s)$ *as* K *increases from 0 to* ∞. Most open-loop transfer functions $F(s)$ encountered in practical applications have more poles than zeros (namely $n>m$), which implies $n-m$ branches of the root locus end at $n-m$ imaginary zeros at infinity. With above preliminary preparation, we can formally present the root locus method.

Step 1: Locate real axis segments that belong to the root locus. For this purpose, take advantage of a useful fact that *each point of these real axis segments is always to the left of an odd number of poles and zeros of the open-loop transfer function* $F(s)$.

Proof For each root locus point located on the real axis, consider the *phase angles* of its zero terms and pole terms of the open-loop transfer function $F(s)$:

$$\sum_{i=1}^{m}\angle(s+z_i)-\sum_{j=1}^{n}\angle(s+p_j)=\angle\frac{\prod_{i=1}^{m}(s+z_i)}{\prod_{j=1}^{n}(s+p_j)}=\angle-\frac{1}{K}=\pi+2k\pi, \tag{5.37}$$

which describes the **phase angle criterion** of the open-loop transfer function $F(s)$.

If the root locus point is to the right of a real zero $-z_i$ or a real pole $-p_j$, then the phase angle of the zero term $s+z_i$ or the pole term $s+p_j$ namely $\angle(s+z_i)$ or $\angle(s+p_j)$ is 0. In contrast, if the root locus point is to the left of a real zero $-z_i$ or a real pole $-p_j$, then $\angle(s+z_i)$ or $\angle(s+p_j)$ is π.

Similarly, if the root locus point is to the right of a pair of conjugate zeros or conjugate poles, then the average phase angle of the conjugate zero terms or the conjugate pole terms is 0. In contrast, if the root locus point is to the left of a pair of conjugate zeros or conjugate poles, then the average phase angle of the conjugate zero terms or the conjugate pole terms is π.

For each root locus point located on the real axis, since the phase angles of all its zero terms and pole terms add[3] to $(2k + 1)\pi$, it must be to the left of an odd number of poles and zeros of the open-loop transfer function $F(s)$. □

It is worth noting another fact that *root locus part not located on the real axis must be symmetrical with respect to the real axis* because complex roots of a real-coefficient polynomial must appear in conjugate pairs.

Step 2: Compute the center c_A and angles ϕ_A of *asymptotes* along which $n - m$ branches of the root locus proceed to the $n - m$ imaginary zeros at infinity. The common center c_A of these asymptotes is located on the real axis, given by

$$c_A = \frac{\sum \text{Poles of } F(s) - \sum \text{Zeros of } F(s)}{n - m} = \frac{\sum_{j=1}^{n}(-p_j) - \sum_{i=1}^{m}(-z_i)}{n - m}. \tag{5.38}$$

The angles of the $n - m$ asymptotes with respect to the real axis are given by

$$\phi_{A(k)} = \frac{2k + 1}{n - m}\pi, \quad k = 0, 1, 2, \ldots, n - m - 1. \tag{5.39}$$

Equations (5.38) and (5.39) are useful conclusions that facilitate sketching of the root locus.

Proof Consider the *phase angles* of zero terms and pole terms of the open-loop transfer function $F(s)$ as s proceeds to infinity along the asymptotes:

$$\pi + 2k\pi = \angle - \frac{1}{K} = \angle \frac{\prod_{i=1}^{m}(s + z_i)}{\prod_{j=1}^{n}(s + p_j)} \to \angle \frac{s^m}{s^n} \to (m - n)\phi_A,$$

from which we can easily conclude (5.39).

Denote an asymptote generically as $s = \lambda d_A + c_A$ ($\lambda \in \mathbb{R}^+$) where $d_A = e^{j\phi_A}$ ($d_A^{n-m} = -1$). Consider a variation Δs on the asymptote as $\lambda \to \infty$ such that

[3] When $\angle(s + p_j)$ is 0 or π, $-\angle(s + p_j) = \angle(s + p_j) \mod 2\pi$. This is why we say the phase angles "add" to $(2k + 1)\pi$.

$$1 + K \frac{\prod_{i=1}^{m} (\lambda d_A + c_A + \Delta s + z_i)}{\prod_{j=1}^{n} (\lambda d_A + c_A + \Delta s + p_j)} = 0.$$

We have

$$1 + \frac{K/\lambda^{n-m}}{d_A^{n-m}} \frac{1 + (mc_A + m\Delta s + \sum z_i)\frac{1}{\lambda d_A} + o\left(\frac{1}{\lambda d_A}\right)}{1 + (nc_A + n\Delta s + \sum p_j)\frac{1}{\lambda d_A} + o\left(\frac{1}{\lambda d_A}\right)} = 0$$

$$\Longleftrightarrow \frac{\lambda^{n-m}}{K} = \frac{1 + (mc_A + m\Delta s + \sum z_i)\frac{1}{\lambda d_A} + o\left(\frac{1}{\lambda d_A}\right)}{1 + (nc_A + n\Delta s + \sum p_j)\frac{1}{\lambda d_A} + o\left(\frac{1}{\lambda d_A}\right)}$$

and

$$\begin{aligned} 2k\pi = \angle\frac{\lambda^{n-m}}{K} &= \angle\frac{1 + (mc_A + m\Delta s + \sum z_i)\frac{1}{\lambda d_A} + o\left(\frac{1}{\lambda d_A}\right)}{1 + (nc_A + n\Delta s + \sum p_j)\frac{1}{\lambda d_A} + o\left(\frac{1}{\lambda d_A}\right)} \\ &\to \angle\left[1 + \left(mc_A + m\Delta s + \sum z_i - nc_A - n\Delta s - \sum p_j\right)\frac{1}{\lambda d_A}\right] \\ &= \angle\left[1 + \left(\sum(-p_j) - \sum(-z_i) - (n-m)c_A - (n-m)\Delta s\right)\frac{1}{\lambda d_A}\right], \end{aligned}$$

which implies

$$\sum(-p_j) - \sum(-z_i) - (n-m)c_A - (n-m)\Delta s \to 0.$$

Note $\Delta s \to 0$ as $\lambda \to \infty$, so we have

$$\sum(-p_j) - \sum(-z_i) - (n-m)c_A \to 0,$$

which holds if and only if (5.38) holds. □

Step 3: Locate where the root locus crosses the imaginary axis (if it does so) via the Routh-Hurwitz criterion presented in Sect. 4.3 in Chap. 4. More specifically, for the characteristic polynomial with parametrized coefficients in terms of K, establish its associated parametrized Routh array. Compute values of K that incur zeros in the first column of the Routh array, and solve their corresponding pure imaginary roots.

Step 4: Locate the *breakaway point* on the real axis (if existing). *The root locus breaks away from the real axis where there are multiple roots (typically double roots) and where $K = -1/F(s)$ achieves the maximum for real s.* To understand this first point, note that the breakaway point is where multiple branches of the root locus converge and each of them is associated with one root. To understand this second point, imagine that before K achieves the maximum for real s, there is always some real s that enables $-1/F(s) = K$. However, once K goes beyond such maximum, there is no real s that enables $-1/F(s) = K$, and the root locus must break into complex (non-real) s-domain to enable $-1/F(s) = K$. Therefore, the breakaway point is exactly the point where $K = -1/F(s)$ achieves the maximum for real s.

Besides, the root locus breaks away from the real axis (if it does so) into branches that are equally spaced over 2π.

Proof Suppose that the breakaway point on the real axis is $-s_b$ with multiplicity l and its associated $K = K_b$, then denote

$$\prod_{j=1}^{n}(s+p_j) + K_b \prod_{i=1}^{m}(s+z_i) \equiv (s+s_b)^l H(s) \quad (l \geqslant 2).$$

Consider a positive infinitesimal variation ΔK on K_b and its associated Δs on $-s_b$ of the root locus. As $\Delta K \to 0^+$ and hence $\Delta s \to 0$, we have

$$\begin{aligned} 0 &= \prod_{j=1}^{n}(s+p_j) + (K_b + \Delta K)\prod_{i=1}^{m}(s+z_i) \\ &= (s+s_b)^l H(s) + \Delta K \prod_{i=1}^{m}(s+z_i) \\ &\to \Delta s^l H(-s_b) + \Delta K \prod_{i=1}^{m}(-s_b+z_i) \end{aligned}$$

and

$$\lim_{\Delta s \to 0} \angle \Delta s^l = \angle\left(-\frac{\Delta K \prod_{i=1}^{m}(-s_b+z_i)}{H(-s_b)}\right) \quad \text{namely a constant,}$$

from which we can conclude that the root locus breaks away from the point $-s_b$ into l branches that are equally spaced over 2π. □

To obtain the maximum of K for real s, one way is to differentiate $K = -1/F(s)$, set the differentiated fractional polynomial to zero, and determine real roots analytically or numerically. Another way is to apply numeric optimization methods (Bonnans et al. 2006) such as the golden section method directly on $K = -1/F(s)$.

Step 5: Compute the angles of departure of the root locus from poles and the angles of arrival of the root locus at zeros, according to the *phase angle criterion* described in (5.37). The root locus departure angle ϕ_{-p_k} of the pole $-p_k$ is given by

$$\begin{aligned}\phi_{-p_k} &= \lim_{s \to -p_k} \angle(s + p_k) \\ &= \pi + \sum_{i=1}^{m} \angle(-p_k + z_i) - \sum_{j=1, j \neq k}^{n} \angle(-p_k + p_j) \mod 2\pi.\end{aligned} \tag{5.40}$$

The root locus arrival angle ϕ_{-z_k} of the zero $-z_k$ is given by

$$\begin{aligned}\phi_{-z_k} &= \lim_{s \to -z_k} \angle - (s + z_k) = \pi + \lim_{s \to -z_k} \angle(s + z_k) \\ &= \sum_{j=1}^{n} \angle(-z_k + p_j) - \sum_{i=1, i \neq k}^{m} \angle(-z_k + z_i) \mod 2\pi.\end{aligned} \tag{5.41}$$

Equations (5.40) and (5.41) are particularly helpful for sketching root locus branches associated with complex poles and zeros.

Step 6: Complete the root locus sketch by connecting all segments of the root locus not covered in previous steps. For a point on a branch of the root locus, we can determine the branch direction at the point via following inference:

$$0 = \mathrm{d}(1 + KF(s)) = F(s)\mathrm{d}K + KF'(s)\mathrm{d}s,$$

then we have

$$\mathrm{d}s = -\frac{F(s)}{F'(s)} \frac{\mathrm{d}K}{K} = \frac{1}{F'(s)} \frac{\mathrm{d}K}{K^2} \propto \frac{1}{F'(s)} \tag{5.42}$$

and

$$\angle \mathrm{d}s = \angle \frac{1}{F'(s)} = -\angle F'(s). \tag{5.43}$$

Equations (5.42) and (5.43) generally hold except at breakaway points which incur singularity. In cases of singularity, techniques described at **Step 4** can be applied to determine directions of root locus branches. Equations (5.42) and (5.43) provide us a valuable guide for completing the root locus sketch, especially when we are confused about how branches of the root locus extend to infinity.

In the light of the root locus sketch, we can conveniently determine desirable dominant control system poles. Suppose a chosen dominant pole is s_d, then its corresponding parameter K can be computed as

$$K(s_\mathrm{d}) = \frac{\prod_{j=1}^{n} |s_\mathrm{d} + p_j|}{\prod_{i=1}^{m} |s_\mathrm{d} + z_i|}. \tag{5.44}$$

5.4.3 Root Locus Example

To facilitate understanding of the root locus method, we may consider the root locus for the characteristic equation example described in (5.45) as K varies from 0 to ∞:

$$1 + \frac{K}{s^4 + 12\,s^3 + 64\,s^2 + 128\,s} = 0, \tag{5.45}$$

where $s^4 + 12s^3 + 64s^2 + 128s \equiv D(s)$.

Step 0: Determine zeros and poles of the open-loop transfer function $1/D(s)$:

$$\frac{1}{D(s)} = \frac{1}{s(s+4)(s+4+4\mathrm{j})(s+4-4\mathrm{j})},$$

which has four poles namely 0, -4, $-4-4\mathrm{j}$, and $-4+4\mathrm{j}$ but has no zero. So the root locus has four separate branches that proceed to four imaginary zeros at infinity. The root locus is symmetrical with respect to the real axis.

Step 1: The interval $(-4, 0)$ is the real axis segment that is to the left of an odd number of poles and zeros of the open-loop transfer function $1/D(s)$. So a segment of the root locus exists on the real axis between -4 and 0.

Step 2: Compute the center c_A and angles ϕ_A of asymptotes along which the four branches of the root locus proceed to infinity. The common center c_A of these asymptotes is located on the real axis, given by

$$c_A = \frac{(-4) + (-4-4\mathrm{j}) + (-4+4\mathrm{j}) + 0}{4} = -3.$$

The angles of the four asymptotes with respect to the real axis are given by

$$\phi_{A(k)} = \frac{2k+1}{4}\pi, \quad k = 0, 1, 2, 3$$
$$\phi_A = \frac{\pi}{4}, \frac{3\pi}{4}, \frac{5\pi}{4}, \frac{7\pi}{4}.$$

The four asymptotes are illustrated in Fig. 5.14.

Step 3: Locate where the root locus crosses the imaginary axis via the Routh-Hurwitz criterion. The characteristic polynomial with parametrized coefficients in terms of K is

$$s^4 + 12\,s^3 + 64\,s^2 + 128\,s + K,$$

for which a parametrized Routh array is established as

s^4	1	64 K
s^3	12	128
s^2	160/3	K
s^1	$128 - 9K/40$	
s^0	K	

Pure imaginary roots exist only if zero exists in the first column of the Routh array. Let $128 - 9K/40 = 0$ and obtain $K = 5\,120/9$. The concrete Routh array with $K = 5\,120/9$ is

s^4	1	64	5 120/9
s^3	12	128	
s^2	160/3	5 120/9	
s^1	0		
s^0	5 120/9		

The concrete characteristic polynomial with $K = 5\,120/9$ is factorized as

$$s^4 + 12\,s^3 + 64\,s^2 + 128\,s + \frac{5\,120}{9} = \left(\frac{160}{3}s^2 + \frac{5\,120}{9}\right)\left(\frac{3}{160}s^2 + \frac{9}{40}s + 1\right),$$

where the factor $\frac{160}{3}s^2 + \frac{5\,120}{9}$ associated with pure imaginary roots coincides with the polynomial associated with the row of s^2 in the Routh array. This reflects how the Routh-Hurwitz criterion together with the Routh array can help us quickly locate pure imaginary roots of the characteristic polynomial. When $K = 5\,120/9$, the four roots of the characteristic polynomial are $\pm\sqrt{32/3}\mathrm{j}$ and $-6 \pm \sqrt{52/3}\mathrm{j}$.

Step 4: Locate the breakaway point on the real axis, namely the point where $K = -D(s)$ achieves the maximum for real s:

$$\frac{\mathrm{d}}{\mathrm{d}s}K = -D'(s) = -(4\,s^3 + 36\,s^2 + 128\,s + 128) = 0,$$

which can be solved numerically and the real root is approximately -1.5767. Its corresponding $K \approx 83.5704$. Two branches that break away from the real axis point $s_B \approx -1.5767$ are equally spaced over 2π.

Step 5: Since there is no zero, we only compute the angles of departure of the root locus from poles according to the *phase angle criterion* described in (5.37):

$$\phi_{-4+4\mathrm{j}} = \pi - \left(\frac{\pi}{2} + \frac{\pi}{2} + \frac{3\pi}{4}\right) = \frac{5\pi}{4} \mod 2\pi,$$

$$\phi_{-4-4\mathrm{j}} = \pi - \left(\frac{3\pi}{2} + \frac{3\pi}{2} + \frac{5\pi}{4}\right) = \frac{3\pi}{4} \mod 2\pi,$$

whereas the departure angles of poles -4 and 0 are apparently known.

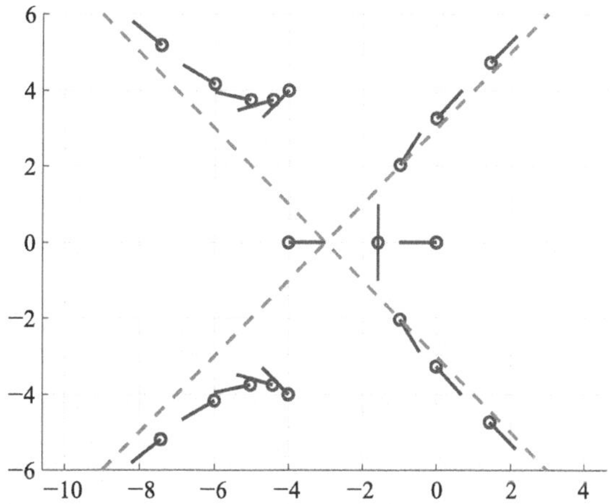

Fig. 5.14 Root locus control points for (5.45)

Step 6: If analysis covered in previous steps is still not clear enough to readers to show how the four branches of the root locus extend to infinity, we may add a few *control points* to better interpolate the contour of root locus branches. For example, we further examine two extra values of K, namely 200 and 2 000. For each of the two extra K values, we solve the roots of its corresponding characteristic polynomial numerically and then determine the root locus branch directions at the roots via (5.42):

$$0 = \mathrm{d}(D(s) + K) = D'(s)\mathrm{d}s + \mathrm{d}K \implies \mathrm{d}s = -\frac{1}{D'(s)}\mathrm{d}K \propto -\frac{1}{D'(s)},$$
$$\angle \mathrm{d}s = \pi - \angle D'(s).$$

The control points that we have by so far include 1) poles of the open-loop transfer function $1/D(s)$, 2) a real axis breakaway point (double roots) and its two other counterpart complex roots, 3) imaginary axis cross points and their two other counterpart complex roots, and 4) control points associated with the two extra K values. The control points together with their root locus branch directions are demonstrated in Fig. 5.14, which is rather clear to show the contour of root locus branches. Then complete the root locus sketch by smooth connection, as demonstrated in Fig. 5.15.

We can resort to modern computational resources to facilitate root locus sketching. Matlab code for sketching the example of root locus control points and their root locus branch directions demonstrated in Fig. 5.14 is given as follows.

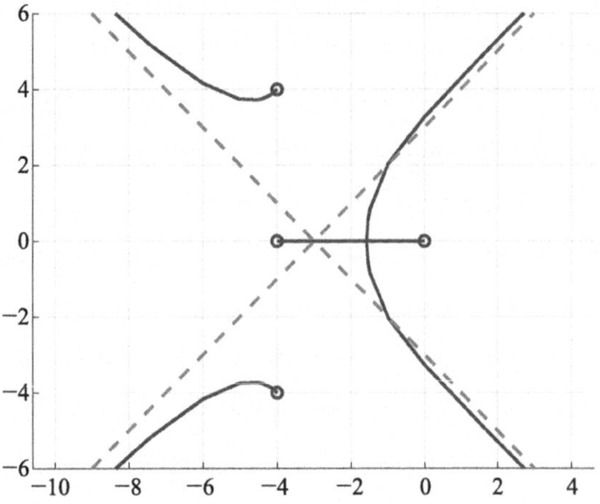

Fig. 5.15 Root locus for (5.45)

RootLocusSketchExample.m

```
syms D(s) dD(s)
D(s) = s^4 + 12*s^3 + 64*s^2 + 128*s;  % 1 + K/D(s) = 0
dD(s) = 4*s^3 + 36*s^2 + 128*s + 128;  % dD(s) = D'(s)

n=4; r2D = zeros(2*n,1); dr2D = r2D; figure(1), hold on;
for K = [0, 83.5704,  200, 5120/9, 2000, 10000]
    % Control system poles for a specific K
    r = vpasolve(D(s) + K == 0, s);
    r2D(1:2:end) = real(r); r2D(2:2:end) = imag(r);
    % Directions of root locus branches
    dr = -1./dD(r); dr = dr./abs(dr);
    dr2D(1:2:end) = real(dr); dr2D(2:2:end) = imag(dr);

    for i = 1:2:2*n
        plot(r2D(i), r2D(i+1), 'ob', 'LineWidth', 2);
        line([r2D(i), r2D(i)+dr2D(i)], [r2D(i+1),
             r2D(i+1)+dr2D(i+1)], 'Color', 'b','LineWidth', 2);
    end
end
plot([-10,4],[-7,7],'--r', [-10,4],[7,-7],'--r', 'LineWidth'2);
grid on; axis equal; ylim([-6, 6]); hold off;
```

Above Matlab code can be easily adapted for tight interpolation even with thousands of control points. With mature numeric computation methods (Quarteroni 2007), modern computational resources, even daily-life computational resources such as personal computers, are sufficient to finish such tight interpolation in moderate time. Since we have a much more convenient way for root locus sketching nowadays, readers may wonder what is the use of the root locus method.

First, it is worth noting that in history the root locus method was developed in years when computational resources were very limited and precious. Root locus sketching by tight interpolation via numeric computation was just a dream probably beyond humans' imagination in those years. Second, even when we have much better computational resources, theoretical analysis techniques reflected in the root locus method can still inspire us in theoretical analysis of more complicated tasks.

In the light of the root locus sketch demonstrated in Fig. 5.15, we can conveniently determine desirable dominant control system poles and its associated parameter K for the parametrized control system modelled by (5.45). As we can know from the root locus sketch, the parameter K must be smaller than the boundary value $5\,120/9$ for the control system to be stable. Otherwise, dominant control system poles will be on or to the right of the imaginary axis and hence incur control system unstability.

In fact, the parameter K should be considerably smaller than the boundary value $5\,120/9$ for dominant control system poles to have comparatively small imaginary parts and hence for the control system to have moderate oscillatory behaviour. For example, to determine the boundary value of K for dominant control system poles to have their imaginary parts smaller than their real parts, draw a line from the origin with phase angle $3\pi/4$ and locate its intersection point with the root locus graphically. It is unnecessary to compute the intersection point s_I, just measure its distances to the four poles graphically and then compute the associated K as

$$K_\text{I} = |s_\text{I}||s_\text{I} + 4||s_\text{I} + 4 - 4\text{j}||s_\text{I} + 4 + 4\text{j}| \approx 126.$$

So the parameter K should be no larger than the approximate boundary value 126 for dominant control system poles to have their imaginary parts smaller than their real parts, as demonstrated in Fig. 5.16.

We can also know from the root locus sketch that the dominant control system pole (or poles) is always on or to the right of the real axis breakaway point. Therefore, to achieve the highest control system responsiveness, we may set the parameter to 84 which corresponds approximately to the real axis breakaway point on the root locus.

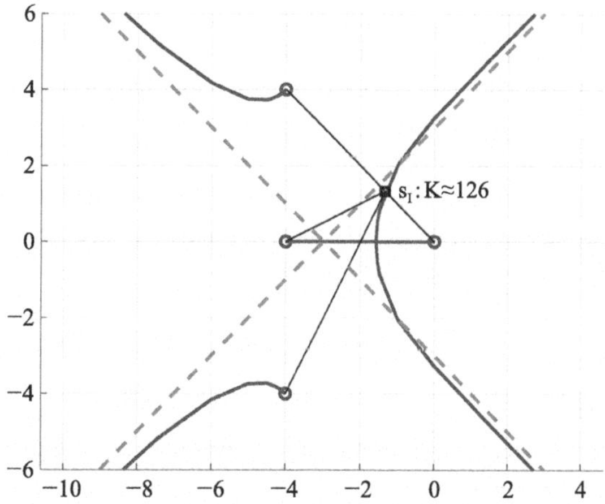

Fig. 5.16 Special intersection point of root locus for (5.45)

With above concrete example, we believe readers can better understand the spirit of the root locus method for *controller-unique-parameter design*. The same spirit can be extended for *controller-multiple-parameter design*: In each round of root locus sketching, vary only one controller parameter while fixing other controller parameters. Alternate the varied parameter in different rounds of root locus sketching. In such iterative way, we can obtain a family of root locus sketches that describe how multiple controller parameters influence locations of control system poles, based on which we then determine a proper set of controller parameters according to concrete applications.

5.4.4 Routh-Hurwitz Criterion Based Axis Shift Method

For the parametrized characteristic polynomial described in (5.34), we can establish its associated Routh array with parametrized elements in terms of controller parameters $\boldsymbol{\theta}$ and then use the Routh-Hurwitz criterion to determine the range of $\boldsymbol{\theta}$ for the control system to be stable, namely for the control system to have all poles to the left of the imaginary axis.

Following the idea of **axis shift**, we can adapt the Routh-Hurwitz criterion to determine the range of $\boldsymbol{\theta}$ for the control system to have all poles to the left of an arbitrary target axis besides the imaginary axis. Supposing that the target axis is $s = -a$ $(a > 0)$, transform (5.34) by axis shift and obtain

$$C_{\mathrm{F}}(s, \boldsymbol{\theta}, a) \equiv 1 + F(s-a, \boldsymbol{\theta})G(s-a) = 0, \tag{5.46}$$

where $C_{\mathrm{F}}(s, \boldsymbol{\theta}, a)$ is a new parametrized characteristic polynomial after axis shift to the left by a. Establish its associated Routh array with parametrized elements in terms of $\boldsymbol{\theta}$ and then use the Routh-Hurwitz criterion to determine the range of $\boldsymbol{\theta}$ for the new characteristic polynomial $C_{\mathrm{F}}(s, \boldsymbol{\theta}, a)$ to have all poles to the left of the imaginary axis. For this purpose, we need to determine the range of $\boldsymbol{\theta}$ that enables all parametrized elements in the first column of the Routh array associated with $C_{\mathrm{F}}(s, \boldsymbol{\theta}, a)$ to be positive. In this way, we can indirectly determine the range of $\boldsymbol{\theta}$ for the original characteristic polynomial $C_{\mathrm{F}}(s, \boldsymbol{\theta})$ to have all poles to the left of the axis $s = -a$.

To facilitate understanding of the Routh-Hurwitz criterion based axis shift method, we still consider the characteristic equation example described in (5.45) which is equivalently transformed into

$$s^4 + 12\,s^3 + 64\,s^2 + 128\,s + K = 0, \tag{5.47}$$

where $s^4 + 12s^3 + 64s^2 + 128s \equiv D(s)$. Perform axis shift to the left by a $(a > 0)$—In practical applications, it is normally meaningless to perform axis shift to the right—The new characteristic polynomial after axis shift is

$$\begin{aligned} D(s-a)+K &= s^4 + (12-4a)s^3 + (64-36a+6a^2)s^2 \\ &\quad + (128-128a+36a^2-4a^3)s \\ &\quad + K - 128a + 64a^2 - 12a^3 + a^4. \end{aligned}$$

If $a = 2$, then the new characteristic polynomial is

$$D(s-2)+K = s^4 + 4s^3 + 16s^2 - 16s + K - 80,$$

for which a parametrized Routh array is established as

$$\begin{array}{c|ccc} s^4 & 1 & 16 & K-80 \\ s^3 & 4 & -16 & \\ s^2 & 20 & K-80 & \\ s^1 & -K/5 & & \\ s^0 & K-80 & & \end{array}$$

Symbolic operation has been incorporated into software such as Matlab. We can take advantage of symbolic operation to conveniently establish a Routh array with parametrized elements. Matlab code for establishing the Routh array with axis shift is given as follows.

RouthHurwitzCriterionAxisShift.m

```
syms s K; a = 2;
D = (s-a)^4 + 12*(s-a)^3 + 64*(s-a)^2 + 128*(s-a) + K;
cs = coeffs(D, s); cs = cs(end:-1:1); n = length(cs)-1;

% Establish the Routh array
gOrg = cs.*s.^[n:-1:0];
% coefficient of s^n
g(n+1) = collect(sum(gOrg(1:2:end)), s); c(n+1) = cs(1);
% coefficient of s^(n-1)
g(n) = collect(sum(gOrg(2:2:end)), s); c(n) = cs(2);
k = n-1;
while (k>=1)
    g(k) = collect(g(k+2) - s*(c(k+2)/c(k+1))*g(k+1), s);
    % coefficient of s^(k-1)
    ctmp = coeffs(g(k), s); c(k) = ctmp(end);
    k = k-1;
end
c = simplify(c(end:-1:1)); % First column of the Routh array
feasibleK = eval(solve(c>0,K))
```

For all elements in the first column of the Routh array to be positive, we require both $-K/5 > 0$ and $K - 80 > 0$. However, these two inequalities can never be satisfied simultaneously. Therefore, no matter for what K, the characteristic polynomial $D(s-2)+K$ can never have all poles to the left of the imaginary axis, which implies the original characteristic polynomial $D(s)+K$ can never have all poles to the left of the axis $s = -2$.

If $a = 1$, then the new characteristic polynomial is

$$D(s-1)+K = s^4 + 8\,s^3 + 34\,s^2 + 32\,s + K - 75,$$

for which a parametrized Routh array is established as

$$\begin{array}{c|ccc} s^4 & 1 & 34 & K-75 \\ s^3 & 8 & 32 & \\ s^2 & 30 & K-75 & \\ s^1 & 52 - 4\,K/15 & & \\ s^0 & K-75 & & \end{array}$$

For all elements in the first column of the Routh array to be positive, we require both $52 - 4K/15 > 0$ and $K - 75 > 0$ namely $75 < K < 195$. This is the range of K for the characteristic polynomial $D(s-1)+K$ to have all poles to the left of the imaginary axis and hence for the original characteristic polynomial $D(s)+K$ to have all poles to the left of the axis $s = -1$. Besides, for the boundary value $K = 195$, note the polynomial $30s^2 + 120$ associated with the row of s^2 in the Routh array, according to which we can conveniently locate pure imaginary poles of $D(s-1)+195$, namely $\pm 2\mathrm{j}$, which implies that the original characteristic polynomial $D(s)+K$ has a pair of conjugate poles $-1 \pm 2\mathrm{j}$ for $K = 195$.

According to the above two aspects of analysis for $a = -2$ and $a = -1$, we can infer that for $75 < K < 195$, dominant control system poles must be located between the axis $s = -2$ and the axis $s = -1$. For $K < 75$ or $K > 195$, dominant control system poles are to the right of the axis $s = -1$. We may apply the Routh-Hurwitz criterion based axis shift method with more tentative axis shift offsets a to acquire more knowledge of the relationship between the parameter K and control system poles. The tentative axis shift offsets may be chosen in *binary section* way. Analysis for more axis shift offsets a is not detailed here, but left to readers for practice.

The Routh-Hurwitz criterion based axis shift method can be naturally extended for *controller-multiple-parameter design*, where a group of inequalities in terms of multiple controller parameters $\boldsymbol{\theta}$ instead of a unique parameter K can be established for each tentative axis shift a. Then solutions to these multi-variate inequalities may be searched via linear programming, quadratic programming, and other kinds of optimization techniques (Bonnans et al. 2006).

5.5 Classical Control Versus Modern Control

By the end of Chap. 5, we have finished presentation of **classical control** theory. We will begin presentation of **modern control** theory since Chap. 6. In fact, there is no strict and distinct categorization of classical control and modern control. Roughly speaking, from the perspective of history, classical control appeared before and flourished during the second world war, whereas modern control made its debut after the second world war especially during 1960s. From the perspective of mathematics, classical control normally involves s-domain analysis based on the Laplace transform, whereas modern control normally involves *state-space analysis* and a much wider range of mathematical techniques. From the perspective of problem complexity, classical control normally applies to *single-input-single-output* control problems which are comparatively simple, whereas modern control can apply to more complicated control problems such as *single-input-multiple-output* and *multiple-input-multiple-output* control problems. We will gradually demonstrate power and wonder of modern control theory in following chapters.

References

J. Bonnans, J. Gilbert, C. Lemarechal, et al. *Numerical optimization: theoretical and practical aspects*. Springer Science & Business Media, 2006.

W. Evans. *Control system dynamics*. McGraw-Hill, 1954.

A. Oppenheim, A. Willsky, S. Nawab. *Signals and systems*. Pearson Education, 1997.

A. Quarteroni, R. Sacco, F. Saleri. *Numerical mathematics*. Springer, 2007.

T. Samad. A survey on industry impact and challenges thereof [technical activities]. *IEEE Control Systems Magazine*, 37(1):17–18, 2017.

Chapter 6
State-Space Analysis and Control

For a control system with certain *target process*,[1] the set of properties that characterizes the target process is called its **process state** (from the perspective of the target process itself), or its **system state** (from the perspective of the control system holistically), or simply its **state**. For example, for a control system that aims at controlling motion of a ball, the position and velocity of the ball may be treated as the state that characterizes motion of the ball.

A classical control system normally has *single-output* that is monitored and controlled in feedback way, where the single-output reflects only part of the process state. In contrast, a modern control system normally has its complete state being estimated and controlled in feedback way, where the state can be regarded as a kind of *multiple-output* of the target process. In other words, modern control aims at handling **multiple-output** control problems. For this purpose, we need **modern control theory**.

6.1 Single-Input-Multiple-Output Control

Multiple-output control problems include *single-input-multiple-output* control problems and *multiple-input-multiple-output* control problems. To facilitate understanding of multiple-output control problems, we may first consider a representative single-input-multiple-output control problem, namely the *inverted pendulum* control problem.

The single inverted pendulum control problem has already been illustrated in Fig. 1.11 in Chap. 1. The single inverted pendulum control system aims at controlling the force exerted on the cart (or equivalently the cart acceleration) so that the cart

[1] If actuator dynamics is not negligible with respect to process dynamics, then the actuator can be incorporated into the target process.

H. Li, *Control Theory for Practical Applications*,
https://doi.org/10.1007/978-981-97-5008-5_6

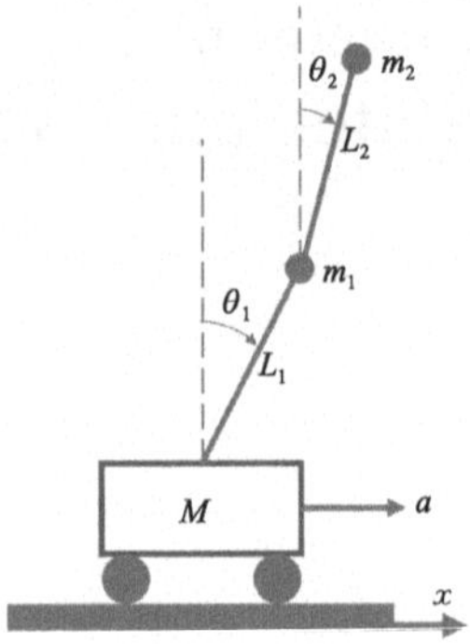

Fig. 6.1 Double inverted pendulum control

is kept at a specific position and the inverted pendulum angle is kept zero. This is a challenging control problem.

If the single inverted pendulum control problem was not challenging enough to readers, we might consider an even more challenging problem, namely the double inverted pendulum control problem, as illustrated in Fig. 6.1. The double inverted pendulum control system aims at controlling the force exerted on the cart so that the cart is kept at a specific position x and both inverted pendulum angles θ_1 and θ_2 are kept zero.

The cart mass is usually much larger than the mass of the two inverted pendulum balls, namely $M \gg m_1, m_2$. So controlling the force exerted on the cart is equivalent to controlling the cart acceleration. For demonstration simplicity, suppose that the control input of the double inverted pendulum is the cart acceleration. Matlab code for simulation of double inverted pendulum control is given as follows. We suggest that readers try the Matlab simulation code and try to design a control method such that the cart position x and both inverted pendulum angles θ_1 and θ_2 are all controlled to zero and kept zero.

DoubleInvertedPendulum.m

```
%% Double inverte pendulum parameters
m1 = 1; m2 = 1; L1 = 1; L2 = 1; g = 10;
%% Simulation preliminary configuration
dt = 0.001; % Numerical computation step
tSpan = 0:dt:3; % Simulation time span
x = 0.2; dx = 0; % Cart position and its velocity
% Inverted pendulum angle theta-1 and its angular velocity
y1 = 0.2; dy1 = 0;
% Inverted pendulum angle theta-2 and its angular velocity
y2 = 0; dy2 = 0;
stt = [y1; dy1; y2; dy2; x; dx]; % Double inverted pendulum
    state
% Record states in simulation
sttAll = zeros(length(stt), length(tSpan)); k = 0;
% Expected equilibrium status
xExpected = 0; y1Expected = 0; y2Expected = 0;
```

```
SimConfig = [m1, m2, L1, L2, g, dt];

%% Simulation of double inverted pendulum control
for t = tSpan
    %% Control method
    acc = 0; % Control law of the cart acceleration
    %% Double inverted pendulum dynamics
    stt = DynamicsDIP(SimConfig, stt, acc);
    sttC = num2cell(stt); [y1, dy1, y2, dy2, x, dx] = sttC{:};
    if (abs(y1)>=pi/2 && abs(y2)>=pi/2)
        fprintf('Control failure!\n'); break;
    end
    k = k+1; sttAll(:,k) = stt;
    %% Double inverted pendulum visualization
    if (rem(k,20) == 0)
        DisplayDIP(x, y1, y2, L1, L2); pause(20*dt);
    end
end
```

The visualization code **DisplayDIP.m** and the double inverted pendulum dynamics code **DynamicsDIP.m** in above code are given as follows.

DisplayDIP.m

```
%% Double inverted pendulum visualization
function DisplayDIP(x, y1, y2, L1, L2)
    Mx = x; My = 0;
    m1x = Mx + L1*sin(y1); m1y = My + L1*cos(y1);
    m2x = m1x + L2*sin(y2); m2y = m1y + L2*cos(y2);
    figure(1); cla; hold on;
    plot(Mx, My, 'or', 'MarkerSize', 16, 'LineWidth', 2);
    plot(m1x, m1y, 'ob', 'MarkerSize', 8, 'LineWidth', 2);
    plot(m2x, m2y, 'ob', 'MarkerSize', 8, 'LineWidth', 2);
    line([Mx,m1x], [My,m1y], 'Color', 'b', 'LineWidth', 2);
    line([m1x,m2x], [m1y,m2y], 'Color', 'b', 'LineWidth', 2);
    axis equal; grid on; xlim([-2, 2]); ylim([0, 3]); hold off;
end
```

DynamicsDIP.m

```
%% Double inverted pendulum dynamics
function stt = DynamicsDIP(SimConfig, sttIn, acc, acc1)
    if (nargin<4) acc1 = 0; end
    SC = num2cell(SimConfig); [m1, m2, L1, L2, g, dt] = SC{:};
    sttC = num2cell(sttIn); [y1, dy1, y2, dy2, x, dx] = sttC{:};
    %% State evolution
    x = x + dx*dt;
    dx = dx + acc*dt;
    y1old = y1; y2old = y2; y12 = y1old - y2old;
    y1 = y1 + dy1*dt;
    dy1 = dy1 + ((sin(y1old) + (m2/m1)*sin(y12)*cos(y2old))*g/L1
        + (-cos(y1old) + (m2/m1)*sin(y12)*sin(y2old))
        *acc/L1 + acc1)*dt;
```

```
    y2 = y2 + dy2*dt;
    dy2 = dy2 + (-sin(y12)*(cos(y1old)+(m2/m1)*cos(y2old)*cos(
    y12))*g/L2 + sin(y12)*(-sin(y1old)+(m2/m1)*sin(y2old)
        *cos(y12))*acc/L2 - cos(y12)*(L1/L2)*acc1)*dt;
    stt = [y1; dy1; y2; dy2; x; dx];
end
```

After trials with the Matlab simulation code, readers may have intuitive understanding of the difficulty of double inverted pendulum control. If we focused only on single-output such as the cart position x, we believe readers would follow classical control theory and easily design a control method to achieve the expected cart position. For example, the proportional-derivative method as follows can achieve the expected cart position.

```
%% Proportional-derivative control method
    acc = - 100*x - 10*dx; % Control law of the cart
    acceleration
```

If the proportional-derivative control method is adopted, the expected cart position will indeed be achieved. However, both inverted pendulums will fall down. The whole objective of double inverted pendulum control is still not achieved. In fact, the difficulty of double inverted pendulum control consists right in the necessity of using single-input to control *multiple-output* which includes the cart position x, the first inverted pendulum angle θ_1, and the second inverted pendulum angle θ_2.

The single inverted pendulum control problem and the double inverted pendulum control problem are representative examples of **multiple-output** control problems especially **single-input-multiple-output** control problems. To realize control systems that handle such difficult control problems, we can no longer resort to classical control theory. Instead, we need **modern control theory**, fundamentals of which will be presented throughout Chap. 6. By default, control systems are assumed time-invariant.

6.2 State-Space Modelling

6.2.1 *State Differential Equation*

Denote $\Delta\theta = \theta_1 - \theta_2$. By physical law, we can establish differential equations for dynamics of the double inverted pendulum control system as

$$\begin{cases} m_1 L_1 \frac{\mathrm{d}^2}{\mathrm{d}t^2}\theta_1 & = m_1(g \sin\theta_1 - a\cos\theta_1) + m_2(g\cos\theta_2 + a\sin\theta_2)\sin\Delta\theta \\ m_2 L_2 \frac{\mathrm{d}^2}{\mathrm{d}t^2}\theta_2 & = m_2(g\sin\theta_2 - a\cos\theta_2) - m_2\cos\Delta\theta L_1 \frac{\mathrm{d}^2}{\mathrm{d}t^2}\theta_1 \end{cases}$$

which can be transformed into

$$\frac{d^2}{dt^2}\theta_1 = \left(\sin\theta_1 + \frac{m_2}{m_1}\sin\Delta\theta\cos\theta_2\right)\frac{g}{L_1} - \left(\cos\theta_1 - \frac{m_2}{m_1}\sin\Delta\theta\sin\theta_2\right)\frac{a}{L_1},$$
$$\frac{d^2}{dt^2}\theta_2 = -\sin\Delta\theta\left[\left(\cos\theta_1 + \frac{m_2}{m_1}\cos\Delta\theta\cos\theta_2\right)\frac{g}{L_2} + \left(\sin\theta_1 - \frac{m_2}{m_1}\cos\Delta\theta\sin\theta_2\right)\frac{a}{L_2}\right].$$

Denote $\boldsymbol{x} \equiv \left[\theta_1 \ \frac{d\theta_1}{dt} \ \theta_2 \ \frac{d\theta_2}{dt} \ x \ \frac{dx}{dt}\right]^T$ and obtain

$$\begin{cases} \frac{d}{dt}\theta_1 = \frac{d\theta_1}{dt} \\ \frac{d}{dt}\frac{d\theta_1}{dt} = \left(\sin\theta_1 + \frac{m_2}{m_1}\sin\Delta\theta\cos\theta_2\right)\frac{g}{L_1} - \left(\cos\theta_1 - \frac{m_2}{m_1}\sin\Delta\theta\sin\theta_2\right)\frac{a}{L_1} \\ \frac{d}{dt}\theta_2 = \frac{d\theta_2}{dt} \\ \frac{d}{dt}\frac{d\theta_2}{dt} = -\sin\Delta\theta\left[\left(\cos\theta_1 + \frac{m_2}{m_1}\cos\Delta\theta\cos\theta_2\right)\frac{g}{L_2}\right. \\ \qquad \left. + \left(\sin\theta_1 - \frac{m_2}{m_1}\cos\Delta\theta\sin\theta_2\right)\frac{a}{L_2}\right] \\ \frac{d}{dt}x = \frac{dx}{dt} \\ \frac{d}{dt}\frac{dx}{dt} = a \end{cases}$$

where the first, third, and fifth equations are trivial equations. The group of differential equations can be expressed compactly by a differential equation as

$$\frac{d}{dt}\boldsymbol{x} \equiv \frac{d}{dt}\begin{bmatrix} \theta_1 \\ \frac{d\theta_1}{dt} \\ \theta_2 \\ \frac{d\theta_2}{dt} \\ x \\ \frac{dx}{dt} \end{bmatrix} = f(\boldsymbol{x}, a),$$

$$f(\boldsymbol{x}, a) \equiv \begin{bmatrix} \frac{d\theta_1}{dt} \\ \left(\sin\theta_1 + \frac{m_2}{m_1}\sin\Delta\theta\cos\theta_2\right)\frac{g}{L_1} - \left(\cos\theta_1 - \frac{m_2}{m_1}\sin\Delta\theta\sin\theta_2\right)\frac{a}{L_1} \\ \frac{d\theta_2}{dt} \\ -\sin\Delta\theta\left[\left(\cos\theta_1 + \frac{m_2}{m_1}\cos\Delta\theta\cos\theta_2\right)\frac{g}{L_2} + \left(\sin\theta_1 - \frac{m_2}{m_1}\cos\Delta\theta\sin\theta_2\right)\frac{a}{L_2}\right] \\ \frac{dx}{dt} \\ a \end{bmatrix}. \tag{6.1}$$

For the double inverted pendulum, the vector $\boldsymbol{x}$ namely the vector of properties composed of its cart position and velocity, its first inverted pendulum angle and angular velocity, and its second inverted pendulum angle and angular velocity, which characterizes its motion status, can be treated as its **state**. The compact differential equation described in (6.1) that models dynamics of the state $\boldsymbol{x}$ is called the **state differential equation** of the double inverted pendulum.

Generally, given a control system with its state denoted as $\boldsymbol{x}$ and its control input to the target process denoted as $\boldsymbol{u}$, dynamics of the state $\boldsymbol{x}$ can be modelled generically by a **state differential equation** as

$$\frac{\mathrm{d}}{\mathrm{d}t}\boldsymbol{x} = f(\boldsymbol{x}, \boldsymbol{u}), \tag{6.2}$$

which reflects the spirit of **state-space modelling** for the control system.

6.2.2 Linear State-Space Modelling in Matrix Form

Nonlinear state differential equations such as that described in (6.1) are usually difficult to handle in practical applications. For analysis and processing convenience, we may consider *linear approximation* of nonlinear state differential equations about representative operation points and establish **linear state-space models** for control systems. The double inverted pendulum control system's representative operation point is the equilibrium state $\boldsymbol{x} = \boldsymbol{0}$. If both inverted pendulum angles θ_1 and θ_2 are close to zero, then the original state differential equation can be fairly linearized about the equilibrium state and simplified into a **linear state differential equation** as

$$\frac{\mathrm{d}}{\mathrm{d}t}\boldsymbol{x} = \begin{bmatrix} 0 & 1 & 0 & 0\,0\,0 \\ \left(1+\frac{m_2}{m_1}\right)\frac{g}{L_1} & 0 & -\frac{m_2}{m_1}\frac{g}{L_1} & 0\,0\,0 \\ 0 & 0 & 0 & 1\,0\,0 \\ -\left(1+\frac{m_2}{m_1}\right)\frac{g}{L_2} & 0 & \left(1+\frac{m_2}{m_1}\right)\frac{g}{L_2} & 0\,0\,0 \\ 0 & 0 & 0 & 0\,0\,1 \\ 0 & 0 & 0 & 0\,0\,0 \end{bmatrix}\boldsymbol{x} + \begin{bmatrix} 0 \\ -\frac{1}{L_1} \\ 0 \\ 0 \\ 0 \\ 1 \end{bmatrix} a \equiv \boldsymbol{A}\boldsymbol{x} + \boldsymbol{B}a. \tag{6.3}$$

Generally, given a control system with its state denoted as $\boldsymbol{x}$ and its control input to the target process denoted as $\boldsymbol{u}$. Suppose that its state $\boldsymbol{x}$ is within a range about certain operation point $\boldsymbol{x}_0$ in which its state differential equation is linear or can be fairly linearized, then dynamics of the state $\boldsymbol{x}$ can be approximately modelled by a **linear state differential equation** as

$$\frac{\mathrm{d}}{\mathrm{d}t}\bar{\boldsymbol{x}} = f(\boldsymbol{x}, \boldsymbol{u}) - f(\boldsymbol{x}_0, \boldsymbol{0}) \approx \boldsymbol{A}\bar{\boldsymbol{x}} + \boldsymbol{B}\boldsymbol{u}, \tag{6.4}$$

where

$$\bar{\boldsymbol{x}} \equiv \boldsymbol{x} - \boldsymbol{x}_0,\quad \boldsymbol{A} = \left.\frac{\partial f(\boldsymbol{x}, \boldsymbol{u})}{\partial \boldsymbol{x}}\right|_{\boldsymbol{x}=\boldsymbol{x}_0, \boldsymbol{u}=\boldsymbol{0}},\quad \boldsymbol{B} = \left.\frac{\partial f(\boldsymbol{x}, \boldsymbol{u})}{\partial \boldsymbol{u}}\right|_{\boldsymbol{x}=\boldsymbol{x}_0, \boldsymbol{u}=\boldsymbol{0}}.$$

For formalism simplicity, we still abuse $\boldsymbol{x}$ to denote $\bar{\boldsymbol{x}} \equiv \boldsymbol{x} - \boldsymbol{x}_0$ in (6.4) and obtain

$$\frac{\mathrm{d}}{\mathrm{d}t}\boldsymbol{x} = \boldsymbol{A}\boldsymbol{x} + \boldsymbol{B}\boldsymbol{u}, \tag{6.5}$$

which is the generic formalism of **linear state-space modelling**. For the linear control system, the square matrix $\boldsymbol{A}$ is called the **state transition matrix** and the matrix $\boldsymbol{B}$ is called the **control input matrix**.

6.2.3 Closed-Loop Feedback State Differential Equation

Feedback is the soul of control science. Most control systems encountered in practical applications are closed-loop feedback control systems. For a closed-loop feedback control system with the state $\boldsymbol{x}$ characterizing the target process, the control input $\boldsymbol{u}$ is normally generated according to feedback of the state $\boldsymbol{x}$ which is compared with certain expected state $\boldsymbol{x}_{\mathrm{E}}$. Therefore, let the feedback control law be generically denoted as $\boldsymbol{u} = g(\boldsymbol{x} - \boldsymbol{x}_{\mathrm{E}})$, substitute it into (6.2) and obtain

$$\frac{\mathrm{d}}{\mathrm{d}t}\boldsymbol{x} = f(\boldsymbol{x}, g(\boldsymbol{x} - \boldsymbol{x}_{\mathrm{E}})) \equiv f_{\mathrm{c}}(\boldsymbol{x} - \boldsymbol{x}_{\mathrm{E}}). \tag{6.6}$$

If $\boldsymbol{x}_{\mathrm{E}} \neq \boldsymbol{0}$, we can shift the state $\boldsymbol{x}$ by an offset of $\boldsymbol{x}_{\mathrm{E}}$ and transform (6.6) into

$$\frac{\mathrm{d}}{\mathrm{d}t}(\boldsymbol{x} - \boldsymbol{x}_{\mathrm{E}}) = \frac{\mathrm{d}}{\mathrm{d}t}\boldsymbol{x} = f_{\mathrm{c}}(\boldsymbol{x} - \boldsymbol{x}_{\mathrm{E}}), \tag{6.7}$$

where $\bar{\boldsymbol{x}} \equiv \boldsymbol{x} - \boldsymbol{x}_{\mathrm{E}}$ can be regarded as a new representation of the state and the expected new state is $\bar{\boldsymbol{x}}_{\mathrm{E}} \equiv \boldsymbol{x}_{\mathrm{E}} - \boldsymbol{x}_{\mathrm{E}} = \boldsymbol{0}$. Therefore, for analysis simplicity yet without influencing analysis essence, we assume $\boldsymbol{x}_{\mathrm{E}} = \boldsymbol{0}$ by default in (6.6) and (6.7) and hence obtain

$$\frac{\mathrm{d}}{\mathrm{d}t}\boldsymbol{x} = f(\boldsymbol{x}, g(\boldsymbol{x})) \equiv f_{\mathrm{c}}(\boldsymbol{x}). \tag{6.8}$$

In other words, once the feedback control law is determined, dynamics of the closed-loop feedback control system is equivalent to dynamics of a **self-evolutionary system** namely a system whose state evolution depends on its own state completely, as modelled by (6.8). The state differential equation described in (6.8) is called the **closed-loop feedback state differential equation** of the control system.

Similarly, for the generic formalism of linear state-space modelling described in (6.5), its corresponding **linear closed-loop feedback state differential equation** is a generic *homogeneous linear state differential equation* as

$$\frac{\mathrm{d}}{\mathrm{d}t}\boldsymbol{x} = \boldsymbol{A}_{\mathrm{c}}\boldsymbol{x}, \tag{6.9}$$

which is equivalent to the state differential equation of a **linear self-evolutionary system**. The square matrix $\boldsymbol{A}_{\mathrm{c}}$ is the **state transition matrix** of the linear self-evolutionary system.

6.2.4 Stability Criterion for Linear Self-Evolutionary Systems

The stability criterion for general self-evolutionary systems especially nonlinear self-evolutionary systems will be discussed in Chap. 7. Here, we focus on the stability criterion for linear self-evolutionary systems which can be generically modelled by the homogeneous linear state differential equation described in (6.9).

The solution to the state differential equation described in (6.9) is given as

$$\boldsymbol{x} = \mathrm{e}^{\boldsymbol{A}_{\mathrm{c}}t}\boldsymbol{x}_0, \tag{6.10}$$

where $\boldsymbol{x}_0 \equiv \boldsymbol{x}(0)$ and the *matrix exponential function* is defined as

$$\mathrm{e}^{\boldsymbol{M}} \equiv \boldsymbol{I} + \boldsymbol{M} + \frac{\boldsymbol{M}^2}{2!} + \frac{\boldsymbol{M}^3}{3!} + \cdots = \sum_{k=0}^{\infty} \frac{\boldsymbol{M}^k}{k!}.$$

If a linear self-evolutionary system is stable, then its state $\boldsymbol{x}$ always converges to the equilibrium state $\boldsymbol{x}_{\mathrm{E}} = \boldsymbol{0}$ no matter for what initial state $\boldsymbol{x}_0$. This is equivalent to the condition

$$\lim_{t\to\infty} \mathrm{e}^{\boldsymbol{A}_{\mathrm{c}}t} = 0. \tag{6.11}$$

In other words, the sufficient and necessary condition for the linear self-evolutionary system to be stable is given in (6.11).

For further analysis of (6.11), we resort to *Jordan canonical decomposition* (Horn and Johnson 2012) of the state transition matrix $\boldsymbol{A}_{\mathrm{c}}$. Suppose that $\boldsymbol{A}_{\mathrm{c}}$ is decomposed as

$$\boldsymbol{A}_{\mathrm{c}} = \boldsymbol{S} \begin{bmatrix} \boldsymbol{J}_{\lambda_1} & \boldsymbol{0} & \cdots & \boldsymbol{0} \\ \boldsymbol{0} & \boldsymbol{J}_{\lambda_2} & \cdots & \boldsymbol{0} \\ \vdots & \vdots & & \vdots \\ \boldsymbol{0} & \boldsymbol{0} & \cdots & \boldsymbol{J}_{\lambda_q} \end{bmatrix} \boldsymbol{S}^{-1} \equiv \boldsymbol{S}\boldsymbol{J}\boldsymbol{S}^{-1}, \tag{6.12}$$

where $\boldsymbol{J}_{\lambda_i}$ $(i = 1, 2, \cdots, q)$ are *Jordan blocks* with each $\boldsymbol{J}_{\lambda_i}$ corresponding to an *eigenvalue* λ_i of $\boldsymbol{A}_{\mathrm{c}}$ in the form as

$$\boldsymbol{J}_{\lambda_i} = \begin{bmatrix} \lambda_i & 1 & & & & \\ & \lambda_i & 1 & & & \\ & & \lambda_i & & & \\ & & & \ddots & & \\ & & & & \lambda_i & 1 \\ & & & & & \lambda_i \end{bmatrix}.$$

Since

$$A_c^k = (SJS^{-1})^k = \underbrace{(SJS^{-1})(SJS^{-1})\cdots(SJS^{-1})}_{k \text{ times}} = SJ^kS^{-1}$$

$$= S\begin{bmatrix} J_{\lambda_1} & & \\ & \ddots & \\ & & J_{\lambda_q} \end{bmatrix}^k S^{-1} = S\begin{bmatrix} J_{\lambda_1}^k & & \\ & \ddots & \\ & & J_{\lambda_q}^k \end{bmatrix} S^{-1},$$

we have

$$e^{A_c t} = \sum_{k=0}^{\infty} \frac{A_c^k t^k}{k!} = \sum_{k=0}^{\infty}\left(S\begin{bmatrix} J_{\lambda_1}^k & & \\ & \ddots & \\ & & J_{\lambda_q}^k \end{bmatrix} \frac{t^k}{k!} S^{-1}\right)$$

$$= S\begin{bmatrix} \sum_{k=0}^{\infty} J_{\lambda_1}^k \frac{t^k}{k!} & & \\ & \ddots & \\ & & \sum_{k=0}^{\infty} J_{\lambda_q}^k \frac{t^k}{k!} \end{bmatrix} S^{-1} = S\begin{bmatrix} e^{J_{\lambda_1} t} & & \\ & \ddots & \\ & & e^{J_{\lambda_q} t} \end{bmatrix} S^{-1}. \quad (6.13)$$

For each Jordan block J_{λ_i}, we have

$$J_{\lambda_i}^k = \begin{bmatrix} \lambda_i^k & C_k^1\lambda_i^{k-1} & C_k^2\lambda_i^{k-2} & \cdots & C_k^{d_i-2}\lambda_i^{k-d_i+2} & C_k^{d_i-1}\lambda_i^{k-d_i+1} \\ & \lambda_i^k & C_k^1\lambda_i^{k-1} & \cdots & C_k^{d_i-3}\lambda_i^{k-d_i+3} & C_k^{d_i-2}\lambda_i^{k-d_i+2} \\ & & \lambda_i^k & \cdots & \ddots & C_k^{d_i-3}\lambda_i^{k-d_i+3} \\ & & & \ddots & \vdots & \vdots \\ & & & & \lambda_i^k & C_k^1\lambda_i^{k-1} \\ & & & & & \lambda_i^k \end{bmatrix},$$

which is an *upper triangular matrix* having the following regularity: elements of the first diagonal line are the same to λ_i^k, elements of the second diagonal line are the same to $C_k^1\lambda_i^{k-1}$, elements of the third diagonal line are the same to $C_k^2\lambda_i^{k-2}$, $\cdots$, elements of the $(d_i - 1)$th diagonal line are the same to $C_k^{d_i-2}\lambda_i^{k-d_i+2}$, and the top right element in the d_ith diagonal line is $C_k^{d_i-1}\lambda_i^{k-d_i+1}$, where d_i denotes the dimension of J_{λ_i}. We further have

$$
\mathrm{e}^{J_{\lambda_i} t} = \sum_{k=0}^{\infty} J_{\lambda_i}^k \frac{t^k}{k!} = \begin{bmatrix} D_{\lambda_i,t,1} & D_{\lambda_i,t,2} & D_{\lambda_i,t,3} & \cdots & D_{\lambda_i,t,d_i-1} & D_{\lambda_i,t,d_i} \\ & D_{\lambda_i,t,1} & D_{\lambda_i,t,2} & \cdots & D_{\lambda_i,t,d_i-2} & D_{\lambda_i,t,d_i-1} \\ & & D_{\lambda_i,t,1} & \cdots & \ddots & D_{\lambda_i,t,d_i-2} \\ & & & \ddots & \vdots & \vdots \\ & & & & D_{\lambda_i,t,1} & D_{\lambda_i,t,2} \\ & & & & & D_{\lambda_i,t,1} \end{bmatrix},
$$

where

$$
\begin{aligned}
D_{\lambda_i,t,1} &= \sum_{k=0}^{\infty} \lambda_i^k \frac{t^k}{k!} = \mathrm{e}^{\lambda_i t}, \\
D_{\lambda_i,t,2} &= \sum_{k=1}^{\infty} C_k^1 \lambda_i^{k-1} \frac{t^k}{k!} = t \sum_{k=1}^{\infty} \lambda_i^{k-1} \frac{t^{k-1}}{(k-1)!} = t\mathrm{e}^{\lambda_i t}, \\
D_{\lambda_i,t,3} &= \sum_{k=2}^{\infty} C_k^2 \lambda_i^{k-2} \frac{t^k}{k!} = \frac{t^2}{2!} \sum_{k=2}^{\infty} \lambda_i^{k-2} \frac{t^{k-2}}{(k-2)!} = \frac{t^2}{2!}\mathrm{e}^{\lambda_i t}, \\
&\vdots \\
D_{\lambda_i,t,d_i} &= \sum_{k=d_i-1}^{\infty} C_k^{d_i-1} \lambda_i^{k-d_i+1} \frac{t^k}{k!} = \frac{t^{d_i-1}}{(d_i-1)!}\mathrm{e}^{\lambda_i t}.
\end{aligned}
$$

Since

$$
\begin{aligned}
\mathrm{Re}(\lambda_i) < 0 &\iff \lim_{t\to\infty} D_{\lambda_i,t,1} = 0, \\
\mathrm{Re}(\lambda_i) < 0 &\iff \lim_{t\to\infty} D_{\lambda_i,t,2} = 0, \\
\mathrm{Re}(\lambda_i) < 0 &\iff \lim_{t\to\infty} D_{\lambda_i,t,3} = 0, \\
&\vdots \\
\mathrm{Re}(\lambda_i) < 0 &\iff \lim_{t\to\infty} D_{\lambda_i,t,d_i} = 0,
\end{aligned}
$$

we have

$$
\mathrm{Re}(\lambda_i) < 0 \iff \lim_{t\to\infty} \mathrm{e}^{J_{\lambda_i} t} = 0. \tag{6.14}
$$

From (6.13) and (6.14) we can infer

$$
\mathrm{Re}(\lambda_1) < 0, \mathrm{Re}(\lambda_2) < 0, \ldots, \mathrm{Re}(\lambda_q) < 0 \iff \lim_{t\to\infty} \mathrm{e}^{A_c t} = 0, \tag{6.15}
$$

which implies that a linear self-evolutionary system is stable if and only if the real parts of all its eigenvalues (namely eigenvalues of its state transition matrix) are negative.

Control system stability criterion: *A linear self-evolutionary system is stable if and only if the real parts of all its eigenvalues are negative.*

A control system that evolves only according to its initial condition is a typical kind of self-evolutionary system. A stable control system's initial response will always fade away no matter what initial condition is given. So a stable control system that evolves only according to its initial condition is also a stable self-evolutionary system. On the other hand, if a control system that evolves only according to its initial condition is a stable self-evolutionary system, then the control system will always have convergent initial response no matter what initial condition is given and hence is stable. Therefore, to judge whether a linear time-invariant control system especially closed-loop feedback control system is stable, we can use above stability criterion to judge whether the control system is a stable linear self-evolutionary system when it evolves only according to its initial condition.

Readers may find that above stability criterion is almost the same to the stability criterion presented in Sect. 4.2 in Chap. 4, except that the term "poles" is replaced by the term "eigenvalues" here. For a linear time-invariant control system, the two stability criteria are essentially the same, because "poles" and "eigenvalues" actually refer to the same thing namely roots of the characteristic polynomial of the control system, yet from different perspectives: "poles" refer to characteristic polynomial roots from the perspective of modelling the control system via transfer function, whereas "eigenvalues" refer to characteristic polynomial roots from the perspective of modelling the control system via state differential equation.

To facilitate understanding of the mutual relationship between system modelling via Laplace transform and linear state-space modelling, we may consider a linear time-invariant control system that evolves only according to its initial condition and consider the mutual transform as follows: On one hand, suppose that the control system adopts linear state-space modelling and dynamics of its initial response is modelled by (6.9). Transform the linear state-space model into a transfer function model by performing the Laplace transform on both sides of (6.9) and obtain

$$\begin{aligned} s\boldsymbol{x}(s) - \boldsymbol{x}_0 &= \boldsymbol{A}_c\boldsymbol{x}(s) \\ \boldsymbol{x}(s) = (s\boldsymbol{I} - \boldsymbol{A}_c)^{-1}\boldsymbol{x}_0 &= \frac{\boldsymbol{C}_{s\boldsymbol{I}-\boldsymbol{A}_c}}{\det(s\boldsymbol{I} - \boldsymbol{A}_c)}\boldsymbol{x}_0. \end{aligned} \tag{6.16}$$

Derivation of (6.16) involves the *Cramer rule* (Golub and Van Loan 1996). The numerator part $\boldsymbol{C}_{s\boldsymbol{I}-\boldsymbol{A}_c}$ denotes the *co-matrix* of $s\boldsymbol{I} - \boldsymbol{A}_c$ that consists of *co-factors*, whereas the denominator part $\det(s\boldsymbol{I} - \boldsymbol{A}_c)$ denotes the determinant of $s\boldsymbol{I} - \boldsymbol{A}_c$. Note that the co-factors are polynomials in terms of the Laplace variable s, so (6.16) implies that $\boldsymbol{x}(s)$ consists of fractional polynomials that share a common denominator namely the polynomial $\det(s\boldsymbol{I} - \boldsymbol{A}_c)$ which is right the characteristic polynomial of the control system. By definition of eigenvalues and poles, we know that eigenvalues of the state transition matrix $\boldsymbol{A}_c$ and control system poles refer to the same thing namely roots of the characteristic polynomial $\det(s\boldsymbol{I} - \boldsymbol{A}_c)$.

On the other hand, suppose that the control system adopts system modelling via Laplace transform and its characteristic polynomial is

$$C_{\mathrm{LT}}(s) \equiv s^n + a_{n-1}s^{n-1} + \cdots + a_0.$$

Dynamics of its initial response is modelled by

$$\begin{aligned} &\frac{\mathrm{d}^n}{\mathrm{d}t^n}y + a_{n-1}\frac{\mathrm{d}^{n-1}}{\mathrm{d}t^{n-1}}y + \cdots + a_0 y = 0, \\ &(s^n + a_{n-1}s^{n-1} + \cdots + a_0)y(s) = A_0(s), \end{aligned} \tag{6.17}$$

where $A_0(s)$ is a constant polynomial determined by initial conditions of the output y.

For the control system, denote its state $\boldsymbol{x}$ as

$$\boldsymbol{x} \equiv \begin{bmatrix} y & \frac{\mathrm{d}y}{\mathrm{d}t} & \frac{\mathrm{d}^2 y}{\mathrm{d}t^2} & \cdots & \frac{\mathrm{d}^{n-1}y}{\mathrm{d}t^{n-1}} \end{bmatrix}^{\mathrm{T}} \tag{6.18}$$

and transform (6.17) into a linear state differential equation as

$$\frac{\mathrm{d}}{\mathrm{d}t}\boldsymbol{x} = \begin{bmatrix} & 1 & & & & \\ & & 1 & & & \\ & & & \ddots & & \\ & & & & 1 & \\ & & & & & 1 \\ -a_0 & -a_1 & -a_2 & \cdots & -a_{n-2} & -a_{n-1} \end{bmatrix} \boldsymbol{x} \equiv \boldsymbol{A}_c\boldsymbol{x}. \tag{6.19}$$

The characteristic polynomial of $\boldsymbol{A}_c$ is

$$\det(s\boldsymbol{I} - \boldsymbol{A}_c) = \begin{bmatrix} s & -1 & & & \\ & s & -1 & & \\ & & s & \ddots & \\ & & & & -1 \\ a_0 & a_1 & a_2 & \cdots & s + a_{n-1} \end{bmatrix}$$

$$= \begin{bmatrix} s & & & & \\ & s & -1 & & \\ & & s & \ddots & \\ & & & & -1 \\ a_0 & a_1 + \frac{a_0}{s} & a_2 & \cdots & s + a_{n-1} \end{bmatrix}$$

$$= \cdots = \begin{bmatrix} s & & & \\ & \ddots & & \\ & & s & \\ a_0 & \cdots & * & s + a_{n-1} + \frac{a_{n-2}}{s} + \cdots + \frac{a_0}{s^{n-1}} \end{bmatrix} = C_{\mathrm{LT}}(s),$$

(cancel -1 in each column gradually)

which implies that poles of the control system whose characteristic polynomial is $C_{\mathrm{LT}}(s)$ coincide with eigenvalues of the state transition matrix $\boldsymbol{A}_c$ whose characteristic polynomial $\det(s\boldsymbol{I} - \boldsymbol{A}_c)$ is right $C_{\mathrm{LT}}(s)$.

6.3 Controllability

Fundamentals of state-space analysis especially the stability criterion presented in Sect. 6.2 provide valuable theoretical guide for handling difficult control problems such as double inverted pendulum control, though more knowledge of modern control theory is still needed. For a control problem, before any tentative design of a control method, a preliminary question arises naturally: Is the target process *controllable*? In other words, is it ever possible to design a control system that can control the target process as desired? Suppose that the target process is characterized by certain state. Then above preliminary question may be posed in another way: Is it ever possible to design a control system that enables the state to evolve as desired especially to achieve any expected state?

Taking double inverted pendulum control as an example, the difficulty of such control problem is easily understood if we imagine that we use our own hand instead of the moving cart to perform double inverted pendulum control. This control problem is so difficult that we might even doubt whether it would ever be possible to succeed in double inverted pendulum control. To avoid blind trials, it is worth making a preliminary theoretical judgement of whether the double inverted pendulum is *controllable*: If the double inverted pendulum is *proved* to be *controllable*, then the intention to design a double inverted pendulum control system tends to make sense.

In contrast, if the double inverted pendulum is *proved* to be *uncontrollable*, then we should waste no time on designing a double inverted pendulum control system.

To design a control system, *it is worth making a preliminary theoretical judgement (if possible) of controllability of the target process*. It is difficult to have a general method to theoretically determine controllability of any arbitrary target process especially severely nonlinear target process. But fortunately, many target processes encountered in practical applications can fairly adopt linear state-space modelling. For example, dynamics of the double inverted pendulum state can be fairly modelled by the linear state differential equation described in (6.3) if both inverted pendulum angles θ_1 and θ_2 are close to zero. For a target process that can fairly adopt linear state-space modelling, we have a systematic method to theoretically determine its controllability.

6.3.1 Solution to Linear State Differential Equation

Given a control system with its state denoted as $\boldsymbol{x}$ and its control input to the target process denoted as $\boldsymbol{u}$, suppose that dynamics of the state $\boldsymbol{x}$ is modelled generically by a linear state differential equation described in (6.5):

$$\frac{\mathrm{d}}{\mathrm{d}t}\boldsymbol{x} = \boldsymbol{A}\boldsymbol{x} + \boldsymbol{B}\boldsymbol{u}.$$

Its homogeneous counterpart equation is

$$\frac{\mathrm{d}}{\mathrm{d}t}\boldsymbol{x} = \boldsymbol{A}\boldsymbol{x},$$

which has the solution as

$$\boldsymbol{x} = \mathrm{e}^{\boldsymbol{A}t}\boldsymbol{x}_0.$$

Definition of the matrix exponential function is already clarified in (6.10).

Replace $\boldsymbol{x}_0 \equiv \boldsymbol{x}(0)$ by an unknown function $\boldsymbol{y}$, let $\boldsymbol{x} = \mathrm{e}^{\boldsymbol{A}t}\boldsymbol{y}$ (apparently $\boldsymbol{y}_0 = \boldsymbol{x}_0$), substitute it into (6.5) and obtain

$$\frac{\mathrm{d}}{\mathrm{d}t}(\mathrm{e}^{\boldsymbol{A}t}\boldsymbol{y}) = \boldsymbol{A}\mathrm{e}^{\boldsymbol{A}t}\boldsymbol{y} + \boldsymbol{B}\boldsymbol{u} \iff \boldsymbol{A}\mathrm{e}^{\boldsymbol{A}t}\boldsymbol{y} + \mathrm{e}^{\boldsymbol{A}t}\frac{\mathrm{d}}{\mathrm{d}t}\boldsymbol{y} = \boldsymbol{A}\mathrm{e}^{\boldsymbol{A}t}\boldsymbol{y} + \boldsymbol{B}\boldsymbol{u}$$

$$\iff \frac{\mathrm{d}}{\mathrm{d}t}\boldsymbol{y} = \mathrm{e}^{-\boldsymbol{A}t}\boldsymbol{B}\boldsymbol{u} \iff \boldsymbol{y} = \boldsymbol{y}_0 + \int_0^t \mathrm{e}^{-\boldsymbol{A}\tau}\boldsymbol{B}\boldsymbol{u}(\tau)\mathrm{d}\tau.$$

Then we have

$$\boldsymbol{x} = \mathrm{e}^{\boldsymbol{A}t}\boldsymbol{y}_0 + \mathrm{e}^{\boldsymbol{A}t}\int_0^t \mathrm{e}^{-\boldsymbol{A}\tau}\boldsymbol{B}\boldsymbol{u}(\tau)\mathrm{d}\tau = \mathrm{e}^{\boldsymbol{A}t}\boldsymbol{x}_0 + \int_0^t \mathrm{e}^{\boldsymbol{A}(t-\tau)}\boldsymbol{B}\boldsymbol{u}(\tau)\mathrm{d}\tau. \tag{6.20}$$

In derivation of (6.20), the *special associative law* of matrix exponential function operation namely $\mathrm{e}^{\boldsymbol{A}t_1}\mathrm{e}^{\boldsymbol{A}t_2} = \mathrm{e}^{\boldsymbol{A}(t_1+t_2)}$ is used.[2] Rearrange (6.20) as

$$\int_0^t \mathrm{e}^{\boldsymbol{A}(t-\tau)}\boldsymbol{B}\boldsymbol{u}(\tau)\mathrm{d}\tau = \boldsymbol{x} - \mathrm{e}^{\boldsymbol{A}t}\boldsymbol{x}_0, \tag{6.21}$$

which can be treated as a *functional equation* in terms of the function $\boldsymbol{u}$ once t is determined. If the target process is controllable, then for arbitrary state $\bar{\boldsymbol{x}}_t \equiv \boldsymbol{x} - \mathrm{e}^{\boldsymbol{A}t}\boldsymbol{x}_0$, the functional equation described in (6.21) must have a solution of $\boldsymbol{u}$ for some t, and *vice versa.*

6.3.2 Controllability Matrix

Transform the left side of the functional equation described in (6.21) as

$$\begin{aligned}\int_0^t \mathrm{e}^{\boldsymbol{A}(t-\tau)}\boldsymbol{B}\boldsymbol{u}(\tau)\mathrm{d}\tau &= \int_0^t \sum_{k=0}^{\infty}\boldsymbol{A}^k\frac{(t-\tau)^k}{k!}\boldsymbol{B}\boldsymbol{u}(\tau)\mathrm{d}\tau \\ &= \sum_{k=0}^{\infty}\boldsymbol{A}^k\boldsymbol{B}\int_0^t \frac{(t-\tau)^k}{k!}\boldsymbol{u}(\tau)\mathrm{d}\tau = \boldsymbol{C}_{\boldsymbol{A},\boldsymbol{B}}\boldsymbol{U}_t,\end{aligned}$$

where

[2] Readers had better not take this *special associative law* for granted. It needs to be proved: $\forall t_1, t_2$, define the function $\phi(t) = \mathrm{e}^{\boldsymbol{A}(t_1-t)}\mathrm{e}^{\boldsymbol{A}(t+t_2)}$. We have

$$\frac{\mathrm{d}}{\mathrm{d}t}\phi(t) = \frac{\mathrm{d}}{\mathrm{d}t}\mathrm{e}^{\boldsymbol{A}(t_1-t)}\mathrm{e}^{\boldsymbol{A}(t+t_2)} + \mathrm{e}^{\boldsymbol{A}(t_1-t)}\frac{\mathrm{d}}{\mathrm{d}t}\mathrm{e}^{\boldsymbol{A}(t+t_2)} = -\boldsymbol{A}\phi(t) + \boldsymbol{A}\phi(t) = \boldsymbol{0},$$

which implies that the function $\phi(t)$ is a constant matrix regardless of t. So

$$\mathrm{e}^{\boldsymbol{A}t_1}\mathrm{e}^{\boldsymbol{A}t_2} = \phi(0) = \phi(t_1) = \mathrm{e}^{\boldsymbol{A}(t_1+t_2)}.$$

$$C_{A,B} = [B \ AB \ A^2B \ \cdots \ A^{n-1}B \ \cdots],$$

$$U_t = \begin{bmatrix} \int_0^t u(\tau)\mathrm{d}\tau \\ \int_0^t (t-\tau)u(\tau)\mathrm{d}\tau \\ \int_0^t \frac{(t-\tau)^2}{2!}u(\tau)\mathrm{d}\tau \\ \vdots \\ \int_0^t \frac{(t-\tau)^{n-1}}{(n-1)!}u(\tau)\mathrm{d}\tau \\ \vdots \end{bmatrix}.$$

Thus the functional equation described in (6.21) can be transformed into a matrix equation form as

$$C_{A,B}U_t = \bar{x}_t, \tag{6.22}$$

where $C_{A,B}$ is called the **controllability matrix** of the control system.

The *necessary and sufficient condition* for the matrix equation described in (6.22) to always have a solution of U_t no matter given what $\bar{x}_t$ is that *the controllability matrix* $C_{A,B}$ *is of full rank (by default in terms of column vectors)* or in other words *the rank of* $C_{A,B}$ *equals the state dimension*. If the controllability matrix $C_{A,B}$ is rank-deficient, there must be $\bar{x}_t$ such that the matrix equation described in (6.22) has no solution of U_t, which implies that the target process is uncontrollable. If the controllability matrix $C_{A,B}$ is of full rank, the matrix equation described in (6.22) has infinite solutions of U_t no matter given what $\bar{x}_t$. For each specific solution of U_t, there is u from which the specific solution of U_t can be obtained,[3] which implies that the target process is controllable.

Denote the characteristic polynomial of A namely $\det(sI - A)$ as

$$C_A(s) = s^n + a_{n-1}s^{n-1} + \cdots + a_0,$$

[3] Given a specific U_t, since $1, t-\tau, (t-\tau)^2, \ldots, (t-\tau)^{n-1}, \ldots$ are *linearly independent*, the following group of linear functional equations in terms of u

$$\begin{bmatrix} \int_0^t u(\tau)\mathrm{d}\tau \\ \int_0^t (t-\tau)u(\tau)\mathrm{d}\tau \\ \int_0^t \frac{(t-\tau)^2}{2!}u(\tau)\mathrm{d}\tau \\ \vdots \end{bmatrix} = U_t$$

are also linearly independent. Besides, the group of equations have *countable* equations, whereas u has *uncountable* dimensions or degrees of freedom. So the group of equations definitely have solutions of u (in fact, infinite solutions of u).

where n is the dimension of the state $\boldsymbol{x}$. The state transition matrix $\boldsymbol{A}$ satisfies

$$C_{A}(\boldsymbol{A}) = \boldsymbol{A}^{n} + a_{n-1}\boldsymbol{A}^{n-1} + \cdots + a_0\boldsymbol{I} = \boldsymbol{0}, \tag{6.23}$$

which is called the *Hamilton-Cayley theorem*.

Proof By *Jordan canonical decomposition*, suppose that $\boldsymbol{A}$ is decomposed as

$$\boldsymbol{A} = \boldsymbol{S}\begin{bmatrix} \boldsymbol{J}_{\lambda_1} & & \\ & \ddots & \\ & & \boldsymbol{J}_{\lambda_q} \end{bmatrix}\boldsymbol{S}^{-1} = \boldsymbol{S}\boldsymbol{J}\boldsymbol{S}^{-1},$$

where Jordan blocks $\boldsymbol{J}_{\lambda_i}$ have dimensions d_i respectively $(i = 1, 2, \ldots, q)$. The eigenvalues λ_i are allowed to be the same. The characteristic polynomial $C_A(s)$ is factorized as

$$C_A(s) = (s - \lambda_1)^{d_1} \cdots (s - \lambda_q)^{d_q},$$

so

$$\begin{aligned} C_A(\boldsymbol{A}) &= (\boldsymbol{S}\boldsymbol{J}\boldsymbol{S}^{-1} - \lambda_1\boldsymbol{I})^{d_1} \cdots (\boldsymbol{S}\boldsymbol{J}\boldsymbol{S}^{-1} - \lambda_q\boldsymbol{I})^{d_q} \\ &= \boldsymbol{S}(\boldsymbol{J} - \lambda_1\boldsymbol{I})^{d_1} \cdots (\boldsymbol{J} - \lambda_q\boldsymbol{I})^{d_q}\boldsymbol{S}^{-1} \\ &= \boldsymbol{S}\begin{bmatrix} (\boldsymbol{J}_{\lambda_1} - \lambda_1\boldsymbol{I})^{d_1} & \\ & \ddots \end{bmatrix} \cdots \begin{bmatrix} \ddots & \\ & (\boldsymbol{J}_{\lambda_q} - \lambda_q\boldsymbol{I})^{d_q} \end{bmatrix}\boldsymbol{S}^{-1} \\ &= \boldsymbol{S}\begin{bmatrix} \boldsymbol{0} & \\ & \ddots \end{bmatrix} \cdots \begin{bmatrix} \ddots & \\ & \boldsymbol{0} \end{bmatrix}\boldsymbol{S}^{-1} = \boldsymbol{S} \;\; \boldsymbol{0} \;\; \boldsymbol{S}^{-1} = \boldsymbol{0}. \end{aligned}$$

The proof is done. □

The Hamilton-Cayley theorem described by (6.23) implies each element after $\boldsymbol{A}^{n-1}\boldsymbol{B}$ in the controllability matrix $\boldsymbol{C}_{A,B}$ namely each $\boldsymbol{A}^{k}\boldsymbol{B}$ for $k \geqslant n$ can be transformed into a linear combination of $\boldsymbol{B}$, $\boldsymbol{A}\boldsymbol{B}$, $\boldsymbol{A}^2\boldsymbol{B}, \ldots, \boldsymbol{A}^{n-1}\boldsymbol{B}$. So we have

$$\begin{aligned} &\text{rank of } [\boldsymbol{B} \;\; \boldsymbol{A}\boldsymbol{B} \;\; \boldsymbol{A}^2\boldsymbol{B} \;\; \cdots \;\; \boldsymbol{A}^{n-1}\boldsymbol{B} \;\; \cdots] \\ = &\text{rank of } [\boldsymbol{B} \;\; \boldsymbol{A}\boldsymbol{B} \;\; \boldsymbol{A}^2\boldsymbol{B} \;\; \cdots \;\; \boldsymbol{A}^{n-1}\boldsymbol{B}]. \end{aligned}$$

Then we can define the **controllability matrix** simply as

$$\boldsymbol{C}_{A,B} = [\boldsymbol{B} \;\; \boldsymbol{A}\boldsymbol{B} \;\; \boldsymbol{A}^2\boldsymbol{B} \;\; \cdots \;\; \boldsymbol{A}^{n-1}\boldsymbol{B}]. \tag{6.24}$$

If the controllability matrix $\boldsymbol{C}_{A,B}$ described in (6.24) is rank-deficient, then the target process is uncontrollable. If the controllability matrix $\boldsymbol{C}_{A,B}$ described in (6.24) is of full rank, then the target process is controllable.

> **Control system controllability criterion**: *For a linear control system, if its controllability matrix is rank-deficient, then its target process is uncontrollable. If its controllability matrix is of full rank, then its target process is controllable.*

We can apply the controllability criterion to determine controllability of the double inverted pendulum that adopts linear state-space modelling. Matlab code for such purpose is given as follows.

ControllabilityMatrixDIP.m

```
%% Double inverted pendulum parameters
m1 = 1; m2 = 1; L1 = 1; L2 = 1; g = 10;
A = [0, 1, 0, 0, 0, 0; ...
    (m1+m2)*g/(m1*L1), 0, -m2*g/(m1*L1), 0, 0, 0; ...
    0, 0, 0, 1, 0, 0; ...
    -(m1+m2)*g/(m1*L2), 0, (m1+m2)*g/(m1*L2), 0, 0, 0; ...
    0, 0, 0, 0, 0, 1; ...
    0, 0, 0, 0, 0, 0];
B = [0; -1/L1; 0; 0; 0; 1];
n = size(A,1); % State dimension

CM = zeros(n); CM(:,1) = B;
for k = 2:n
    CM(:,k) = A*CM(:,k-1);
end % CM = [B, A*B, A^2*B, A^3*B, A^4*B, A^5*B];
fprintf('Controllability matrix: '); CM

%% Check if the controllability matrix is of full rank
if (rank(CM) == n)
    fprintf('The double inverted pendulum is controllable\n');
else
    fprintf('The double inverted pendulum is uncontrollable\n');
end
```

After trying the Matlab code, readers will find that the double inverted pendulum is indeed controllable, at least when both inverted pendulum angles θ_1 and θ_2 are close to zero so that linear state-space modelling can be fairly adopted. Based on this positive preliminary theoretical judgement, we can move forward with confidence to handle the double inverted pendulum control problem.

6.4 Full-State Feedback Control

6.4.1 Control Strategy

For a control system that adopts linear state-space modelling described by (6.5), to theoretically analyse controllability of the target process, we resort to functional equations in terms of the control law $\boldsymbol{u}$, as described in (6.21) and (6.22). However, we hardly rely on solving (6.21) and (6.22) directly to obtain a solution of $\boldsymbol{u}$. In fact, much more realistic and effective control methods are available.

In Sect. 6.1, we have mentioned that if we try the proportional-derivative method for double inverted pendulum control, namely

$$a = -k_x x - k_{\mathrm{d}x} \frac{\mathrm{d}x}{\mathrm{d}t},$$

then the double inverted pendulum control system can achieve the expected cart position $x_{\mathrm{E}} = 0$, but cannot achieve expected inverted pendulum angles $\theta_{1\mathrm{E}} = 0$ and $\theta_{2\mathrm{E}} = 0$.

It might not be difficult for readers to understand why the proportional-derivative method with feedback of cart position fails in double inverted pendulum control: The proportional-derivative method resorts to feedback of cart position x only, without utilizing any feedback of inverted pendulum angles θ_1 and θ_2. No wonder the proportional-derivative method can only succeed in cart position control but has no guarantee of control performance with respect to both inverted pendulum angles θ_1 and θ_2.

Inspired by above analysis of the failure of the proportional-derivative method, readers may have a heuristic idea: what if the proportional-derivative method is extended by incorporating feedback of both inverted pendulum angles θ_1 and θ_2? For example, what if we use an extended proportional-derivative method as

$$a = -k_{\theta_1}\theta_1 - k_{\mathrm{d}\theta_1}\frac{\mathrm{d}\theta_1}{\mathrm{d}t} - k_{\theta_2}\theta_2 - k_{\mathrm{d}\theta_2}\frac{\mathrm{d}\theta_2}{\mathrm{d}t} - k_x x - k_{\mathrm{d}x}\frac{\mathrm{d}x}{\mathrm{d}t}? \quad (6.25)$$

The formula described in (6.25) is expressed compactly as

$$a = -\boldsymbol{K}^{\mathrm{T}}\boldsymbol{x}, \quad (6.26)$$

where $\boldsymbol{x} \equiv \begin{bmatrix}\theta_1 & \frac{\mathrm{d}\theta_1}{\mathrm{d}t} & \theta_2 & \frac{\mathrm{d}\theta_2}{\mathrm{d}t} & x & \frac{\mathrm{d}x}{\mathrm{d}t}\end{bmatrix}^{\mathrm{T}}$ is the double inverted pendulum state and $\boldsymbol{K} \equiv \begin{bmatrix}k_{\theta_1} & k_{\mathrm{d}\theta_1} & k_{\theta_2} & k_{\mathrm{d}\theta_2} & k_x & k_{\mathrm{d}x}\end{bmatrix}$ is the vector of coefficients corresponding to state terms respectively.

The generic formalism described in (6.26) is indeed adopted for design of our intended double inverted pendulum control method, though more knowledge is still needed to instantiate the generic formalism into a concrete and effective control method.

Generally, given a control system that adopts linear state-space modelling described by (6.5), with its state denoted as $\boldsymbol{x}$ and its control input to the target process denoted as $\boldsymbol{u}$:

$$\frac{\mathrm{d}}{\mathrm{d}t}\boldsymbol{x} = \boldsymbol{A}\boldsymbol{x} + \boldsymbol{B}\boldsymbol{u}.$$

Let $\boldsymbol{x}_{\mathrm{E}} = \boldsymbol{0}$ be the expected state. Suppose that the **full-state feedback control method** formalized as

$$\boldsymbol{u} = -\boldsymbol{K}^{\mathrm{T}}\boldsymbol{x} \tag{6.27}$$

is applied, where $\boldsymbol{K}$ is called the **gain matrix** of the full-state feedback control law.

Substitute (6.27) into (6.5) to obtain the closed-loop feedback state differential equation of the control system:

$$\frac{\mathrm{d}}{\mathrm{d}t}\boldsymbol{x} = \boldsymbol{A}\boldsymbol{x} + \boldsymbol{B}(-\boldsymbol{K}^{\mathrm{T}}\boldsymbol{x}) = (\boldsymbol{A} - \boldsymbol{B}\boldsymbol{K}^{\mathrm{T}})\boldsymbol{x}, \tag{6.28}$$

which is equivalent to the state differential equation of a linear self-evolutionary system with the state transition matrix $\boldsymbol{A} - \boldsymbol{B}\boldsymbol{K}^{\mathrm{T}}$.

According to the stability criterion presented in Sect. 6.2.4, the closed-loop feedback control system is stable, or in other words the state $\boldsymbol{x}$ can always converge to the expected state $\boldsymbol{x}_{\mathrm{E}} = \boldsymbol{0}$, if and only if the real parts of all eigenvalues of the state transition matrix $\boldsymbol{A} - \boldsymbol{B}\boldsymbol{K}^{\mathrm{T}}$ are negative. Therefore, a promising control strategy for the control system is to adopt the control law specified in (6.27) and set certain gain matrix $\boldsymbol{K}$ such that the real parts of all eigenvalues of $\boldsymbol{A} - \boldsymbol{B}\boldsymbol{K}^{\mathrm{T}}$ are negative.

Full-state feedback control strategy: *Set certain gain matrix* $\boldsymbol{K}$ *such that the real parts of all eigenvalues of* $\boldsymbol{A} - \boldsymbol{B}\boldsymbol{K}^{\mathrm{T}}$ *are negative, and then adopt the full-state feedback control method* $\boldsymbol{u} = -\boldsymbol{K}^{\mathrm{T}}\boldsymbol{x}$.

6.4.2 Design the Gain Matrix: Expedient Method

A key issue of the full-state feedback control strategy is design of the gain matrix $\boldsymbol{K}$, at least to find a gain matrix $\boldsymbol{K}$ such that the real parts of all eigenvalues of $\boldsymbol{A} - \boldsymbol{B}\boldsymbol{K}^{\mathrm{T}}$ are negative and hence the closed-loop feedback control system is stable.

Sometimes, a proper gain matrix $\boldsymbol{K}$ may be found by *empirical tuning* with little theoretical analysis. Take lateral control of low-speed autonomous vehicle navigation illustrated in Fig. 6.2 as an example.

In low-speed autonomous vehicle navigation, vehicle dynamics at a constant speed v can be modelled by differential equations as

$$\begin{cases} \frac{\mathrm{d}}{\mathrm{d}t}x &= v\cos\theta \\ \frac{\mathrm{d}}{\mathrm{d}t}y &= v\sin\theta \\ \frac{\mathrm{d}}{\mathrm{d}t}\theta &= \frac{v}{L}\tan\beta \end{cases}$$

which is called the *bicycle kinematics model*. In the model, x denotes the vehicle longitudinal position, y denotes the vehicle lateral position, θ denotes the vehicle orientation or heading angle, L denotes the vehicle wheel-base, and β denotes the vehicle steering angle. The vehicle *lateral dynamics* is extracted and approximated as

$$\frac{\mathrm{d}}{\mathrm{d}t}\begin{bmatrix} y \\ \theta \end{bmatrix} = \begin{bmatrix} 0 & v \\ 0 & 0 \end{bmatrix}\begin{bmatrix} y \\ \theta \end{bmatrix} + \begin{bmatrix} 0 \\ \frac{v}{L} \end{bmatrix}\beta \equiv \boldsymbol{A}\begin{bmatrix} y \\ \theta \end{bmatrix} + \boldsymbol{B}\beta, \tag{6.29}$$

where $[y \quad \theta]^{\mathrm{T}}$ denotes the *vehicle lateral state* and the steering angle β serves as control input to vehicle lateral control.

The controllability matrix of the vehicle lateral control system modelled by (6.29) is

$$\boldsymbol{C}_{\boldsymbol{A},\boldsymbol{B}} = [\boldsymbol{B} \quad \boldsymbol{AB}] = \begin{bmatrix} 0 & \frac{v^2}{L} \\ \frac{v}{L} & 0 \end{bmatrix},$$

which is of full rank. So the target process of vehicle lateral control is controllable, which is consistent with our obvious daily-life experience.

Let $\boldsymbol{K} = [k_y \quad k_\theta]^{\mathrm{T}}$ and $\beta = -k_y y - k_\theta \theta$, and the characteristic polynomial of the closed-loop feedback vehicle lateral control system is

$$\det[s\boldsymbol{I} - (\boldsymbol{A} - \boldsymbol{BK}^{\mathrm{T}})] = \det\begin{bmatrix} s & -v \\ \frac{v}{L}k_y & s + \frac{v}{L}k_\theta \end{bmatrix} = s^2 + \frac{v}{L}k_\theta s + \frac{v^2}{L}k_y,$$

for which we can easily set $\boldsymbol{K} = [k_y \quad k_\theta]^{\mathrm{T}}$ by empirical tuning to make the two eigenvalues have negative real parts. For example, set $k_y = L$ and $k_\theta = 2L$ (if $v > 0$) or $k_\theta = -2L$ (if $v < 0$).

However, empirical tuning is no longer applicable to difficult control problems such as double inverted pendulum control. When the Matlab simulation code **DoubleInvertedPendulum.m** was presented in Sect. 6.1 for the first time, readers might have no idea of how to handle the problem of double inverted pendulum control.

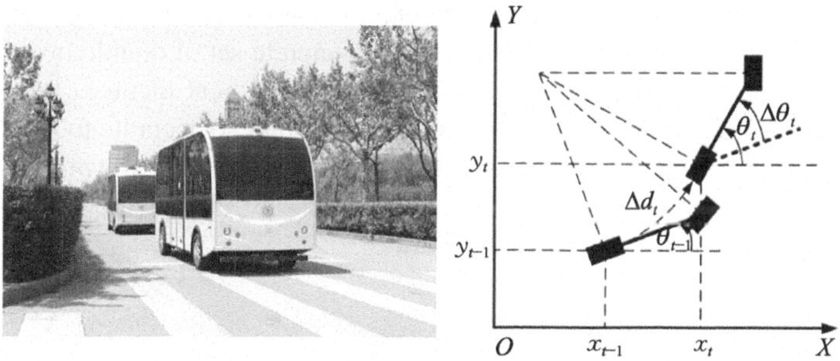

Fig. 6.2 Low-speed autonomous vehicle navigation

In contrast, now the strategy to handle the control problem, namely the *full-state feedback control strategy*, is clear to readers. Readers can try again with the Matlab simulation code **DoubleInvertedPendulum.m**. Yet we believe that after further trials with empirical tuning of the gain matrix $\boldsymbol{K}$ in (6.26), readers would still find it difficult to succeed in double inverted pendulum control.

In fact, design of the gain matrix for difficult control problems is not a trivial issue, but involves systematic knowledge. An expedient method of designing the gain matrix is based on **symbolic operation** that is incorporated into software such as Matlab. The idea of the expedient method is as follows: compute the closed-loop characteristic polynomial $\det[s\boldsymbol{I} - (\boldsymbol{A} - \boldsymbol{B}\boldsymbol{K}^{\mathrm{T}})]$ via symbolic operation and express it as a characteristic polynomial with parametrized coefficients in terms of the gain matrix $\boldsymbol{K}$. Denote

$$\det[s\boldsymbol{I} - (\boldsymbol{A} - \boldsymbol{B}\boldsymbol{K}^{\mathrm{T}})] = s^n + a_{n-1}(\boldsymbol{K})s^{n-1} + \cdots + a_0(\boldsymbol{K}). \tag{6.30}$$

Suppose that an expected closed-loop characteristic polynomial is

$$C_{\mathrm{E}}(s) = s^n + a^{\mathrm{E}}_{n-1}s^{n-1} + \cdots + a^{\mathrm{E}}_0. \tag{6.31}$$

Compare corresponding s exponential terms between (6.30) and (6.31) to establish a group of equations in terms of the gain matrix $\boldsymbol{K}$ as

$$\begin{cases} a_{n-1}(\boldsymbol{K}) & = a^{\mathrm{E}}_{n-1} \\ a_{n-2}(\boldsymbol{K}) & = a^{\mathrm{E}}_{n-2} \\ & \vdots \\ a_1(\boldsymbol{K}) & = a^{\mathrm{E}}_1 \\ a_0(\boldsymbol{K}) & = a^{\mathrm{E}}_0 \end{cases} \tag{6.32}$$

from which elements of the gain matrix $\boldsymbol{K}$ are solved.

Double inverted pendulum control

We apply above expedient method to design the gain matrix for double inverted pendulum control. The double inverted pendulum control system adopts linear state-space modelling described by (6.3). We take a concrete set of double inverted pendulum parameters as example to demonstrate the expedient method. Let $m_1 = 1$, $m_2 = 1$, $L_1 = 1$, $L_2 = 1$, and $g = 10$—It is obvious not so accurate to set the earth gravitational constant as $g = 10$, yet this has no essential influence on presentation of control theory. For numeric simplicity, we set it as $g = 10$—Then the state transition matrix $\boldsymbol{A}$ and the control input matrix $\boldsymbol{B}$ are

$$
\boldsymbol{A} = \begin{bmatrix} 0 & 1 & 0 & 0 & 0 & 0 \\ 20 & 0 & -10 & 0 & 0 & 0 \\ 0 & 0 & 0 & 1 & 0 & 0 \\ -20 & 0 & 20 & 0 & 0 & 0 \\ 0 & 0 & 0 & 0 & 0 & 1 \\ 0 & 0 & 0 & 0 & 0 & 0 \end{bmatrix}, \quad \boldsymbol{B} = \begin{bmatrix} 0 \\ -1 \\ 0 \\ 0 \\ 0 \\ 1 \end{bmatrix}.
$$

Let the gain matrix $\boldsymbol{K} \equiv [k_1\ k_2\ k_3\ k_4\ k_5\ k_6]^{\mathrm{T}}$. Compute the closed-loop characteristic polynomial $\det[s\boldsymbol{I} - (\boldsymbol{A} - \boldsymbol{B}\boldsymbol{K}^{\mathrm{T}})]$ via symbolic operation and obtain

$$
\begin{aligned}
&\det[s\boldsymbol{I} - (\boldsymbol{A} - \boldsymbol{B}\boldsymbol{K}^{\mathrm{T}})] = s^6 + (k_6 - k_2)s^5 + \\
&(k_5 - k_1 - 40)s^4 + (20k_2 + 20k_4 - 40k_6)s^3 + \\
&(20k_1 + 20k_3 - 40k_5 + 200)s^2 + 200k_6 s + 200k_5.
\end{aligned}
$$

Set the expected closed-loop characteristic polynomial as

$$
C_{\mathrm{E}}(s) = (s+4)^6 = s^6 + 24\,s^5 + 240\,s^4 + 1\,280\,s^3 + 3\,840\,s^2 + 6\,144\,s + 4\,096.
$$

This is only an example of expected closed-loop characteristic polynomial, whereas other expected closed-loop characteristic polynomials can also be adopted. We can follow knowledge presented in Chap. 5 to design expected locations of control system eigenvalues namely control system poles (equivalence between eigenvalues and poles is explained in Sect. 6.2.4).

Compare corresponding s exponential terms between the closed-loop characteristic polynomial $\det[s\boldsymbol{I} - (\boldsymbol{A} - \boldsymbol{B}\boldsymbol{K}^{\mathrm{T}})]$ and the expected closed-loop characteristic polynomial $C_{\mathrm{E}}(s)$ to establish a group of equations in terms of the gain matrix elements k_1, k_2, k_3, k_4, k_5, and k_6 as

$$
\begin{cases}
k_6 - k_2 & = 24 \\
k_5 - k_1 - 40 & = 240 \\
20k_2 + 20k_4 - 40k_6 & = 1\,280 \\
20k_1 + 20k_3 - 40k_5 + 200 & = 3\,840 \\
200k_6 & = 6\,144 \\
200k_5 & = 4\,096
\end{cases}
$$

from which the gain matrix elements k_1, k_2, k_3, k_4, k_5, and k_6 are solved. The solution is

$$
\boldsymbol{K} \equiv \begin{bmatrix} k_1 \\ k_2 \\ k_3 \\ k_4 \\ k_5 \\ k_6 \end{bmatrix} = \begin{bmatrix} -259.52 \\ 6.72 \\ 482.48 \\ 118.72 \\ 20.48 \\ 30.72 \end{bmatrix}.
$$

Matlab code for the expedient method of designing the gain matrix via symbolic operation is given as follows.

DesignGainMatrixExpedient.m

```
%% Double inverted pendulum parameters
m1 = 1; m2 = 1; L1 = 1; L2 = 1; g = 10;
A = [0, 1, 0, 0, 0, 0; ...
     (m1+m2)*g/(m1*L1), 0, -m2*g/(m1*L1), 0, 0, 0; ...
     0, 0, 0, 1, 0, 0; ...
     -(m1+m2)*g/(m1*L2), 0, (m1+m2)*g/(m1*L2), 0, 0, 0; ...
     0, 0, 0, 0, 0, 1; ...
     0, 0, 0, 0, 0, 0];
B = [0; -1/L1; 0; 0; 0; 1];
n = 6; % State dimension

%% Compute characteristic polynomials via symbolic operation
syms k1 k2 k3 k4 k5 k6 s
csK = collect(det(s*eye(n) - (A - B*[k1, k2, k3, k4, k5, k6])));
fprintf('Characteristic polynomial with parametrized
    coefficients: '); csK
csE = collect((s+4)^6); % Expected characteristic polynomial
fprintf('Expected characteristic polynomial: '); csE

%% Compare characteristic polynomials and solve the gain matrix
kvars = [k1 k2 k3 k4 k5 k6];
csKcoeffs = coeffs(csK, s); csEcoeffs = coeffs(csE, s);
K = vpasolve(csKcoeffs == csEcoeffs, kvars);
fprintf('Gain matrix K: '); sttK = eval([K.k1; K.k2; K.k3; K.k4;
    K.k5; K.k6])
```

Once the gain matrix $\boldsymbol{K}$ is designed, readers can try the concrete full-state feedback control method to see how the double inverted pendulum is controlled to and kept stable at the expected state. Matlab simulation code for complete demonstration of double inverted pendulum control is given as follows.

DoubleInvertedPendulum2.m

```
%% Double inverted pendulum parameters
m1 = 1; m2 = 1; L1 = 1; L2 = 1; g = 10;
%% Simulation preliminary configuration
dt = 0.001; % Numerical computation step
tSpan = 0:dt:3; % Simulation time span
x = 0.2; dx = 0; % Cart position and its velocity
% Inverted pendulum angle theta-1 and its angular velocity
y1 = 0.2; dy1 = 0;
% Inverted pendulum angle theta-2 and its angular velocity
y2 = 0; dy2 = 0;
stt = [y1; dy1; y2; dy2; x; dx]; % Double inverted pendulum
    state
% Record states in simulation
sttAll = zeros(length(stt), length(tSpan)); k = 0;
% Expected equilibrium status
xExpected = 0; y1Expected = 0; y2Expected = 0;
```

```
SimConfig = [m1, m2, L1, L2, g, dt];

%% Design the gain matrix via the expedient method
DesignGainMatrixExpedient % Obtain the gain matrix sttK

%% Simulation of double inverted pendulum control
for t = tSpan
    %% Control method
    acc = -sttK'*stt; % Full-state feedback control of the cart
    acceleration

    %% Double inverted pendulum dynamics
    stt = DynamicsDIP(SimConfig, stt, acc);
    sttC = num2cell(stt); [y1, dy1, y2, dy2, x, dx] = sttC{:};
    if (abs(y1)>=pi/2 && abs(y2)>=pi/2)
        fprintf('Control failure!\n'); break;
    end
    k = k+1; sttAll(:,k) = stt;
    %% Double inverted pendulum visualization
    if (rem(k,20) == 0)
        DisplayDIP(x, y1, y2, L1, L2); pause(20*dt);
    end
end
```

By so far, the difficult problem of double inverted pendulum control is handled. On the other hand, the expedient method of designing the gain matrix via symbolic operation is computational demanding. It is usually limited to control systems that have low-dimensional states and is not generally applicable in practical applications. Besides, software such as Matlab that incorporates symbolic operation may not be available to readers, which also restricts applicability of the expedient method. These limitations of the expedient method are reasons why it is called "expedient". Besides the expedient method, we need a method of designing the gain matrix that is generally applicable in practical applications and can be conveniently implemented in common programming languages.

6.4.3 Controllability Canonical Form of Linear State-Space Modelling

Given a control system that adopts linear state-space modelling described by (6.5), with its state denoted as $\boldsymbol{x}$ (suppose that the state dimension is n) and its control input to the target process denoted as $\boldsymbol{u}$, for the moment, we focus on *single-input-multiple-output* control problems, which implies that the control input $\boldsymbol{u}$ is one-dimensional and the control input matrix $\boldsymbol{B}$ is a n-by-1 matrix namely a n-dimensional vector.

$$\frac{\mathrm{d}}{\mathrm{d}t}\boldsymbol{x} = \boldsymbol{A}\boldsymbol{x} + \boldsymbol{B}\boldsymbol{u}.$$

Its controllability matrix is defined in (6.24) as

$$\boldsymbol{C}_{A,B} = [\boldsymbol{B}\ \ \boldsymbol{A}\boldsymbol{B}\ \ \boldsymbol{A}^2\boldsymbol{B}\ \ \cdots\ \ \boldsymbol{A}^{n-1}\boldsymbol{B}],$$

which is a n-by-n square matrix as $\boldsymbol{B}$ is a n-by-1 matrix. Suppose that the target process is controllable, in other words, suppose that the controllability matrix $\boldsymbol{C}_{A,B}$ is of full rank and hence invertible.

Denote the characteristic polynomial of the state transition matrix $\boldsymbol{A}$ namely the open-loop characteristic polynomial $\det(s\boldsymbol{I} - \boldsymbol{A})$ as

$$C_A(s) \equiv \det(s\boldsymbol{I} - \boldsymbol{A}) = s^n + a_{n-1}s^{n-1} + \cdots + a_0.$$

Define three **auxiliary matrices** $\boldsymbol{M}_A$, $\boldsymbol{P}_{A,B}$, and $\boldsymbol{S}_A$ in (6.33), (6.34), and (6.35) respectively:

$$\boldsymbol{M}_A = \begin{bmatrix} a_1 & a_2 & a_3 & \cdots & a_{n-2} & a_{n-1} & 1 \\ a_2 & a_3 & a_4 & \cdots & a_{n-1} & 1 & \\ a_3 & a_4 & a_5 & \cdots & 1 & & \\ \vdots & \vdots & \vdots & \ddots & & & \\ a_{n-2} & a_{n-1} & 1 & & & & \\ a_{n-1} & 1 & & & & & \\ 1 & & & & & & \end{bmatrix}, \tag{6.33}$$

$$\boldsymbol{P}_{A,B} = \boldsymbol{C}_{A,B}\boldsymbol{M}_A, \tag{6.34}$$

$$\boldsymbol{S}_A = \begin{bmatrix} & 1 & & & & \\ & & 1 & & & \\ & & & \ddots & & \\ & & & & 1 & \\ & & & & & 1 \\ -a_0 & -a_1 & -a_2 & \cdots & -a_{n-2} & -a_{n-1} \end{bmatrix}. \tag{6.35}$$

We have

$$\boldsymbol{P}_{A,B}^{-1}\boldsymbol{A}\boldsymbol{P}_{A,B} = \boldsymbol{S}_A, \tag{6.36}$$

$$\boldsymbol{P}_{A,B}^{-1}\boldsymbol{B} = \boldsymbol{e}_n \equiv [0\ 0\ \ \cdots\ \ 0\ 1]^{\mathrm{T}}. \tag{6.37}$$

Proof Consider (6.37) and have

$$\begin{aligned} \boldsymbol{P}_{A,B}^{-1}\boldsymbol{B} = \boldsymbol{e}_n \iff \boldsymbol{B} = \boldsymbol{P}_{A,B}\boldsymbol{e}_n \iff \boldsymbol{B} = \boldsymbol{C}_{A,B}\boldsymbol{M}_A\boldsymbol{e}_n \\ (\text{note } \boldsymbol{M}_A\boldsymbol{e}_n = \boldsymbol{e}_1 \equiv [1\ \ 0\ \ \cdots\ \ 0\ \ 0]^{\mathrm{T}}) \\ \iff \boldsymbol{B} = \boldsymbol{C}_{A,B}\boldsymbol{e}_1 \\ \iff \boldsymbol{B} = [\boldsymbol{B}\ \ \boldsymbol{A}\boldsymbol{B}\ \ \cdots\ \ \boldsymbol{A}^{n-1}\boldsymbol{B}]\boldsymbol{e}_1 = \boldsymbol{B}, \end{aligned}$$

which implies that (6.37) holds. Since

$$C_{A,B}^{-1}C_{A,B} = C_{A,B}^{-1}[B \;\; AB \;\; \cdots \;\; A^{n-1}B] = I,$$

we have

$$\begin{aligned} C_{A,B}^{-1}B &= e_1, \\ C_{A,B}^{-1}AB &= e_2 \equiv [0 \;\; 1 \;\; \cdots \;\; 0 \;\; 0]^{\mathrm{T}}, \\ &\;\;\vdots \\ C_{A,B}^{-1}A^{n-2}B &= e_{n-1} \equiv [0 \;\; 0 \;\; \cdots \;\; 1 \;\; 0]^{\mathrm{T}}, \\ C_{A,B}^{-1}A^{n-1}B &= e_n, \end{aligned} \tag{6.38}$$

where e_k denotes the k-th column of the identity matrix I, namely the n-dimensional vector whose k-th element is 1 with other elements being 0.

According to (6.23), we also have

$$\begin{aligned} & C_A(A) = A^n + a_{n-1}A^{n-1} + \cdots + a_0 I = 0 \\ \iff & A^n = -a_{n-1}A^{n-1} - a_{n-2}A^{n-2} \cdots - a_1 A - a_0 I, \end{aligned}$$

so

$$\begin{aligned} C_{A,B}^{-1}AC_{A,B} &= C_{A,B}^{-1}[AB \;\; A^2B \;\; \cdots \;\; A^nB] \\ &= \left[e_2 \;\; \cdots \;\; e_n \;\; \sum_{k=1}^{n}(-a_{k-1})e_k \right], \end{aligned}$$

namely

$$C_{A,B}^{-1}AC_{A,B} = \begin{bmatrix} & & & -a_0 \\ 1 & & & -a_1 \\ & 1 & & -a_2 \\ & & \ddots & \vdots \\ & & 1 & -a_{n-1} \end{bmatrix} = S_A^{\mathrm{T}}.$$

Consider (6.36) and have

$$\begin{aligned} P_{A,B}^{-1}AP_{A,B} = S_A &\iff M_A^{-1}C_{A,B}^{-1}AC_{A,B}M_A = S_A \\ &\iff S_A^{\mathrm{T}}M_A = M_A S_A. \end{aligned}$$

Denote

$$M_A^* \equiv \begin{bmatrix} a_2 & a_3 & \cdots & a_{n-1} & 1 \\ a_3 & a_4 & \cdots & 1 & \\ \vdots & \vdots & \ddots & & \\ a_{n-1} & 1 & & & \\ 1 & & & & \end{bmatrix} \quad m = \begin{bmatrix} a_1 \\ a_2 \\ \vdots \\ a_{n-2} \\ a_{n-1} \end{bmatrix}.$$

We have

$$S_A^{\mathrm{T}} M_A = \begin{bmatrix} & & & -a_0 \\ 1 & & & -a_1 \\ & 1 & & -a_2 \\ & & \ddots & \vdots \\ & & 1 & -a_{n-1} \end{bmatrix} \begin{bmatrix} a_1 & a_2 & \cdots & a_{n-1} & 1 \\ a_2 & a_3 & \cdots & 1 & \\ \vdots & \vdots & \ddots & & \\ a_{n-1} & 1 & & & \\ 1 & & & & \end{bmatrix}$$
$$= \begin{bmatrix} \mathbf{0}^{\mathrm{T}} & -a_0 \\ I & -m \end{bmatrix} \begin{bmatrix} m & M_A^* \\ 1 & \mathbf{0}^{\mathrm{T}} \end{bmatrix} = \begin{bmatrix} -a_0 & \mathbf{0}^{\mathrm{T}} \\ \mathbf{0} & M_A^* \end{bmatrix}$$

and

$$M_A S_A = \begin{bmatrix} a_1 & a_2 & \cdots & a_{n-1} & 1 \\ a_2 & a_3 & \cdots & 1 & \\ \vdots & \vdots & \ddots & & \\ a_{n-1} & 1 & & & \\ 1 & & & & \end{bmatrix} \begin{bmatrix} & 1 & & & \\ & & \ddots & & \\ & & & 1 & \\ & & & & 1 \\ -a_0 & -a_1 & \cdots & -a_{n-2} & -a_{n-1} \end{bmatrix}$$
$$= \begin{bmatrix} m^{\mathrm{T}} & 1 \\ M_A^* & \mathbf{0} \end{bmatrix} \begin{bmatrix} \mathbf{0} & I \\ -a_0 & -m^{\mathrm{T}} \end{bmatrix} = \begin{bmatrix} -a_0 & \mathbf{0}^{\mathrm{T}} \\ \mathbf{0} & M_A^* \end{bmatrix}.$$

So we have $S_A^{\mathrm{T}} M_A = M_A S_A$ which implies that (6.36) holds. □

Take **state transform** on (6.5) as

$$\bar{x} = P_{A,B}^{-1} x \iff x = P_{A,B} \bar{x} \tag{6.39}$$

and obtain

$$\frac{\mathrm{d}}{\mathrm{d}t} x = Ax + Bu \iff \frac{\mathrm{d}}{\mathrm{d}t} \bar{x} = P_{A,B}^{-1} A P_{A,B} \bar{x} + P_{A,B}^{-1} B u,$$

where the auxiliary matrix $P_{A,B}$ is the **state transform matrix**. According to (6.36) and (6.37), we have

$$\frac{\mathrm{d}}{\mathrm{d}t} \bar{x} = S_A \bar{x} + e_n u, \tag{6.40}$$

namely

$$\frac{\mathrm{d}}{\mathrm{d}t}\bar{x} = \begin{bmatrix} & 1 & & & \\ & & \ddots & & \\ & & & 1 & \\ & & & & 1 \\ -a_0 & -a_1 & \cdots & -a_{n-2} & -a_{n-1} \end{bmatrix} \bar{x} + \begin{bmatrix} 0 \\ 0 \\ \vdots \\ 0 \\ 1 \end{bmatrix} \boldsymbol{u},$$

which is called the **controllability canonical form** of linear state-space modelling for the control system.

6.4.4 Design the Gain Matrix: General Method

Given a *single-input-multiple-output* control system, based on the *controllability canonical form* described in (6.40), we have a much more effective method of designing the gain matrix for the full-state feedback control strategy.

Step 1: Compute the controllability matrix $\boldsymbol{C}_{\boldsymbol{A},\boldsymbol{B}}$ via (6.24) as

$$\boldsymbol{C}_{\boldsymbol{A},\boldsymbol{B}} = [\boldsymbol{B} \;\; \boldsymbol{A}\boldsymbol{B} \;\; \boldsymbol{A}^2\boldsymbol{B} \;\; \cdots \;\; \boldsymbol{A}^{n-1}\boldsymbol{B}].$$

Step 2: Obtain the characteristic polynomial of the state transition matrix $\boldsymbol{A}$ namely the the open-loop characteristic polynomial $\det(s\boldsymbol{I} - \boldsymbol{A})$:

$$C_{\boldsymbol{A}}(s) \equiv \det(s\boldsymbol{I} - \boldsymbol{A}) = s^n + a_{n-1}s^{n-1} + \cdots + a_0.$$

For this purpose, the first way is to compute eigenvalues of $\boldsymbol{A}$ via numeric computation (Quarteroni et al. 2007). Denote the computed eigenvalues as $\lambda_1, \lambda_2, \cdots, \lambda_n$ and denote

$$\begin{aligned}
C_{\boldsymbol{A},1}(s) &\equiv a_{1,1}s + a_{1,0} = s - \lambda_1, \\
C_{\boldsymbol{A},2}(s) &\equiv a_{2,2}s^2 + a_{2,1}s + a_{2,0} = (s - \lambda_1)(s - \lambda_2), \\
&\vdots \\
C_{\boldsymbol{A},k}(s) &\equiv a_{k,k}s^k + \sum_{i=0}^{k-1} a_{k,i}s^i = \prod_{i=1}^{k}(s - \lambda_i), \\
&\vdots \\
C_{\boldsymbol{A},n}(s) &\equiv a_{n,n}s^n + a_{n,n-1}s^{n-1} + \cdots + a_{n,0} = \prod_{i=1}^{n}(s - \lambda_i),
\end{aligned}$$

where the last polynomial $C_{\boldsymbol{A},n}(s)$ is the characteristic polynomial $C_{\boldsymbol{A}}(s) \equiv \det(s\boldsymbol{I} - \boldsymbol{A})$ and its coefficients $a_{n,0}, a_{n,1}, \ldots, a_{n,n-1}$ are the characteristic polynomial coefficients $a_0, a_1, \ldots, a_{n-1}$ respectively. Apparently $a_{1,1} = 1, a_{2,2} = 1, \ldots, a_{n,n} = 1$, whereas other characteristic polynomial coefficients can be computed iteratively with eigenvalues $\lambda_1, \lambda_2, \ldots, \lambda_n$ as (define $a_{k,-1} \equiv 0$ for $k = 1, \ldots, n$)

$$\begin{aligned}
a_{1,0} &= -\lambda_1, \\
a_{2,0} &= a_{1,-1} + a_{1,0}(-\lambda_2), \quad a_{2,1} = a_{1,0} + a_{1,1}(-\lambda_2), \\
&\vdots \\
a_{k,i} &= a_{k-1,i-1} + a_{k-1,i}(-\lambda_k) \text{ for } i = 0, \ldots, k-1, \\
&\vdots \\
a_{n,i} &= a_{n-1,i-1} + a_{n-1,i}(-\lambda_n) \text{ for } i = 0, \ldots, n-1, \\
a_0 &= a_{n,0}, \quad a_1 = a_{n,1}, \quad \ldots, \quad a_{n-1} = a_{n,n-1}.
\end{aligned} \tag{6.41}$$

The second way is to solve the characteristic polynomial coefficients $a_0, a_1, \cdots,$ a_{n-1} directly as

$$\boldsymbol{C}_{\boldsymbol{A},\boldsymbol{B}} \boldsymbol{X}_a = \boldsymbol{A}^n \boldsymbol{B} \iff \boldsymbol{X}_a = \boldsymbol{C}_{\boldsymbol{A},\boldsymbol{B}}^{-1} \boldsymbol{A}^n \boldsymbol{B}, \tag{6.42}$$

where $\boldsymbol{X}_a \equiv [-a_0 \ \ -a_1 \ \ \cdots \ \ -a_{n-1}]^{\mathrm{T}}$. This is because according to (6.23), we have

$$\begin{aligned}
& \boldsymbol{A}^n + a_{n-1}\boldsymbol{A}^{n-1} + \cdots + a_0 \boldsymbol{I} = \boldsymbol{0} \\
\iff & -a_0 \boldsymbol{I} - a_1 \boldsymbol{A} \cdots - a_{n-1}\boldsymbol{A}^{n-1} = \boldsymbol{A}^n \\
\Longrightarrow & -a_0 \boldsymbol{B} - a_1 \boldsymbol{A}\boldsymbol{B} \cdots - a_{n-1}\boldsymbol{A}^{n-1}\boldsymbol{B} = \boldsymbol{A}^n \boldsymbol{B},
\end{aligned}$$

which implies that (6.42) holds. Compared with the first way, the second way is computationally more desirable because it is exempt from numeric computation of eigenvalues of the state transition matrix $\boldsymbol{A}$ and only involves basic matrix operations.

Step 3: Generate the auxiliary matrix $\boldsymbol{M}_{\boldsymbol{A}}$ via (6.33) as

$$\boldsymbol{M}_{\boldsymbol{A}} = \begin{bmatrix} a_1 & a_2 & \cdots & a_{n-1} & 1 \\ a_2 & a_3 & \cdots & 1 & \\ \vdots & \vdots & \ddots & & \\ a_{n-1} & 1 & & & \\ 1 & & & & \end{bmatrix},$$

and compute the state transform matrix $\boldsymbol{P}_{\boldsymbol{A},\boldsymbol{B}}$ via (6.34) as

$$\boldsymbol{P}_{\boldsymbol{A},\boldsymbol{B}} = \boldsymbol{C}_{\boldsymbol{A},\boldsymbol{B}} \boldsymbol{M}_{\boldsymbol{A}}.$$

If we take the state transform matrix $\boldsymbol{P}_{\boldsymbol{A},\boldsymbol{B}}$ and take state transform on (6.5) according to (6.39) as

$$\bar{\boldsymbol{x}} = \boldsymbol{P}_{A,B}^{-1}\boldsymbol{x} \iff \boldsymbol{x} = \boldsymbol{P}_{A,B}\bar{\boldsymbol{x}},$$

then we have the controllability canonical form described in (6.40) as

$$\frac{\mathrm{d}}{\mathrm{d}t}\bar{\boldsymbol{x}} = \boldsymbol{S}_A\bar{\boldsymbol{x}} + \boldsymbol{e}_n\boldsymbol{u}$$

namely

$$\frac{\mathrm{d}}{\mathrm{d}t}\bar{\boldsymbol{x}} = \begin{bmatrix} & 1 & & & \\ & & \ddots & & \\ & & & 1 & \\ & & & & 1 \\ -a_0 & -a_1 & \cdots & -a_{n-2} & -a_{n-1} \end{bmatrix}\bar{\boldsymbol{x}} + \begin{bmatrix} 0 \\ 0 \\ \vdots \\ 0 \\ 1 \end{bmatrix}\boldsymbol{u}.$$

In fact, for gain matrix design, we do not need to really perform any state transform. Instead, we only need to bear in mind that the state transform $\bar{\boldsymbol{x}} = \boldsymbol{P}_{A,B}^{-1}\boldsymbol{x}$ or $\boldsymbol{x} = \boldsymbol{P}_{A,B}\bar{\boldsymbol{x}}$ with the state transform matrix $\boldsymbol{P}_{A,B}$ enables (6.5) to transform into the controllability canonical form described in (6.40).

Step 4: Design the transformed gain matrix $\bar{\boldsymbol{K}} \equiv \boldsymbol{P}_{A,B}^{\mathrm{T}}\boldsymbol{K}$. More specifically, substitute the state transform $\boldsymbol{x} = \boldsymbol{P}_{A,B}\bar{\boldsymbol{x}}$ into the full-state feedback control law described in (6.27) and obtain

$$\boldsymbol{u} = -\boldsymbol{K}^{\mathrm{T}}\boldsymbol{x} = -\boldsymbol{K}^{\mathrm{T}}\boldsymbol{P}_{A,B}\bar{\boldsymbol{x}} = -\bar{\boldsymbol{K}}^{\mathrm{T}}\bar{\boldsymbol{x}}.$$

Denote $\bar{\boldsymbol{K}} \equiv [\bar{k}_1\ \bar{k}_2\ \cdots\ \bar{k}_n]^{\mathrm{T}}$, substitute $\boldsymbol{u} = -\bar{\boldsymbol{K}}^{\mathrm{T}}\bar{\boldsymbol{x}}$ into the controllability canonical form described in (6.40) and obtain

$$\begin{aligned}\frac{\mathrm{d}}{\mathrm{d}t}\bar{\boldsymbol{x}} &= \boldsymbol{S}_A\bar{\boldsymbol{x}} + \boldsymbol{e}_n\boldsymbol{u} = \left(\boldsymbol{S}_A - \boldsymbol{e}_n\bar{\boldsymbol{K}}^{\mathrm{T}}\right)\bar{\boldsymbol{x}} \\ &= \begin{bmatrix} & 1 & & & \\ & & \ddots & & \\ & & & 1 & \\ & & & & 1 \\ -a_0-\bar{k}_1 & -a_1-\bar{k}_2 & \cdots & -a_{n-2}-\bar{k}_{n-1} & -a_{n-1}-\bar{k}_n \end{bmatrix}\bar{\boldsymbol{x}}.\end{aligned}$$

Review the state differential equation described in (6.19) and derivation of the characteristic polynomial of the state transition matrix $\boldsymbol{A}_c$. By analogue, we can know that the characteristic polynomial of the state transition matrix $\boldsymbol{S}_A - \boldsymbol{e}_n\bar{\boldsymbol{K}}^{\mathrm{T}}$ is

$$C_{\boldsymbol{S}_A-\boldsymbol{e}_n\bar{\boldsymbol{K}}^{\mathrm{T}}}(s) = s^n + (a_{n-1}+\bar{k}_n)s^{n-1} + \cdots + (a_1+\bar{k}_2)s + (a_0+\bar{k}_1).$$

Suppose that an expected closed-loop characteristic polynomial is

$$C_{\mathrm{E}}(s) = s^n + a_{n-1}^{\mathrm{E}} s^{n-1} + \cdots + a_0^{\mathrm{E}}.$$

To achieve $C_{S_A - e_n \bar{K}^{\mathrm{T}}}(s) = C_{\mathrm{E}}(s)$, we can easily set $\bar{k}_1, \bar{k}_2, \cdots, \bar{k}_n$ as

$$\begin{cases} \bar{k}_1 & = a_0^{\mathrm{E}} - a_0 \\ \bar{k}_2 & = a_1^{\mathrm{E}} - a_1 \\ & \vdots \\ \bar{k}_{n-1} & = a_{n-2}^{\mathrm{E}} - a_{n-2} \\ \bar{k}_n & = a_{n-1}^{\mathrm{E}} - a_{n-1} \end{cases} \tag{6.43}$$

Step 5: Recover the gain matrix $\boldsymbol{K}$ from the transformed gain matrix $\bar{\boldsymbol{K}}$ via

$$\boldsymbol{K} = \boldsymbol{P}_{A,B}^{-\mathrm{T}} \bar{\boldsymbol{K}}. \tag{6.44}$$

Double inverted pendulum control

We apply above general method to design the gain matrix for double inverted pendulum control. The double inverted pendulum control system adopts linear state-space modelling described by (6.3). As in Sect. 6.4.2, we take the same set of double inverted pendulum parameters as example to demonstrate the general method. Let $m_1 = 1$, $m_2 = 1$, $L_1 = 1$, $L_2 = 1$, and $g = 10$. Then the state transition matrix $\boldsymbol{A}$ and the control input matrix $\boldsymbol{B}$ are

$$\boldsymbol{A} = \begin{bmatrix} 0 & 1 & 0 & 0 & 0 & 0 \\ 20 & 0 & -10 & 0 & 0 & 0 \\ 0 & 0 & 0 & 1 & 0 & 0 \\ -20 & 0 & 20 & 0 & 0 & 0 \\ 0 & 0 & 0 & 0 & 0 & 1 \\ 0 & 0 & 0 & 0 & 0 & 0 \end{bmatrix}, \quad \boldsymbol{B} = \begin{bmatrix} 0 \\ -1 \\ 0 \\ 0 \\ 0 \\ 1 \end{bmatrix}.$$

Step 1: Compute the controllability matrix $\boldsymbol{C}_{A,B}$ via (6.24) as

$$\boldsymbol{C}_{A,B} = \begin{bmatrix} \boldsymbol{B} & \boldsymbol{AB} & \boldsymbol{A}^2\boldsymbol{B} & \boldsymbol{A}^3\boldsymbol{B} & \boldsymbol{A}^4\boldsymbol{B} & \boldsymbol{A}^5\boldsymbol{B} \end{bmatrix} = \begin{bmatrix} 0 & -1 & 0 & -20 & 0 & -600 \\ -1 & 0 & -20 & 0 & -600 & 0 \\ 0 & 0 & 0 & 20 & 0 & 800 \\ 0 & 0 & 20 & 0 & 800 & 0 \\ 0 & 1 & 0 & 0 & 0 & 0 \\ 1 & 0 & 0 & 0 & 0 & 0 \end{bmatrix}.$$

Step 2: If by the first way, compute eigenvalues of $\boldsymbol{A}$ numerically and obtain $\lambda_1 = -5.843\,1$, $\lambda_2 = -2.420\,3$, $\lambda_3 = 5.843\,1$, $\lambda_4 = 2.420\,3$, and $\lambda_5 = \lambda_6 = 0$. Compute characteristic polynomial coefficients iteratively via (6.41) and obtain $a_0 = a_1 = 0$, $a_2 = 200$, $a_3 = 0$, $a_4 = -40$, and $a_5 = 0$, which implies that

$$C_A(s) \equiv \det(s\boldsymbol{I} - \boldsymbol{A}) = s^6 - 40\,s^4 + 200\,s^2.$$

If by the second way, compute $\boldsymbol{A}^n\boldsymbol{B}$ by multiplying $\boldsymbol{A}$ with the already computed last column of the controllability matrix $\boldsymbol{C}_{\boldsymbol{A},\boldsymbol{B}}$:

$$\boldsymbol{A}^n\boldsymbol{B} = [0 \;\; -20\,000 \;\; 0 \;\; 28\,000 \;\; 0 \;\; 0]^{\mathrm{T}}.$$

Then solve the equation described in (6.42) and obtain

$$\begin{bmatrix} -a_0 \\ -a_1 \\ -a_2 \\ -a_3 \\ -a_4 \\ -a_5 \end{bmatrix} = \begin{bmatrix} 0 & -1 & 0 & -20 & 0 & -600 \\ -1 & 0 & -20 & 0 & -600 & 0 \\ 0 & 0 & 0 & 20 & 0 & 800 \\ 0 & 0 & 20 & 0 & 800 & 0 \\ 0 & 1 & 0 & 0 & 0 & 0 \\ 1 & 0 & 0 & 0 & 0 & 0 \end{bmatrix}^{-1} \begin{bmatrix} 0 \\ -20\,000 \\ 0 \\ 28\,000 \\ 0 \\ 0 \end{bmatrix} = \begin{bmatrix} 0 \\ 0 \\ -200 \\ 0 \\ 40 \\ 0 \end{bmatrix},$$

which also implies

$$C_A(s) \equiv \det(s\boldsymbol{I} - \boldsymbol{A}) = s^6 - 40\,s^4 + 200\,s^2.$$

Step 3: Generate the auxiliary matrix $\boldsymbol{M}_A$ via (6.33) as

$$\boldsymbol{M}_A = \begin{bmatrix} & 200 & & -40 & & 1 \\ 200 & & -40 & & 1 & \\ & -40 & & 1 & & \\ -40 & & 1 & & & \\ & 1 & & & & \\ 1 & & & & & \end{bmatrix}$$

and compute the state transform matrix $\boldsymbol{P}_{\boldsymbol{A},\boldsymbol{B}}$ via (6.34) as

$$\boldsymbol{P}_{\boldsymbol{A},\boldsymbol{B}} = \boldsymbol{C}_{\boldsymbol{A},\boldsymbol{B}}\boldsymbol{M}_A = \begin{bmatrix} 0 & 0 & 20 & 0 & -1 & 0 \\ 0 & 0 & 0 & 20 & 0 & -1 \\ 0 & 0 & 20 & 0 & 0 & 0 \\ 0 & 0 & 0 & 20 & 0 & 0 \\ 200 & 0 & -40 & 0 & 1 & 0 \\ 0 & 200 & 0 & -40 & 0 & 1 \end{bmatrix}.$$

Note that the state transform $\bar{\boldsymbol{x}} = \boldsymbol{P}_{\boldsymbol{A},\boldsymbol{B}}^{-1}\boldsymbol{x}$ or $\boldsymbol{x} = \boldsymbol{P}_{\boldsymbol{A},\boldsymbol{B}}\bar{\boldsymbol{x}}$ with the state transform matrix $\boldsymbol{P}_{\boldsymbol{A},\boldsymbol{B}}$ enables (6.3) w le inverted pendulum parameters to transform into the controllability canonical form as

$$\frac{\mathrm{d}}{\mathrm{d}t}\bar{x} = \begin{bmatrix} & 1 & & & & \\ & & 1 & & & \\ & & & 1 & & \\ & & & & 1 & \\ & & & & & 1 \\ 0 & 0 & -200 & 0 & 40 & 0 \end{bmatrix} \bar{x} + \begin{bmatrix} 0 \\ 0 \\ 0 \\ 0 \\ 0 \\ 1 \end{bmatrix} a.$$

Step 4: Set the expected closed-loop characteristic polynomial as

$$C_{\mathrm{E}}(s) = (s+4)^6 = s^6 + 24\,s^5 + 240\,s^4 + 1\,280\,s^3 + 3\,840\,s^2 + 6\,144\,s + 4\,096,$$

which is the same to the example of expected closed-loop characteristic polynomial used in Sect. 6.4.2. Compute the transformed gain matrix $\bar{K} \equiv P_{A,B}^{\mathrm{T}} K$ via (6.43) as

$$\begin{cases} \bar{k}_1 = a_0^{\mathrm{E}} - a_0 = 4\,096 - 0 = 4\,096 \\ \bar{k}_2 = a_1^{\mathrm{E}} - a_1 = 6\,144 - 0 = 6\,144 \\ \bar{k}_3 = a_2^{\mathrm{E}} - a_2 = 3\,840 - 200 = 3\,640 \\ \bar{k}_4 = a_3^{\mathrm{E}} - a_3 = 1\,280 - 0 = 1\,280 \\ \bar{k}_5 = a_4^{\mathrm{E}} - a_4 = 240 - (-40) = 280 \\ \bar{k}_6 = a_5^{\mathrm{E}} - a_5 = 24 \end{cases} \qquad \bar{K} \equiv \begin{bmatrix} \bar{k}_1 \\ \bar{k}_2 \\ \bar{k}_3 \\ \bar{k}_4 \\ \bar{k}_5 \\ \bar{k}_6 \end{bmatrix} = \begin{bmatrix} 4\,096 \\ 6\,144 \\ 3\,640 \\ 1\,280 \\ 280 \\ 24 \end{bmatrix}.$$

Step 5: Recover the gain matrix K from the transformed gain matrix $\bar{K}$ via (6.44):

$$K = P_{A,B}^{-\mathrm{T}} \bar{K} = \begin{bmatrix} 0 & 0 & 20 & 0 & -1 & 0 \\ 0 & 0 & 0 & 20 & 0 & -1 \\ 0 & 0 & 20 & 0 & 0 & 0 \\ 0 & 0 & 0 & 20 & 0 & 0 \\ 200 & 0 & -40 & 0 & 1 & 0 \\ 0 & 200 & 0 & -40 & 0 & 1 \end{bmatrix}^{-\mathrm{T}} \begin{bmatrix} 4\,096 \\ 6\,144 \\ 3\,640 \\ 1\,280 \\ 280 \\ 24 \end{bmatrix} = \begin{bmatrix} -259.52 \\ 6.72 \\ 482.48 \\ 118.72 \\ 20.48 \\ 30.72 \end{bmatrix}.$$

Matlab code for the general method of designing the gain matrix is given as follows.

DesignGainMatrixGeneral.m

```
%% Double inverted pendulum parameters
m1 = 1; m2 = 1; L1 = 1; L2 = 1; g = 10;
A = [0, 1, 0, 0, 0, 0; ...
     (m1+m2)*g/(m1*L1), 0, -m2*g/(m1*L1), 0, 0, 0; ...
     0, 0, 0, 1, 0, 0; ...
     -(m1+m2)*g/(m1*L2), 0, (m1+m2)*g/(m1*L2), 0, 0, 0; ...
     0, 0, 0, 0, 0, 1; ...
     0, 0, 0, 0, 0, 0];
B = [0; -1/L1; 0; 0; 0; 1];
n = 6; % State dimension

%% Compute the controllability matrix
```

```
CM = zeros(n); CM(:,1) = B;
for k = 2:n
    CM(:,k) = A*CM(:,k-1);
end % CM = [B, A*B, A^2*B, A^3*B, A^4*B, A^5*B];
fprintf('Controllability matrix: '); CM

%% Compute the characteristic polynomial of A
if (0)
    fprintf('First way: compute coefficients iteratively from
    eigenvalues\n');
    lambda = eig(A); % Eigenvalues of A
    aA = GetCharacteristicPolynomialCoefficients(lambda);
else
    fprintf('Second way: solve equation [B, ..., A^(n-1)*B]*X =
    A^n*B\n');
    aA = CM\(A*CM(:,n)); aA = [-aA', 1];
end
fprintf('Characteristic polynomial coefficients (a0, a1, ..., an
    ): '); aA

%% Compute the state transform matrix
M = zeros(n, n);
for k = 1:n % Auxiliary matrix M
    M(k,1:(n+1-k)) = aA((k+1):end);
end
fprintf('Auxiliary matrix M: '); M
P = CM*M;
fprintf('State transform matrix P: '); P

%% Design the transformed gain matrix
% Expected characteristic polynomial (s+4)^6
lambdaE = [-4; -4; -4; -4; -4; -4];
aE = GetCharacteristicPolynomialCoefficients(lambdaE);
fprintf('Expected characteristic polynomial coefficients (a0, a1
    , ..., an): '); aE
transformedK = transpose(aE(1:n) - aA(1:n));
fprintf('Transformed gain matrix K: '); transformedK

%% Recover the gain matrix
sttK = inv(transpose(P))*transformedK;
fprintf('Gain matrix K: '); sttK

%% Compute the characteristic polynomial coefficients from
    eigenvalues
function a = GetCharacteristicPolynomialCoefficients(lambda)
    n = length(lambda);
    % Characteristic polynomial coefficients a(-1), a0, ..., an
    a = zeros(1,n+2);
    a(2) = 1;
    for k = 1:n
        aOld = a;
        for i = 2:(k+1)
            a(i) = aOld(i-1) + aOld(i)*(-lambda(k));
```

```
        end
        a(k+2) = 1;
    end
    % Characteristic polynomial coefficients a0, a1, ..., an
    a = a(2:end);
end
```

Matlab simulation code for complete demonstration of double inverted pendulum control is given as follows.

DoubleInvertedPendulum3.m

```
%% Double inverted pendulum parameters
m1 = 1; m2 = 1; L1 = 1; L2 = 1; g = 10;
%% Simulation preliminary configuration
dt = 0.001; % Numerical computation step
tSpan = 0:dt:3; % Simulation time span
x = 0.2; dx = 0; % Cart position and its velocity
% Inverted pendulum angle theta-1 and its angular velocity
y1 = 0.2; dy1 = 0;
% Inverted pendulum angle theta-2 and its angular velocity
y2 = 0; dy2 = 0;
stt = [y1; dy1; y2; dy2; x; dx]; % Double inverted pendulum
    state
% Record states in simulation
sttAll = zeros(length(stt), length(tSpan)); k = 0;
% Expected equilibrium status
xExpected = 0; y1Expected = 0; y2Expected = 0;
SimConfig = [m1, m2, L1, L2, g, dt];

%% Design the gain matrix via the general method
DesignGainMatrixGeneral % Obtain the gain matrix sttK

%% Simulation of double inverted pendulum control
for t = tSpan
    %% Control method
    acc = -sttK'*stt; % Full-state feedback control of the cart
    acceleration

    %% Double inverted pendulum dynamics
    stt = DynamicsDIP(SimConfig, stt, acc);
    sttC = num2cell(stt); [y1, dy1, y2, dy2, x, dx] = sttC{:};
    if (abs(y1)>=pi/2 && abs(y2)>=pi/2)
        fprintf('Control failure!\n'); break;
    end
    k = k+1; sttAll(:,k) = stt;
    %% Double inverted pendulum visualization
    if (rem(k,20) == 0)
        DisplayDIP(x, y1, y2, L1, L2); pause(20*dt);
    end
end
```

The general method of designing the gain matrix does not resort to symbolic operation and is computationally much more effective than the expedient method presented in Sect. 6.4.2. This general method is generally applicable in practical applications and can be conveniently implemented in common programming languages.

6.4.5 Ackermann Formula

For the general method of designing the gain matrix, we have another formalism to describe the obtained gain matrix $\boldsymbol{K}$. Supposing that all mathematical notations follow those presented in Sect. 6.4.4, then we have

$$\boldsymbol{K} = \boldsymbol{P}_{\boldsymbol{A},\boldsymbol{B}}^{-\mathrm{T}} \bar{\boldsymbol{K}} \iff \boldsymbol{K}^{\mathrm{T}} = \bar{\boldsymbol{K}}^{\mathrm{T}} \boldsymbol{P}_{\boldsymbol{A},\boldsymbol{B}}^{-1} = \boldsymbol{e}_n^{\mathrm{T}} \boldsymbol{C}_{\boldsymbol{A},\boldsymbol{B}}^{-1} C_{\mathrm{E}}(\boldsymbol{A}), \tag{6.45}$$

which is called the **Ackermann formula** (Ackermann 1985).

Proof We have

$$\begin{aligned} \bar{\boldsymbol{K}}^{\mathrm{T}} \boldsymbol{P}_{\boldsymbol{A},\boldsymbol{B}}^{-1} = \boldsymbol{e}_n^{\mathrm{T}} \boldsymbol{C}_{\boldsymbol{A},\boldsymbol{B}}^{-1} C_{\mathrm{E}}(\boldsymbol{A}) &\iff \bar{\boldsymbol{K}}^{\mathrm{T}} = \boldsymbol{e}_n^{\mathrm{T}} \boldsymbol{C}_{\boldsymbol{A},\boldsymbol{B}}^{-1} C_{\mathrm{E}}(\boldsymbol{A}) \boldsymbol{P}_{\boldsymbol{A},\boldsymbol{B}} \\ &\iff \bar{\boldsymbol{K}}^{\mathrm{T}} = \boldsymbol{e}_n^{\mathrm{T}} \boldsymbol{C}_{\boldsymbol{A},\boldsymbol{B}}^{-1} C_{\mathrm{E}}(\boldsymbol{A}) \boldsymbol{C}_{\boldsymbol{A},\boldsymbol{B}} \boldsymbol{M}_{\boldsymbol{A}} \end{aligned}$$

namely

$$\bar{\boldsymbol{K}}^{\mathrm{T}} \equiv [\bar{k}_1 \ \bar{k}_2 \ \cdots \ \bar{k}_n] = \boldsymbol{e}_n^{\mathrm{T}} \boldsymbol{C}_{\boldsymbol{A},\boldsymbol{B}}^{-1} C_{\mathrm{E}}(\boldsymbol{A}) \boldsymbol{C}_{\boldsymbol{A},\boldsymbol{B}} \boldsymbol{M}_{\boldsymbol{A}}. \tag{6.46}$$

Denote $a_n \equiv a_n^{\mathrm{E}} \equiv 1$, so the characteristic polynomial of $\boldsymbol{A}$ and the expected closed-loop characteristic polynomial are represented respectively as

$$\begin{aligned} C_{\boldsymbol{A}}(s) &= s^n + a_{n-1}s^{n-1} + \cdots + a_0 = \sum_{i=0}^{n} a_i s^i, \\ C_{\mathrm{E}}(s) &= s^n + a_{n-1}^{\mathrm{E}} s^{n-1} + \cdots + a_0^{\mathrm{E}} = \sum_{i=0}^{n} a_i^{\mathrm{E}} s^i. \end{aligned}$$

According to (6.23), we have

$$C_{\boldsymbol{A}}(\boldsymbol{A}) = \boldsymbol{A}^n + a_{n-1}\boldsymbol{A}^{n-1} + \cdots + a_0 \boldsymbol{I} = \sum_{i=0}^{n} a_i \boldsymbol{A}^i = \boldsymbol{0}$$

and

$$C_{\mathrm{E}}(\boldsymbol{A}) = \sum_{i=0}^{n} a_i^{\mathrm{E}} \boldsymbol{A}^i - \sum_{i=0}^{n} a_i \boldsymbol{A}^i = \sum_{i=0}^{n-1} (a_i^{\mathrm{E}} - a_i) \boldsymbol{A}^i = \sum_{i=1}^{n} \bar{k}_i \boldsymbol{A}^{i-1}. \tag{6.47}$$

We also have

$$\begin{aligned} \boldsymbol{C}_{\boldsymbol{A},\boldsymbol{B}} \boldsymbol{M}_{\boldsymbol{A}} &= \left[\sum_{j=1}^{n} a_j \boldsymbol{A}^{j-1} \boldsymbol{B} \ \sum_{j=2}^{n} a_j \boldsymbol{A}^{j-2} \boldsymbol{B} \ \cdots \ \sum_{j=n}^{n} a_j \boldsymbol{A}^{j-n} \boldsymbol{B} \right] \\ &\equiv \left[\sum_{j=t}^{n} a_j \boldsymbol{A}^{j-t} \boldsymbol{B} \right]_{t=1,\cdots,n}. \end{aligned} \tag{6.48}$$

Substitute (6.47) and (6.48) into the right side of (6.46) and obtain

$$\begin{aligned} \boldsymbol{e}_n^{\mathrm{T}} \boldsymbol{C}_{\boldsymbol{A},\boldsymbol{B}}^{-1} C_{\mathrm{E}}(\boldsymbol{A}) \boldsymbol{C}_{\boldsymbol{A},\boldsymbol{B}} \boldsymbol{M}_{\boldsymbol{A}} &= \boldsymbol{e}_n^{\mathrm{T}} \boldsymbol{C}_{\boldsymbol{A},\boldsymbol{B}}^{-1} \sum_{i=1}^{n} \bar{k}_i \boldsymbol{A}^{i-1} \left[\sum_{j=t}^{n} a_j \boldsymbol{A}^{j-t} \boldsymbol{B} \right]_{t=1,\cdots,n} \\ &= \sum_{i=1}^{n} \bar{k}_i \boldsymbol{e}_n^{\mathrm{T}} \boldsymbol{C}_{\boldsymbol{A},\boldsymbol{B}}^{-1} \left[\sum_{j=t}^{n} a_j \boldsymbol{A}^{i+j-t-1} \boldsymbol{B} \right]_{t=1,\cdots,n}. \end{aligned}$$

Thus if we have

$$\boldsymbol{e}_n^{\mathrm{T}} \boldsymbol{C}_{\boldsymbol{A},\boldsymbol{B}}^{-1} \left[\sum_{j=t}^{n} a_j \boldsymbol{A}^{i+j-t-1} \boldsymbol{B} \right]_{t=1,\cdots,n} = \boldsymbol{e}_i^{\mathrm{T}} \tag{6.49}$$

for each $i = 1, \ldots, n$, then we can derive (6.46).

Review (6.38):

$$\begin{aligned} \boldsymbol{C}_{\boldsymbol{A},\boldsymbol{B}}^{-1} \boldsymbol{B} &= \boldsymbol{e}_1, \\ &\vdots \\ \boldsymbol{C}_{\boldsymbol{A},\boldsymbol{B}}^{-1} \boldsymbol{A}^{i-1} \boldsymbol{B} &= \boldsymbol{e}_i, \\ &\vdots \\ \boldsymbol{C}_{\boldsymbol{A},\boldsymbol{B}}^{-1} \boldsymbol{A}^{n-1} \boldsymbol{B} &= \boldsymbol{e}_n. \end{aligned}$$

If $t < i$, we have

$$\boldsymbol{e}_n^{\mathrm{T}}\boldsymbol{C}_{\boldsymbol{A},\boldsymbol{B}}^{-1}\sum_{j=t}^{n} a_j \boldsymbol{A}^{i+j-t-1}\boldsymbol{B} = \boldsymbol{e}_n^{\mathrm{T}}\boldsymbol{C}_{\boldsymbol{A},\boldsymbol{B}}^{-1}\left(\sum_{j=t}^{n} a_j \boldsymbol{A}^{j}\right)\boldsymbol{A}^{i-t-1}\boldsymbol{B}$$

$$= \boldsymbol{e}_n^{\mathrm{T}}\boldsymbol{C}_{\boldsymbol{A},\boldsymbol{B}}^{-1}\left(-\sum_{j=0}^{t-1} a_j \boldsymbol{A}^{j}\right)\boldsymbol{A}^{i-t-1}\boldsymbol{B} = -\sum_{j=0}^{t-1} a_j \boldsymbol{e}_n^{\mathrm{T}}\boldsymbol{C}_{\boldsymbol{A},\boldsymbol{B}}^{-1}\boldsymbol{A}^{i+j-t-1}\boldsymbol{B}$$

$$= -\sum_{j=0}^{t-1} a_j \boldsymbol{e}_n^{\mathrm{T}}\boldsymbol{e}_{i+j-t} = 0 \quad (\text{note } i + j - t < i \leqslant n).$$

If $t > i$, we have

$$\boldsymbol{e}_n^{\mathrm{T}}\boldsymbol{C}_{\boldsymbol{A},\boldsymbol{B}}^{-1}\sum_{j=t}^{n} a_j \boldsymbol{A}^{i+j-t-1}\boldsymbol{B} = \sum_{j=t}^{n} a_j \boldsymbol{e}_n^{\mathrm{T}}\boldsymbol{C}_{\boldsymbol{A},\boldsymbol{B}}^{-1}\boldsymbol{A}^{j+i-t-1}\boldsymbol{B}$$

$$= \sum_{j=t}^{n} a_j \boldsymbol{e}_n^{\mathrm{T}}\boldsymbol{e}_{j+i-t} = 0 \quad (\text{note } j + i - t < j \leqslant n).$$

If $t = i$, we have

$$\boldsymbol{e}_n^{\mathrm{T}}\boldsymbol{C}_{\boldsymbol{A},\boldsymbol{B}}^{-1}\sum_{j=t}^{n} a_j \boldsymbol{A}^{i+j-t-1}\boldsymbol{B} = \sum_{j=t}^{n} a_j \boldsymbol{e}_n^{\mathrm{T}}\boldsymbol{C}_{\boldsymbol{A},\boldsymbol{B}}^{-1}\boldsymbol{A}^{j-1}\boldsymbol{B} = \sum_{j=t}^{n} a_j \boldsymbol{e}_n^{\mathrm{T}}\boldsymbol{e}_j = a_n = 1.$$

Above three cases imply (6.49) holds and hence (6.46) and (6.45) also hold. □

We apply the Ackermann formula to design the gain matrix for double inverted pendulum control. The double inverted pendulum control system adopts linear state-space modelling described by (6.3). As in Sects. 6.4.2 and 6.4.4, we take the same set of double inverted pendulum parameters as example to demonstrate the Ackermann formula. Let $m_1 = 1, m_2 = 1, L_1 = 1, L_2 = 1$, and $g = 10$. Then the state transition matrix $\boldsymbol{A}$ and the control input matrix $\boldsymbol{B}$ are

$$\boldsymbol{A} = \begin{bmatrix} 0 & 1 & 0 & 0 & 0 & 0 \\ 20 & 0 & -10 & 0 & 0 & 0 \\ 0 & 0 & 0 & 1 & 0 & 0 \\ -20 & 0 & 20 & 0 & 0 & 0 \\ 0 & 0 & 0 & 0 & 0 & 1 \\ 0 & 0 & 0 & 0 & 0 & 0 \end{bmatrix}, \quad \boldsymbol{B} = \begin{bmatrix} 0 \\ -1 \\ 0 \\ 0 \\ 0 \\ 1 \end{bmatrix}.$$

Compute the controllability matrix $\boldsymbol{C}_{\boldsymbol{A},\boldsymbol{B}}$ via (6.24) as

$$\boldsymbol{C}_{\boldsymbol{A},\boldsymbol{B}} = \begin{bmatrix} 0 & -1 & 0 & -20 & 0 & -600 \\ -1 & 0 & -20 & 0 & -600 & 0 \\ 0 & 0 & 0 & 20 & 0 & 800 \\ 0 & 0 & 20 & 0 & 800 & 0 \\ 0 & 1 & 0 & 0 & 0 & 0 \\ 1 & 0 & 0 & 0 & 0 & 0 \end{bmatrix}.$$

Set the expected closed-loop characteristic polynomial as

$$C_{\mathrm{E}}(s) = (s+4)^6 = s^6 + 24\,s^5 + 240\,s^4 + 1\,280\,s^3 + 3\,840\,s^2 + 6\,144\,s + 4\,096,$$

which is the same to the example of expected closed-loop characteristic polynomial used in Sects. 6.4.2 and 6.4.4. For numeric efficiency, do not compute $C_{\mathrm{E}}(A)$ directly as

$$C_{\mathrm{E}}(\boldsymbol{A}) = \sum_{i=0}^{n} a_i^{\mathrm{E}} \boldsymbol{A}^i,$$

but compute $C_{\mathrm{E}}(A)$ iteratively as

$$C_{\mathrm{E}}(\boldsymbol{A}) = (\cdots(((a_n^{\mathrm{E}}\boldsymbol{A} + a_{n-1}^{\mathrm{E}}\boldsymbol{I})\boldsymbol{A} + a_{n-2}^{\mathrm{E}}\boldsymbol{I})\boldsymbol{A} + a_{n-3}^{\mathrm{E}}\boldsymbol{I})\cdots + a_1^{\mathrm{E}}\boldsymbol{I})\boldsymbol{A} + a_0^{\mathrm{E}}\boldsymbol{I},$$

namely

$$\begin{aligned} C_{\mathrm{E},n}^{\boldsymbol{A}} &= a_n^{\mathrm{E}}\boldsymbol{I}, \\ C_{\mathrm{E},n-1}^{\boldsymbol{A}} &= C_{\mathrm{E},n}^{\boldsymbol{A}}\boldsymbol{A} + a_{n-1}^{\mathrm{E}}\boldsymbol{I}, \\ C_{\mathrm{E},n-2}^{\boldsymbol{A}} &= C_{\mathrm{E},n-1}^{\boldsymbol{A}}\boldsymbol{A} + a_{n-2}^{\mathrm{E}}\boldsymbol{I}, \\ &\vdots \\ C_{\mathrm{E},1}^{\boldsymbol{A}} &= C_{\mathrm{E},2}^{\boldsymbol{A}}\boldsymbol{A} + a_1^{\mathrm{E}}\boldsymbol{I}, \\ C_{\mathrm{E}}(\boldsymbol{A}) = C_{\mathrm{E},0}^{\boldsymbol{A}} &= C_{\mathrm{E},1}^{\boldsymbol{A}}\boldsymbol{A} + a_0^{\mathrm{E}}\boldsymbol{I}. \end{aligned}$$

Take numbers of the concrete example and compute

$$C_{\mathrm{E}}(\boldsymbol{A}) = (((((\boldsymbol{A}+24\boldsymbol{I})\boldsymbol{A}+240\boldsymbol{I})\boldsymbol{A}+1\,280\boldsymbol{I})\boldsymbol{A}+3\,840\boldsymbol{I})\boldsymbol{A}+6\,144\boldsymbol{I})\boldsymbol{A}+4\,096\boldsymbol{I}$$

$$
= \begin{bmatrix}
244\,896 & 46\,144 & -148\,400 & -22\,400 & & \\
1\,370\,880 & 244\,896 & -909\,440 & -148\,400 & & \\
-296\,800 & -44\,800 & 244\,896 & 46\,144 & & \\
-1\,818\,880 & -296\,800 & 1\,370\,880 & 244\,896 & & \\
 & & & & 4\,096 & 6\,144 \\
 & & & & & 4\,096
\end{bmatrix}.
$$

Finally, obtain the gain matrix $\boldsymbol{K}$ via (6.45) as

$$
\boldsymbol{K} = \left[\boldsymbol{e}_n^{\mathrm{T}} \boldsymbol{C}_{\boldsymbol{A},\boldsymbol{B}}^{-1} C_{\mathrm{E}}(\boldsymbol{A})\right]^{\mathrm{T}} = \begin{bmatrix} -259.52 \\ 6.72 \\ 482.48 \\ 118.72 \\ 20.48 \\ 30.72 \end{bmatrix}.
$$

Matlab code for realizing the Ackermann formula is given as follows.

AckermannFormula.m

```
%% Double inverted pendulum parameters
m1 = 1; m2 = 1; L1 = 1; L2 = 1; g = 10;
A = [0, 1, 0, 0, 0, 0; ...
     (m1+m2)*g/(m1*L1), 0, -m2*g/(m1*L1), 0, 0, 0; ...
     0, 0, 0, 1, 0, 0; ...
     -(m1+m2)*g/(m1*L2), 0, (m1+m2)*g/(m1*L2), 0, 0, 0; ...
     0, 0, 0, 0, 0, 1; ...
     0, 0, 0, 0, 0, 0];
B = [0; -1/L1; 0; 0; 0; 1];
n = 6; % State dimension

%% Ackermann formula
CM = zeros(n); CM(:,1) = B;
for k = 2:n
    CM(:,k) = A*CM(:,k-1);
end % CM = [B, A*B, A^2*B, A^3*B, A^4*B, A^5*B];
fprintf('Controllability matrix: '); CM
% Expected characteristic polynomial (s+4)^6
lambdaE = [-4; -4; -4; -4; -4; -4];
aE = GetCharacteristicPolynomialCoefficients(lambdaE);
fprintf('Expected characteristic polynomial coefficients (a0, a1
    , ..., an): '); aE
CEA = aE(n+1)*eye(n); % CE(A)
for k = n:-1:1
    CEA = CEA*A + aE(k)*eye(n);
end
fprintf('Expected characteristic polynomial with argument A
    namely CE(A): '); CEA
enT = zeros(1,n); enT(n) = 1;
sttK = transpose(enT*inv(CM)*CEA);
fprintf('Gain matrix K: '); sttK
```

```
%% Compute the characteristic polynomial coefficients from
    eigenvalues
function a = GetCharacteristicPolynomialCoefficients(lambda)
    n = length(lambda);
    % Characteristic polynomial coefficients a(-1), a0, ..., an
    a = zeros(1,n+2);
    a(2) = 1;
    for k = 1:n
        aOld = a;
        for i = 2:(k+1)
            a(i) = aOld(i-1) + aOld(i)*(-lambda(k));
        end
        a(k+2) = 1;
    end
    % Characteristic polynomial coefficients a0, a1, ..., an
    a = a(2:end);
end
```

As (6.45) conveys, the *Ackermann formula* is another formalism for the general method of designing the gain matrix. The Ackermann formula is also computationally desirable as it only involves basic matrix operations and can be conveniently implemented in common programming languages.

By so far, not only has the difficult problem of double inverted pendulum control been handled, but also a general methodology based on the full-state feedback control strategy together with effective methods of designing the gain matrix is available. Although we focus on *single-input-multiple-output* control systems for presenting the general method of designing the gain matrix in Sects. 6.4.4 and 6.4.5, such fundamental methodology can be extended to *multiple-input-multiple-output* control systems.

6.5 Multiple-Input-Multiple-Output Control

Consider a variant of the double inverted pendulum control problem, as illustrated in Fig. 6.3. Compared with the original double inverted pendulum control problem presented in previous sections, the variant of the double inverted pendulum control problem shares all configurations except that its control input to the process or target process is no longer the *single-input* of cart acceleration a but the *multiple-input* of both cart acceleration a and first inverted pendulum angular acceleration a_1. The variant of the double inverted pendulum control system aims at controlling the cart acceleration a and the first inverted pendulum angular acceleration a_1 simultaneously so that the cart is kept at a specific position x and both inverted pendulum angles θ_1 and θ_2 are kept zero.

Denote $\Delta\theta = \theta_1 - \theta_2$. By physical law, we can establish differential equations for dynamics of the variant of the double inverted pendulum control system as

$$\begin{cases} m_1 L_1 \frac{\mathrm{d}^2}{\mathrm{d}t^2}\theta_1 = m_1(g\sin\theta_1 - a\cos\theta_1) + m_2(g\cos\theta_2 + a\sin\theta_2)\sin\Delta\theta + m_1 L_1 a_1 \\ m_2 L_2 \frac{\mathrm{d}^2}{\mathrm{d}t^2}\theta_2 = m_2(g\sin\theta_2 - a\cos\theta_2) - m_2\cos\Delta\theta L_1 \frac{\mathrm{d}^2}{\mathrm{d}t^2}\theta_1 \end{cases}$$

Denote $\boldsymbol{x} \equiv \left[\theta_1 \ \frac{\mathrm{d}\theta_1}{\mathrm{d}t} \ \theta_2 \ \frac{\mathrm{d}\theta_2}{\mathrm{d}t} \ x \ \frac{\mathrm{d}x}{\mathrm{d}t}\right]^{\mathrm{T}}$ and obtain

$$\frac{\mathrm{d}}{\mathrm{d}t}\boldsymbol{x} = f(\boldsymbol{x}) + g(\boldsymbol{x})a + g_1(\boldsymbol{x})a_1, \tag{6.50}$$

where

$$f(\boldsymbol{x}) \equiv \begin{bmatrix} \frac{\mathrm{d}\theta_1}{\mathrm{d}t} \\ \left(\sin\theta_1 + \frac{m_2}{m_1}\sin\Delta\theta\cos\theta_2\right)\frac{g}{L_1} \\ \frac{\mathrm{d}\theta_2}{\mathrm{d}t} \\ -\sin\Delta\theta\left(\cos\theta_1 + \frac{m_2}{m_1}\cos\Delta\theta\cos\theta_2\right)\frac{g}{L_2} \\ \frac{\mathrm{d}x}{\mathrm{d}t} \\ 0 \end{bmatrix},$$

$$g(\boldsymbol{x}) \equiv \begin{bmatrix} 0 \\ -\left(\cos\theta_1 - \frac{m_2}{m_1}\sin\Delta\theta\sin\theta_2\right)\frac{1}{L_1} \\ 0 \\ -\sin\Delta\theta\left(\sin\theta_1 - \frac{m_2}{m_1}\cos\Delta\theta\sin\theta_2\right)\frac{1}{L_2} \\ 0 \\ 1 \end{bmatrix}, \quad g_1(\boldsymbol{x}) \equiv \begin{bmatrix} 0 \\ 1 \\ 0 \\ -\cos\Delta\theta\frac{L_1}{L_2} \\ 0 \\ 0 \end{bmatrix}.$$

If both inverted pendulum angles θ_1 and θ_2 are close to zero, then the state differential equation described in (6.50) can be fairly linearized about the equilibrium state and simplified into a linear state differential equation as

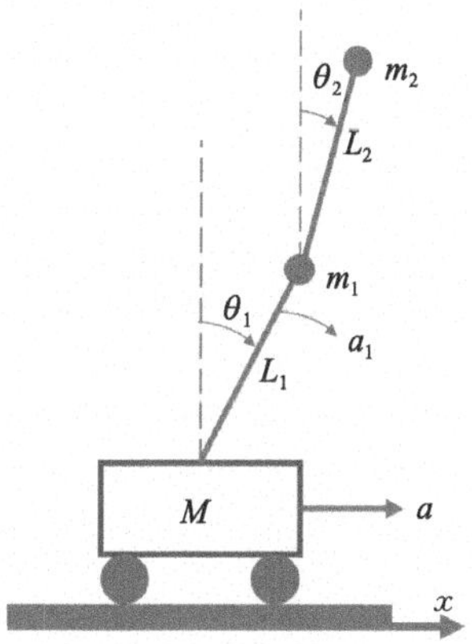

Fig. 6.3 Variant of double inverted pendulum control

$$\frac{\mathrm{d}}{\mathrm{d}t}\boldsymbol{x} = \begin{bmatrix} 0 & 1 & 0 & 0\,0\,0 \\ \left(1+\frac{m_2}{m_1}\right)\frac{g}{L_1} & 0 & -\frac{m_2}{m_1}\frac{g}{L_1} & 0\,0\,0 \\ 0 & 0 & 0 & 1\,0\,0 \\ -\left(1+\frac{m_2}{m_1}\right)\frac{g}{L_2} & 0 & \left(1+\frac{m_2}{m_1}\right)\frac{g}{L_2} & 0\,0\,0 \\ 0 & 0 & 0 & 0\,0\,1 \\ 0 & 0 & 0 & 0\,0\,0 \end{bmatrix}\boldsymbol{x} + \begin{bmatrix} 0 & 0 \\ -\frac{1}{L_1} & 1 \\ 0 & 0 \\ 0 & -\frac{L_1}{L_2} \\ 0 & 0 \\ 1 & 0 \end{bmatrix}\begin{bmatrix} a \\ a_1 \end{bmatrix}$$
$$\equiv \boldsymbol{Ax} + \boldsymbol{Bu}. \tag{6.51}$$

It is worth noting that although the variant of the double inverted pendulum control problem and the original double inverted pendulum control problem share the same process entity namely the double inverted pendulum and the same state (the vector of properties composed of the cart position and velocity, the first inverted pendulum angle and angular velocity, the second inverted pendulum angle and angular velocity), *their processes or target processes are different because they have different control input to the process respectively*: The control input for the original double inverted pendulum control problem is the *single-input* of cart acceleration a, whereas the control input for the variant of the double inverted pendulum control problem is the *multiple-input* of cart acceleration a and first inverted pendulum angular acceleration a_1.

The variant of double inverted pendulum control is a representative example of **multiple-input-multiple-output** control. As demonstrated in Sect. 6.4, the double inverted pendulum can be well controlled by the *single-input* of cart acceleration a, then reader may wonder what the use of the *multiple-input* with first inverted pendulum angular acceleration a_1 as redundant control input is. Reasons can be various, yet simply speaking, compared with a *single-input* control system, a *multiple-input* control system enjoys enhanced flexibility and has the potential to achieve the same expected state in better way.

Besides the variant of double inverted pendulum control, cooperative longitudinal control of vehicle platooning is another representative example of multiple-input-multiple-output control. For simplicity, suppose that there are two vehicles and consider their longitudinal motion. The process of the cooperative longitudinal control system is longitudinal spatial relationship between the two vehicles. Its absolute state consists of the inter-vehicle position $\Delta x = x_1 - x_2$, the first vehicle velocity $v_1 \equiv \frac{\mathrm{d}x_1}{\mathrm{d}t}$, and the second vehicle velocity $v_2 \equiv \frac{\mathrm{d}x_2}{\mathrm{d}t}$. The control input to the target process is the multiple-input of first vehicle acceleration a_1 and second vehicle acceleration a_2. The cooperative longitudinal control system aims at controlling the multiple-input of a_1 and a_2 simultaneously to achieve certain expected inter-vehicle position Δx_E and expected vehicle velocity v_E.

Denote the state $\boldsymbol{x} \equiv [\Delta x - \Delta x_\mathrm{E} \;\; v_1 - v_\mathrm{E} \;\; v_2 - v_\mathrm{E}]^\mathrm{T}$. Dynamics of the state $\boldsymbol{x}$ can be modelled by a state differential equation as

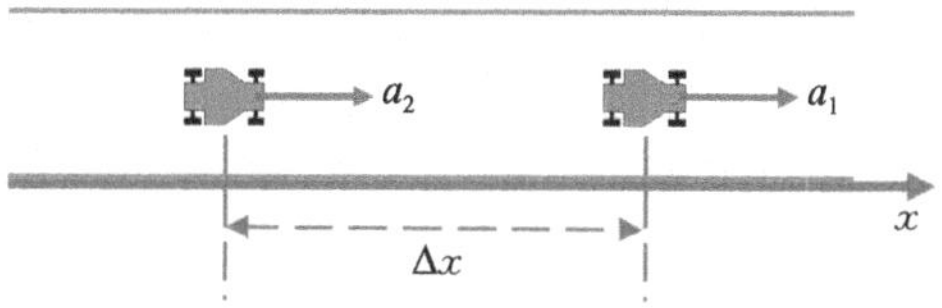

Fig. 6.4 Cooperative longitudinal control of vehicle platooning

$$\frac{\mathrm{d}}{\mathrm{d}t}\boldsymbol{x} = \begin{bmatrix} 0 & 1 & -1 \\ 0 & 0 & 0 \\ 0 & 0 & 0 \end{bmatrix} \boldsymbol{x} + \begin{bmatrix} 0 & 0 \\ 1 & 0 \\ 0 & 1 \end{bmatrix} \begin{bmatrix} a_1 \\ a_2 \end{bmatrix} \equiv \boldsymbol{A}\boldsymbol{x} + \boldsymbol{B}\boldsymbol{u}. \tag{6.52}$$

The control objective is to control the state $\boldsymbol{x}$ to the equilibrium state $\boldsymbol{x}_\mathrm{E} = \boldsymbol{0}$. Unlike the variant of the double inverted pendulum control system, the cooperative longitudinal control system has no redundant control input part. Both vehicle accelerations a_1 and a_2 are necessary to achieve the control objective.

For a multiple-input-multiple-output control system, no matter whether it has redundant control input part or not, it is always worth analysing first how each part of control input can influence the state or control the state to what extent. The need of such analysis leads to the topic of *partial controllability* of the control system.

6.5.1 Partial Controllability

Given a control system that adopts linear state-space modelling described by (6.5), with its state denoted as $\boldsymbol{x}$ (suppose that the state dimension is n) and its control input to the target process denoted as $\boldsymbol{u}$, for the moment, we still focus on single-input-multiple-output control, but suppose that the control system's controllability matrix

$$\boldsymbol{C}_{\boldsymbol{A},\boldsymbol{B}} = [\boldsymbol{B} \;\; \boldsymbol{A}\boldsymbol{B} \;\; \boldsymbol{A}^2\boldsymbol{B} \;\; \cdots \;\; \boldsymbol{A}^{n-1}\boldsymbol{B}]$$

is rank-deficient and hence the target process is uncontrollable. Although the entire state $\boldsymbol{x}$ is uncontrollable, we intend to analyse to what extent the state $\boldsymbol{x}$ is partially controllable. In other words, we intend to analyse **partial controllability** of the control system. Knowledge on this point will be used later in extension of the fundamental methodology of *single-input-multiple-output* control to *multiple-input-multiple-output* control.

Suppose that the rank of the controllability matrix $\boldsymbol{C}_{\boldsymbol{A},\boldsymbol{B}}$ is $m < n$, and then we have

$$\text{rank of } [\boldsymbol{B} \;\; \boldsymbol{A}\boldsymbol{B} \;\; \cdots \;\; \boldsymbol{A}^{m-1}\boldsymbol{B}] = m. \tag{6.53}$$

Proof Let $\bar{m}$ be the maximum index such that $\boldsymbol{B}, \boldsymbol{A}\boldsymbol{B}, \cdots, \boldsymbol{A}^{\bar{m}-1}\boldsymbol{B}$ are linearly independent. In other words, $\bar{m}$ is defined as

$$\bar{m} = \max\{k|\text{rank of } [\boldsymbol{B} \;\; \boldsymbol{AB} \;\; \cdots \;\; \boldsymbol{A}^{k-1}\boldsymbol{B}] = k\}.$$

On one hand, $\bar{m} \leqslant m$ holds obviously. On the other hand, by the maximality of $\bar{m}$, we have

$$\boldsymbol{A}^{\bar{m}}\boldsymbol{B} \in \langle \boldsymbol{B}, \boldsymbol{AB}, \ldots, \boldsymbol{A}^{\bar{m}-1}\boldsymbol{B}\rangle,$$

where $\langle \boldsymbol{B}, \boldsymbol{AB}, \cdots, \boldsymbol{A}^{\bar{m}-1}\boldsymbol{B}\rangle$ denotes the span of the $\bar{m}$ vectors $\{\boldsymbol{B}, \boldsymbol{AB}, \cdots, \boldsymbol{A}^{\bar{m}-1}\boldsymbol{B}\}$. So we further have

$$\begin{aligned}
\boldsymbol{A}^{\bar{m}+1}\boldsymbol{B} = \boldsymbol{A}\boldsymbol{A}^{\bar{m}}\boldsymbol{B} &\in \boldsymbol{A}\langle \boldsymbol{B}, \boldsymbol{AB}, \ldots, \boldsymbol{A}^{\bar{m}-1}\boldsymbol{B}\rangle \\
&= \langle\langle \boldsymbol{AB}, \ldots, \boldsymbol{A}^{\bar{m}-1}\boldsymbol{B}\rangle, \boldsymbol{A}^{\bar{m}}\boldsymbol{B}\rangle \\
&\subset \langle\langle \boldsymbol{B}, \ldots, \boldsymbol{A}^{\bar{m}-1}\boldsymbol{B}\rangle, \langle \boldsymbol{B}, \ldots, \boldsymbol{A}^{\bar{m}-1}\boldsymbol{B}\rangle\rangle \\
&= \langle \boldsymbol{B}, \boldsymbol{AB}, \ldots, \boldsymbol{A}^{\bar{m}-1}\boldsymbol{B}\rangle
\end{aligned}$$

and similarly

$$\begin{aligned}
\boldsymbol{A}^{\bar{m}+2}\boldsymbol{B} &= \boldsymbol{A}\boldsymbol{A}^{\bar{m}+1}\boldsymbol{B} \in \langle \boldsymbol{B}, \boldsymbol{AB}, \ldots, \boldsymbol{A}^{\bar{m}-1}\boldsymbol{B}\rangle, \\
\boldsymbol{A}^{\bar{m}+3}\boldsymbol{B} &= \boldsymbol{A}\boldsymbol{A}^{\bar{m}+2}\boldsymbol{B} \in \langle \boldsymbol{B}, \boldsymbol{AB}, \ldots, \boldsymbol{A}^{\bar{m}-1}\boldsymbol{B}\rangle, \\
&\vdots \\
\boldsymbol{A}^{n-1}\boldsymbol{B} &= \boldsymbol{A}\boldsymbol{A}^{n-2}\boldsymbol{B} \in \langle \boldsymbol{B}, \boldsymbol{AB}, \ldots, \boldsymbol{A}^{\bar{m}-1}\boldsymbol{B}\rangle,
\end{aligned}$$

which implies $m \leqslant \bar{m}$. Thus we have $\bar{m} = m$ and (6.53) holds. □

Denote the **partial controllability matrix $\bar{\boldsymbol{C}}_{A,B}$** as

$$\bar{\boldsymbol{C}}_{A,B} \equiv [\boldsymbol{B} \;\; \boldsymbol{AB} \;\; \cdots \;\; \boldsymbol{A}^{m-1}\boldsymbol{B}], \tag{6.54}$$

where m is the rank of the controllability matrix $\boldsymbol{C}_{A,B}$. Denote $\bar{\boldsymbol{C}}^{+}_{A,B}$ as the n-by-$(n-m)$ complement matrix of $\bar{\boldsymbol{C}}_{A,B}$ such that the **complemented controllability matrix** of $\bar{\boldsymbol{C}}_{A,B}$ given as

$$\bar{\boldsymbol{C}}^{*}_{A,B} \equiv [\bar{\boldsymbol{C}}_{A,B} \;\; \bar{\boldsymbol{C}}^{+}_{A,B}] \tag{6.55}$$

is a n-by-n invertible square matrix.

When the controllability matrix $\boldsymbol{C}_{A,B}$ is of full rank, the complement matrix $\bar{\boldsymbol{C}}^{+}_{A,B}$ is empty and the complemented controllability matrix $\bar{\boldsymbol{C}}^{*}_{A,B}$ is the same to the controllability matrix $\boldsymbol{C}_{A,B}$. When the controllability matrix $\boldsymbol{C}_{A,B}$ is rank-deficient, the complement matrix $\bar{\boldsymbol{C}}^{+}_{A,B}$ can be set as follows: Perform *Gaussian pivot elimination* on columns of $\bar{\boldsymbol{C}}_{A,B}$ to find m rows of $\bar{\boldsymbol{C}}_{A,B}$ that are linearly independent, then set rows of $\bar{\boldsymbol{C}}_{\boldsymbol{A},\boldsymbol{B}}^{+}$ at these linearly independent rows to $\mathbf{0}$ while setting other rows of $\bar{\boldsymbol{C}}_{\boldsymbol{A},\boldsymbol{B}}^{+}$ to the $n-m$ row vectors of the $(n-m)$-dimensional identity matrix respectively.

Suppose $\boldsymbol{A}^m\boldsymbol{B}+\bar{a}_{m-1}\boldsymbol{A}^{m-1}\boldsymbol{B}+\cdots+\bar{a}_1\boldsymbol{A}\boldsymbol{B}+\bar{a}_0\boldsymbol{B}=\mathbf{0}$. The coefficients $\bar{a}_0,\bar{a}_1,\ldots,\bar{a}_{m-1}$ can be uniquely determined by solving the following equation:

$$\bar{\boldsymbol{C}}_{\boldsymbol{A},\boldsymbol{B}}\boldsymbol{X}_{\bar{a}}=\boldsymbol{A}^m\boldsymbol{B} \tag{6.56}$$

namely

$$[\boldsymbol{B}\ \ \boldsymbol{A}\boldsymbol{B}\ \ \cdots\ \ \boldsymbol{A}^{m-1}\boldsymbol{B}]\begin{bmatrix}-\bar{a}_0\\-\bar{a}_1\\\vdots\\-\bar{a}_{m-1}\end{bmatrix}=\boldsymbol{A}^m\boldsymbol{B},$$

where $\boldsymbol{X}_{\bar{a}}\equiv[-\bar{a}_0\ \ -\bar{a}_1\ \ \cdots\ \ -\bar{a}_{m-1}]^{\mathrm{T}}$ is solved as

$$\boldsymbol{X}_{\bar{a}}=(\bar{\boldsymbol{C}}_{\boldsymbol{A},\boldsymbol{B}}^{\mathrm{T}}\bar{\boldsymbol{C}}_{\boldsymbol{A},\boldsymbol{B}})^{-1}\bar{\boldsymbol{C}}_{\boldsymbol{A},\boldsymbol{B}}^{\mathrm{T}}\boldsymbol{A}^m\boldsymbol{B}. \tag{6.57}$$

Define three **auxiliary matrices** $\bar{\boldsymbol{M}}_{\boldsymbol{A},\boldsymbol{B}}$, $\bar{\boldsymbol{P}}_{\boldsymbol{A},\boldsymbol{B}}$, and $\bar{\boldsymbol{S}}_{\boldsymbol{A},\boldsymbol{B}}$ in (6.58), (6.59), and (6.60) respectively:

$$\bar{\boldsymbol{M}}_{\boldsymbol{A},\boldsymbol{B}}=\begin{bmatrix}\bar{a}_1 & \bar{a}_2 & \bar{a}_3 & \cdots & \bar{a}_{m-2} & \bar{a}_{m-1} & 1\\ \bar{a}_2 & \bar{a}_3 & \bar{a}_4 & \cdots & \bar{a}_{m-1} & 1 & \\ \bar{a}_3 & \bar{a}_4 & \bar{a}_5 & \cdots & 1 & & \\ \vdots & \vdots & \vdots & \ddots & & & \\ \bar{a}_{m-2} & \bar{a}_{m-1} & 1 & & & & \\ \bar{a}_{m-1} & 1 & & & & & \\ 1 & & & & & & \end{bmatrix}, \tag{6.58}$$

$$\bar{\boldsymbol{P}}_{\boldsymbol{A},\boldsymbol{B}}=\bar{\boldsymbol{C}}_{\boldsymbol{A},\boldsymbol{B}}^{*}\begin{bmatrix}\bar{\boldsymbol{M}}_{\boldsymbol{A},\boldsymbol{B}} & \mathbf{0}\\ \mathbf{0} & \boldsymbol{I}\end{bmatrix}=\begin{bmatrix}\bar{\boldsymbol{C}}_{\boldsymbol{A},\boldsymbol{B}}\bar{\boldsymbol{M}}_{\boldsymbol{A},\boldsymbol{B}} & \bar{\boldsymbol{C}}_{\boldsymbol{A},\boldsymbol{B}}^{+}\end{bmatrix}, \tag{6.59}$$

$$\bar{\boldsymbol{S}}_{\boldsymbol{A},\boldsymbol{B}}=\begin{bmatrix} & 1 & & & & \\ & & 1 & & & \\ & & & \ddots & & \\ & & & & 1 & \\ & & & & & 1\\ -\bar{a}_0 & -\bar{a}_1 & -\bar{a}_2 & \cdots & -\bar{a}_{m-2} & -\bar{a}_{m-1}\end{bmatrix}. \tag{6.60}$$

Then we have

$$\bar{\boldsymbol{P}}_{\boldsymbol{A},\boldsymbol{B}}^{-1}\boldsymbol{A}\bar{\boldsymbol{P}}_{\boldsymbol{A},\boldsymbol{B}}=\begin{bmatrix}\bar{\boldsymbol{S}}_{\boldsymbol{A},\boldsymbol{B}} & *\\ \mathbf{0} & *\end{bmatrix} \tag{6.61}$$

in the *upper triangular block* form and also have

$$\bar{P}_{A,B}^{-1}B = e_m = \begin{bmatrix} \bar{e}_m \\ 0 \end{bmatrix}, \tag{6.62}$$

where $\bar{e}_m$ denotes the m-dimensional vector $[0 \ \cdots \ 0 \ 1]$ namely the first m-element block of e_m.

Proof Denote

$$\bar{M}_{2:} \equiv \begin{bmatrix} \bar{a}_2 & \bar{a}_3 & \cdots & \bar{a}_{m-1} & 1 \\ \bar{a}_3 & \bar{a}_4 & \cdots & 1 & \\ \vdots & \vdots & \ddots & & \\ \bar{a}_{m-1} & 1 & & & \\ 1 & & & & \end{bmatrix}, \quad \bar{a}_{1:} = \begin{bmatrix} \bar{a}_1 \\ \bar{a}_2 \\ \vdots \\ \bar{a}_{m-2} \\ \bar{a}_{m-1} \end{bmatrix}.$$

We have

$$\bar{M}_{A,B}^{-1} = \begin{bmatrix} \bar{a}_{1:}^{\mathrm{T}} & 1 \\ \bar{M}_{2:} & 0 \end{bmatrix}^{-1} = \begin{bmatrix} 0 & \bar{M}_{2:}^{-1} \\ 1 & -\bar{a}_{1:}^{\mathrm{T}}\bar{M}_{2:}^{-1} \end{bmatrix}. \tag{6.63}$$

Also note that

$$\begin{aligned} I &= \bar{C}_{A,B}^{*-1}\bar{C}_{A,B}^{*} = \bar{C}_{A,B}^{*-1}[\bar{C}_{A,B} \ \ \bar{C}_{A,B}^{+}] \\ &= \bar{C}_{A,B}^{*-1}[B \ \ AB \ \ \cdots \ \ A^{m-1}B \ \ \bar{C}_{A,B}^{+}], \end{aligned}$$

which implies

$$\begin{aligned} \bar{C}_{A,B}^{*-1}B &= e_1 = \begin{bmatrix} \bar{e}_1 \\ 0 \end{bmatrix}, \\ \bar{C}_{A,B}^{*-1}AB &= e_2 = \begin{bmatrix} \bar{e}_2 \\ 0 \end{bmatrix}, \\ &\vdots \\ \bar{C}_{A,B}^{*-1}A^{m-1}B &= e_m = \begin{bmatrix} \bar{e}_m \\ 0 \end{bmatrix}, \\ \bar{C}_{A,B}^{*-1}A^{m}B &= \sum_{i=0}^{m-1}(-\bar{a}_i)\bar{C}_{A,B}^{*-1}A^{i}B = \begin{bmatrix} X_{\bar{a}} \\ 0 \end{bmatrix}, \end{aligned} \tag{6.64}$$

where $\bar{e}_1, \bar{e}_2, \ldots, \bar{e}_m$ denote the first m-element blocks of $e_1, e_2, \ldots, e_m$ respectively and $X_{\bar{a}}$ is defined in (6.57).

Consider (6.62) and have

$$\bar{\boldsymbol{P}}_{A,B}^{-1}\boldsymbol{B} = \begin{bmatrix} \bar{\boldsymbol{M}}_{A,B} & \boldsymbol{0} \\ \boldsymbol{0} & \boldsymbol{I} \end{bmatrix}^{-1} \bar{\boldsymbol{C}}_{A,B}^{*-1}\boldsymbol{B} = \begin{bmatrix} \bar{\boldsymbol{M}}_{A,B}^{-1} & \boldsymbol{0} \\ \boldsymbol{0} & \boldsymbol{I} \end{bmatrix} \boldsymbol{e}_1 = \begin{bmatrix} \bar{\boldsymbol{M}}_{A,B}^{-1}\bar{\boldsymbol{e}}_1 \\ \boldsymbol{0} \end{bmatrix}.$$

Substitute (6.63) into above equation and obtain

$$\bar{\boldsymbol{M}}_{A,B}^{-1}\bar{\boldsymbol{e}}_1 = \begin{bmatrix} \boldsymbol{0} & \bar{\boldsymbol{M}}_{2:}^{-1} \\ 1 & -\bar{\boldsymbol{a}}_{1:}^{\mathrm{T}}\bar{\boldsymbol{M}}_{2:}^{-1} \end{bmatrix} \bar{\boldsymbol{e}}_1 = \begin{bmatrix} \boldsymbol{0} \\ 1 \end{bmatrix} = \bar{\boldsymbol{e}}_m,$$

which implies that (6.62) holds. Note (6.64) and

$$\begin{aligned}
\bar{\boldsymbol{C}}_{A,B}^{*-1}\boldsymbol{A}\bar{\boldsymbol{C}}_{A,B}^{*} &= \bar{\boldsymbol{C}}_{A,B}^{*-1}\boldsymbol{A}[\boldsymbol{B}\ \ \boldsymbol{AB}\ \ \cdots\ \ \boldsymbol{A}^{m-1}\boldsymbol{B}\ \ \bar{\boldsymbol{C}}_{A,B}^{+}] \\
&= \bar{\boldsymbol{C}}_{A,B}^{*-1}[\boldsymbol{AB}\ \ \cdots\ \ \boldsymbol{A}^{m-1}\boldsymbol{B}\ \ \boldsymbol{A}^{m}\boldsymbol{B}\ \ *] \\
&= \begin{bmatrix} \bar{\boldsymbol{e}}_2 & \cdots & \bar{\boldsymbol{e}}_m & \boldsymbol{X}_{\bar{a}} & * \\ \boldsymbol{0} & \cdots & \boldsymbol{0} & \boldsymbol{0} & * \end{bmatrix} = \begin{bmatrix} \bar{\boldsymbol{S}}_{A,B}^{\mathrm{T}} & * \\ \boldsymbol{0} & * \end{bmatrix}.
\end{aligned}$$

Consider (6.61) and have

$$\begin{aligned}
\bar{\boldsymbol{P}}_{A,B}^{-1}\boldsymbol{A}\bar{\boldsymbol{P}}_{A,B} &= \begin{bmatrix} \bar{\boldsymbol{M}}_{A,B} & \boldsymbol{0} \\ \boldsymbol{0} & \boldsymbol{I} \end{bmatrix}^{-1} \bar{\boldsymbol{C}}_{A,B}^{*-1}\boldsymbol{A}\bar{\boldsymbol{C}}_{A,B}^{*} \begin{bmatrix} \bar{\boldsymbol{M}}_{A,B} & \boldsymbol{0} \\ \boldsymbol{0} & \boldsymbol{I} \end{bmatrix} \\
&= \begin{bmatrix} \bar{\boldsymbol{M}}_{A,B}^{-1} & \boldsymbol{0} \\ \boldsymbol{0} & \boldsymbol{I} \end{bmatrix} \begin{bmatrix} \bar{\boldsymbol{S}}_{A,B}^{\mathrm{T}} & * \\ \boldsymbol{0} & * \end{bmatrix} \begin{bmatrix} \bar{\boldsymbol{M}}_{A,B} & \boldsymbol{0} \\ \boldsymbol{0} & \boldsymbol{I} \end{bmatrix} \\
&= \begin{bmatrix} \bar{\boldsymbol{M}}_{A,B}^{-1}\bar{\boldsymbol{S}}_{A,B}^{\mathrm{T}}\bar{\boldsymbol{M}}_{A,B} & * \\ \boldsymbol{0} & * \end{bmatrix},
\end{aligned}$$

where

$$\begin{aligned}
\bar{\boldsymbol{M}}_{A,B}^{-1}\bar{\boldsymbol{S}}_{A,B}^{\mathrm{T}}\bar{\boldsymbol{M}}_{A,B} &= \begin{bmatrix} \boldsymbol{0} & \bar{\boldsymbol{M}}_{2:}^{-1} \\ 1 & -\bar{\boldsymbol{a}}_{1:}^{\mathrm{T}}\bar{\boldsymbol{M}}_{2:}^{-1} \end{bmatrix} \begin{bmatrix} \boldsymbol{0}^{\mathrm{T}} & -\bar{a}_0 \\ \boldsymbol{I} & -\bar{\boldsymbol{a}}_{1:} \end{bmatrix} \begin{bmatrix} \bar{\boldsymbol{a}}_{1:} & \bar{\boldsymbol{M}}_{2:} \\ 1 & \boldsymbol{0} \end{bmatrix} \\
&= \begin{bmatrix} \bar{\boldsymbol{M}}_{2:}^{-1} & -\bar{\boldsymbol{M}}_{2:}^{-1}\bar{\boldsymbol{a}}_{1:} \\ -\bar{\boldsymbol{a}}_{1:}^{\mathrm{T}}\bar{\boldsymbol{M}}_{2:}^{-1} & -\bar{a}_0 + \bar{\boldsymbol{a}}_{1:}^{\mathrm{T}}\bar{\boldsymbol{M}}_{2:}^{-1}\bar{\boldsymbol{a}}_{1:} \end{bmatrix} \begin{bmatrix} \bar{\boldsymbol{a}}_{1:} & \bar{\boldsymbol{M}}_{2:} \\ 1 & \boldsymbol{0} \end{bmatrix} \\
&= \begin{bmatrix} \boldsymbol{0} & \boldsymbol{I} \\ -\bar{a}_0 & -\bar{\boldsymbol{a}}_{1:}^{\mathrm{T}} \end{bmatrix} = \bar{\boldsymbol{S}}_{A,B}.
\end{aligned}$$

So (6.61) holds. □

Choose the auxiliary matrix $\bar{\boldsymbol{P}}_{A,B}$ as the **state transform matrix** and take **state transform** on (6.5) as

$$\bar{\boldsymbol{x}} = \bar{\boldsymbol{P}}_{A,B}^{-1}\boldsymbol{x} \iff \boldsymbol{x} = \bar{\boldsymbol{P}}_{A,B}\bar{\boldsymbol{x}}, \tag{6.65}$$

via which we can obtain

$$\frac{\mathrm{d}}{\mathrm{d}t}\boldsymbol{x} = \boldsymbol{A}\boldsymbol{x} + \boldsymbol{B}\boldsymbol{u} \iff \frac{\mathrm{d}}{\mathrm{d}t}\bar{\boldsymbol{x}} = \bar{\boldsymbol{P}}_{\boldsymbol{A},\boldsymbol{B}}^{-1}\boldsymbol{A}\bar{\boldsymbol{P}}_{\boldsymbol{A},\boldsymbol{B}}\bar{\boldsymbol{x}} + \bar{\boldsymbol{P}}_{\boldsymbol{A},\boldsymbol{B}}^{-1}\boldsymbol{B}\boldsymbol{u}.$$

According to (6.61) and (6.62), we have

$$\frac{\mathrm{d}}{\mathrm{d}t}\bar{\boldsymbol{x}} = \begin{bmatrix} \bar{\boldsymbol{S}}_{\boldsymbol{A},\boldsymbol{B}} & * \\ \boldsymbol{0} & * \end{bmatrix}\bar{\boldsymbol{x}} + \begin{bmatrix} \bar{\boldsymbol{e}}_m \\ \boldsymbol{0} \end{bmatrix}\boldsymbol{u}, \tag{6.66}$$

which is called the **partial controllability canonical form** of linear state-space modelling for the control system.

We can decompose the transformed state $\bar{\boldsymbol{x}}$ into two blocks as

$$\bar{\boldsymbol{x}} = \begin{bmatrix} \bar{\boldsymbol{x}}_c \\ \bar{\boldsymbol{x}}_s \end{bmatrix}, \tag{6.67}$$

where $\bar{\boldsymbol{x}}_c$ is the first m-element block of $\bar{\boldsymbol{x}}$ and $\bar{\boldsymbol{x}}_s$ is the second $(n-m)$-element block of $\bar{\boldsymbol{x}}$. From (6.66) we can see that dynamics of $\bar{\boldsymbol{x}}_s$ is equivalent to dynamics of a self-evolutionary system which cannot be influenced by control input $\boldsymbol{u}$:

$$\frac{\mathrm{d}}{\mathrm{d}t}\bar{\boldsymbol{x}}_s = [*]\bar{\boldsymbol{x}}_s.$$

The partial controllability canonical form described in (6.66) shows explicitly that the transformed state $\bar{\boldsymbol{x}}$ has uncontrollable part namely the part $\bar{\boldsymbol{x}}_s$ and hence the target process is obviously uncontrollable.

In contrast with the uncontrollable part $\bar{\boldsymbol{x}}_s$, the part $\bar{\boldsymbol{x}}_c$ can be influenced by control input $\boldsymbol{u}$. Dynamics of $\bar{\boldsymbol{x}}_c$ is modelled by the following state differential equation:

$$\frac{\mathrm{d}}{\mathrm{d}t}\bar{\boldsymbol{x}}_c = \bar{\boldsymbol{S}}_{\boldsymbol{A},\boldsymbol{B}}\bar{\boldsymbol{x}}_c + [*]\bar{\boldsymbol{x}}_s + \bar{\boldsymbol{e}}_m\boldsymbol{u}.$$

Especially when $\bar{\boldsymbol{x}}_s$ is already at or can be somehow kept at the equilibrium state $\boldsymbol{0}$, then above equation is reduced to

$$\frac{\mathrm{d}}{\mathrm{d}t}\bar{\boldsymbol{x}}_c = \bar{\boldsymbol{S}}_{\boldsymbol{A},\boldsymbol{B}}\bar{\boldsymbol{x}}_c + \bar{\boldsymbol{e}}_m\boldsymbol{u},$$

which is the controllability canonical form for $\bar{\boldsymbol{x}}_c$ given control input $\boldsymbol{u}$. In this case, $\bar{\boldsymbol{x}}_c$ is controllable and can also be kept at the equilibrium state $\boldsymbol{0}$, for which we can resort to methods presented in Sect. 6.4. So the part $\bar{\boldsymbol{x}}_c$ is partially controllable part of the transformed state $\bar{\boldsymbol{x}}$.

To facilitate understanding of above control theory on **partial controllability**, we may take the variant of the double inverted pendulum control problem as a concrete example. Suppose that both inverted pendulum angles θ_1 and θ_2 are close to zero. Here, we focus on the control input part of first inverted pendulum angular acceleration a_1 only, so the linear state differential equation described in (6.51) is reduced to

$$\frac{\mathrm{d}}{\mathrm{d}t}\boldsymbol{x} = \begin{bmatrix} 0 & 1 & 0 & 0 & 0 & 0 \\ \left(1+\frac{m_2}{m_1}\right)\frac{g}{L_1} & 0 & -\frac{m_2}{m_1}\frac{g}{L_1} & 0 & 0 & 0 \\ 0 & 0 & 0 & 1 & 0 & 0 \\ -\left(1+\frac{m_2}{m_1}\right)\frac{g}{L_2} & 0 & \left(1+\frac{m_2}{m_1}\right)\frac{g}{L_2} & 0 & 0 & 0 \\ 0 & 0 & 0 & 0 & 0 & 1 \\ 0 & 0 & 0 & 0 & 0 & 0 \end{bmatrix}\boldsymbol{x} + \begin{bmatrix} 0 \\ 1 \\ 0 \\ -\frac{L_1}{L_2} \\ 0 \\ 0 \end{bmatrix} a_1$$

$$\equiv \boldsymbol{A}\boldsymbol{x} + \boldsymbol{B}a_1. \tag{6.68}$$

As in Sect. 6.4, we take the same set of double inverted pendulum parameters: Let $m_1 = 1$, $m_2 = 1$, $L_1 = 1$, $L_2 = 1$, and $g = 10$. Then the state transition matrix $\boldsymbol{A}$ and the control input matrix $\boldsymbol{B}$ are

$$\boldsymbol{A} = \begin{bmatrix} 0 & 1 & 0 & 0 & 0 & 0 \\ 20 & 0 & -10 & 0 & 0 & 0 \\ 0 & 0 & 0 & 1 & 0 & 0 \\ -20 & 0 & 20 & 0 & 0 & 0 \\ 0 & 0 & 0 & 0 & 0 & 1 \\ 0 & 0 & 0 & 0 & 0 & 0 \end{bmatrix}, \quad \boldsymbol{B} = \begin{bmatrix} 0 \\ 1 \\ 0 \\ -1 \\ 0 \\ 0 \end{bmatrix}.$$

For such control system example, we can know by intuition that the double inverted pendulum is uncontrollable, because it has no direct or indirect control on the cart position and velocity. On the other hand, we refrain from intuitive judgement but resort to theoretical analysis of its partial controllability.

Compute the controllability matrix $\boldsymbol{C}_{\boldsymbol{A},\boldsymbol{B}}$ via (6.24) as

$$\boldsymbol{C}_{\boldsymbol{A},\boldsymbol{B}} = \begin{bmatrix} 0 & 1 & 0 & 30 & 0 & 1\,000 \\ 1 & 0 & 30 & 0 & 1\,000 & 0 \\ 0 & -1 & 0 & -40 & 0 & -1\,400 \\ -1 & 0 & -40 & 0 & -1\,400 & 0 \\ 0 & 0 & 0 & 0 & 0 & 0 \\ 0 & 0 & 0 & 0 & 0 & 0 \end{bmatrix},$$

which is of rank $m = 4$. Then compute the partial controllability matrix $\bar{\boldsymbol{C}}_{\boldsymbol{A},\boldsymbol{B}}$ via (6.54) as

$$\bar{\boldsymbol{C}}_{\boldsymbol{A},\boldsymbol{B}} \equiv [\boldsymbol{B}\ \ \boldsymbol{A}\boldsymbol{B}\ \ \boldsymbol{A}^2\boldsymbol{B}\ \ \boldsymbol{A}^3\boldsymbol{B}] = \begin{bmatrix} 0 & 1 & 0 & 30 \\ 1 & 0 & 30 & 0 \\ 0 & -1 & 0 & -40 \\ -1 & 0 & -40 & 0 \\ 0 & 0 & 0 & 0 \\ 0 & 0 & 0 & 0 \end{bmatrix}.$$

We can verify via *Gaussian pivot elimination* that the partial controllability matrix $\bar{C}_{A,B}$ is indeed of rank $m = 4$, consistent with (6.53). Its first four rows are linearly independent, so we can generate the complemented controllability matrix $\bar{C}^*_{A,B}$ via (6.55) as

$$\bar{C}^*_{A,B} \equiv [\bar{C}_{A,B}\ \bar{C}^+_{A,B}] = \begin{bmatrix} 0 & 1 & 0 & 30 & 0 & 0 \\ 1 & 0 & 30 & 0 & 0 & 0 \\ 0 & -1 & 0 & -40 & 0 & 0 \\ -1 & 0 & -40 & 0 & 0 & 0 \\ 0 & 0 & 0 & 0 & 1 & 0 \\ 0 & 0 & 0 & 0 & 0 & 1 \end{bmatrix}.$$

Compute the coefficients $\bar{a}_0$, $\bar{a}_1$, $\bar{a}_2$, and $\bar{a}_3$ via (6.57) as

$$\begin{bmatrix} -\bar{a}_0 \\ -\bar{a}_1 \\ -\bar{a}_2 \\ -\bar{a}_3 \end{bmatrix} = (\bar{C}^{\mathrm{T}}_{A,B}\bar{C}_{A,B})^{-1}\bar{C}^{\mathrm{T}}_{A,B}A^m B = \begin{bmatrix} -200 \\ 0 \\ 40 \\ 0 \end{bmatrix}$$

such that $A^4B + \bar{a}_3A^3B + \bar{a}_2A^2B + \bar{a}_1AB + \bar{a}_0B = \mathbf{0}$. Generate the auxiliary matrix $\bar{M}_{A,B}$ via (6.58) as

$$\bar{M}_{A,B} = \begin{bmatrix} & & -40 & 1 \\ & -40 & & 1 & \\ & & 1 & & \\ & 1 & & & \end{bmatrix}$$

and compute the state transform matrix $\bar{P}_{A,B}$ via (6.59) as

$$\bar{P}_{A,B} = \left[\bar{C}_{A,B}\bar{M}_{A,B}\ \bar{C}^+_{A,B}\right] = \begin{bmatrix} -10 & 0 & 1 & 0 & 0 & 0 \\ 0 & -10 & 0 & 1 & 0 & 0 \\ 0 & 0 & -1 & 0 & 0 & 0 \\ 0 & 0 & 0 & -1 & 0 & 0 \\ 0 & 0 & 0 & 0 & 1 & 0 \\ 0 & 0 & 0 & 0 & 0 & 1 \end{bmatrix}.$$

Note that the state transform $\bar{x} = \bar{P}^{-1}_{A,B}x$ or $x = \bar{P}_{A,B}\bar{x}$ with the state transform matrix $\bar{P}_{A,B}$ enables (6.68) with the specific set of double inverted pendulum parameters to transform into the partial controllability canonical form as

$$\frac{\mathrm{d}}{\mathrm{d}t}\bar{\boldsymbol{x}} = \begin{bmatrix} & 1 & & \\ & & 1 & & * \\ & & & 1 & \\ -200 & 0 & 40 & 0 & \\ & \boldsymbol{0} & & & * \end{bmatrix} \bar{\boldsymbol{x}} + \begin{bmatrix} 0 \\ 0 \\ 0 \\ 1 \\ \boldsymbol{0} \end{bmatrix} a_1. \tag{6.69}$$

As in (6.67), decompose the transformed state $\bar{\boldsymbol{x}}$ into the first m-element block namely the first 4-element block $\bar{\boldsymbol{x}}_c$ and the second $(n-m)$-element block namely the second 2-element block $\bar{\boldsymbol{x}}_s$:

$$\bar{\boldsymbol{x}}_c = [\boldsymbol{I}_4\ \ \boldsymbol{0}]\bar{\boldsymbol{P}}_{\boldsymbol{A},\boldsymbol{B}}^{-1}\boldsymbol{x} = \begin{bmatrix} -0.1 & 0 & -0.1 & 0 & \\ 0 & -0.1 & 0 & -0.1 & \boldsymbol{0} \\ 0 & 0 & -1 & 0 & \\ 0 & 0 & 0 & -1 & \end{bmatrix} \boldsymbol{x} = \begin{bmatrix} -0.1(\theta_1+\theta_2) \\ -0.1(\frac{\mathrm{d}\theta_1}{\mathrm{d}t}+\frac{\mathrm{d}\theta_2}{\mathrm{d}t}) \\ -\theta_2 \\ -\frac{\mathrm{d}\theta_2}{\mathrm{d}t} \end{bmatrix},$$

$$\bar{\boldsymbol{x}}_s = [\boldsymbol{0}\ \ \boldsymbol{I}_2]\bar{\boldsymbol{P}}_{\boldsymbol{A},\boldsymbol{B}}^{-1}\boldsymbol{x} = \begin{bmatrix} \boldsymbol{0} & \begin{matrix} 1 & 0 \\ 0 & 1 \end{matrix} \end{bmatrix} \boldsymbol{x} = \begin{bmatrix} x \\ \frac{\mathrm{d}x}{\mathrm{d}t} \end{bmatrix}.$$

The partial controllability canonical form described in (6.69) shows explicitly that $\bar{\boldsymbol{x}}_s$ is uncontrollable part of the the transformed state $\bar{\boldsymbol{x}}$, whereas $\bar{\boldsymbol{x}}_c$ is partially controllable part of the transformed state $\bar{\boldsymbol{x}}$.

Suppose that $\bar{\boldsymbol{x}}_s$ is already at or somehow kept at the equilibrium state $\boldsymbol{0}$. Then we have

$$\frac{\mathrm{d}}{\mathrm{d}t}\bar{\boldsymbol{x}}_c = \begin{bmatrix} & 1 & & \\ & & 1 & \\ & & & 1 \\ -200 & 0 & 40 & 0 \end{bmatrix} \bar{\boldsymbol{x}}_c + \begin{bmatrix} 0 \\ 0 \\ 0 \\ 1 \end{bmatrix} a_1,$$

which is the controllability canonical form for $\bar{\boldsymbol{x}}_c$ given control input a_1. According to methods presented in Sect. 6.4, design the transformed gain matrix $\bar{\boldsymbol{K}}$ of full-state feedback control law of a_1 dedicated to $\bar{\boldsymbol{x}}_c$. For example, set the expected closed-loop characteristic polynomial for $\bar{\boldsymbol{x}}_c$ as

$$\bar{C}_{\mathrm{E}}(s) = (s+4)^4 = s^4 + 16\,s^3 + 96\,s^2 + 256\,s + 256.$$

Compute the transformed gain matrix via (6.43) as

$$\begin{cases} \bar{k}_1 = \bar{a}_0^{\mathrm{E}} - \bar{a}_0 = 256 - 200 = 56 \\ \bar{k}_2 = \bar{a}_1^{\mathrm{E}} - \bar{a}_1 = 256 - 0 = 256 \\ \bar{k}_3 = \bar{a}_2^{\mathrm{E}} - \bar{a}_2 = 96 - (-40) = 136 \\ \bar{k}_4 = \bar{a}_3^{\mathrm{E}} - \bar{a}_3 = 16 - 0 = 16 \end{cases} \qquad \bar{\boldsymbol{K}} \equiv \begin{bmatrix} \bar{k}_1 \\ \bar{k}_2 \\ \bar{k}_3 \\ \bar{k}_4 \end{bmatrix} = \begin{bmatrix} 56 \\ 256 \\ 136 \\ 16 \end{bmatrix}$$

Recover the gain matrix $\boldsymbol{K}$ of full-state feedback control law of a_1 from the transformed gain matrix $\bar{\boldsymbol{K}}$ as

$$K = \bar{P}_{A,B}^{-T}\begin{bmatrix}\bar{K}\\ 0\end{bmatrix} = \begin{bmatrix} -10 & 0 & 1 & 0 & 0 & 0\\ 0 & -10 & 0 & 1 & 0 & 0\\ 0 & 0 & -1 & 0 & 0 & 0\\ 0 & 0 & 0 & -1 & 0 & 0\\ 0 & 0 & 0 & 0 & 1 & 0\\ 0 & 0 & 0 & 0 & 0 & 1 \end{bmatrix}^{-T}\begin{bmatrix}56\\256\\136\\16\\0\\0\end{bmatrix} = \begin{bmatrix}-5.6\\-25.6\\-141.6\\-41.6\\0\\0\end{bmatrix}.$$

So the full-state feedback control law of the first inverted pendulum angular acceleration a_1 is

$$a_1 = -K^T x = 5.6\theta_1 + 25.6\frac{d\theta_1}{dt} + 141.6\theta_2 + 41.6\frac{d\theta_2}{dt}.$$

Matlab code for analysing partial controllability of the variant of double inverted pendulum control by the control input part of first inverted pendulum angular acceleration a_1 is given as follows.

PartialControllabilityMatrixDIP.m

```
%% Double inverted pendulum parameters
m1 = 1; m2 = 1; L1 = 1; L2 = 1; g = 10;
A = [0, 1, 0, 0, 0, 0; ...
    (m1+m2)*g/(m1*L1), 0, -m2*g/(m1*L1), 0, 0, 0; ...
    0, 0, 0, 1, 0, 0; ...
    -(m1+m2)*g/(m1*L2), 0, (m1+m2)*g/(m1*L2), 0, 0, 0; ...
    0, 0, 0, 0, 0, 1; ...
    0, 0, 0, 0, 0, 0];
B = [0; 1; 0; -L1/L2; 0; 0];
n = 6; % State dimension

%% Compute the partial controllability matrix
CM = zeros(n); CM(:,1) = B;
for k = 2:n
    CM(:,k) = A*CM(:,k-1);
end % CM = [B, A*B, A^2*B, A^3*B, A^4*B, A^5*B];
m = rank(CM); PCM = zeros(n,m); PCM(:,1) = B;
for k = 2:m
    PCM(:,k) = A*PCM(:,k-1);
end % CM = [B, A*B, A^2*B, A^3*B];
fprintf('Partial controllability matrix: '); PCMorg = PCM; PCM

%% Gaussian pivot elimination
r = 1; rL = []; k = 1; % Linearly independent rows
while (k<=size(PCM,2))
    pI = -1; pM = 0; % Pivot
    for i = k:size(PCM,2)
        if (abs(PCM(r,i)) > pM) pM = abs(PCM(r,i)); pI = i; end
    end
    if (pM<10*eps) r = r+1; continue; end
    if (k ~= pI) PCM(r:end,[k, pI]) = PCM(r:end,[pI, k]); end
    ck = (k+1):size(PCM,2);
```

```
    PCM(r:end,ck) = PCM(r:end,ck) - PCM(r:end,k)*PCM(r,ck)/PCM(r
    ,k);
    rL = [rL, r]; r = r+1; k = k+1;
end
fprintf('Partial controllability matrix after Gaussian
    elimination: '); PCM
fprintf('Indices of linearly independent rows: '); rL
PCM = PCMorg;

%% Generate the complemented controllability matrix
rC = ones(1,n); rC(rL) = 0; rC = find(rC == 1);
CompM = zeros(n,n-m); CompM(rC,:) = eye(n-m); % Complement
    matrix
CompCM = [PCM, CompM]; % Complemented controllability matrix
fprintf('Complemented controllability matrix: '); CompCM

%% Compute the state transform matrix
aA = PCM\(A*PCM(:,m)); aA = [-aA', 1];
M = zeros(m, m);
for k = 1:m % Auxiliary matrix M
    M(k,1:(m+1-k)) = aA((k+1):(m+1));
end
fprintf('Auxiliary matrix M: '); M
P = [PCM*M, CompM]; % State transform matrix
fprintf('State transform matrix P: '); P

%% State transform
fprintf('Transformed state transition matrix: ');
transformedA = inv(P)*A*P
fprintf('Transformed control input matrix: ');
transformedB = inv(P)*B
fprintf('Partially controllable part: ');
xcM = [eye(m), zeros(m,n-m)]*inv(P)
fprintf('Uncontrollable part: ');
xsM = [zeros(n-m,m), eye(n-m)]*inv(P)

%% Design the transformed gain matrix
lambdaE = -4*ones(1,m);  % Expected characteristic polynomial (s
    +4)^4
aE - GetCharacteristicPolynomialCoefficients(lambdaE);
fprintf('Expected characteristic polynomial coefficients (a0, a1
    , ...): '); aE
transformedK = zeros(n,1); transformedK(1:m) = transpose(aE(1:m)
     - aA(1:m));
fprintf('Transformed gain matrix K: '); transformedK

%% Recover the gain matrix
sttK1 = inv(transpose(P))*transformedK;
fprintf('Gain matrix K: '); sttK1

%% Compute the characteristic polynomial coefficients from
    eigenvalues
function a = GetCharacteristicPolynomialCoefficients(lambda)
```

```
    n = length(lambda);
    % Characteristic polynomial coefficients a(-1), a0, ..., an
    a = zeros(1,n+2);
    a(2) = 1;
    for k = 1:n
        aOld = a;
        for i = 2:(k+1)
            a(i) = aOld(i-1) + aOld(i)*(-lambda(k));
        end
        a(k+2) = 1;
    end
    % Characteristic polynomial coefficients a0, a1, ..., an
    a = a(2:end);
end
```

Matlab simulation code for complete demonstration of double inverted pendulum control by the control input part of first inverted pendulum angular acceleration a_1 is given as follows.

DoubleInvertedPendulumPartial.m

```
%% Double inverted pendulum parameters
m1 = 1; m2 = 1; L1 = 1; L2 = 1; g = 10;
%% Simulation preliminary configuration
dt = 0.001; % Numerical computation step
tSpan = 0:dt:3; % Simulation time span
x = 0.2; dx = 0; % Cart position and its velocity
% Inverted pendulum angle theta-1 and its angular velocity
y1 = 0.2; dy1 = 0;
% Inverted pendulum angle theta-2 and its angular velocity
y2 = 0; dy2 = 0;
stt = [y1; dy1; y2; dy2; x; dx]; % Double inverted pendulum
    state
% Record states in simulation
sttAll = zeros(length(stt), length(tSpan)); k = 0;
% Expected equilibrium status
xExpected = 0; y1Expected = 0; y2Expected = 0;
SimConfig = [m1, m2, L1, L2, g, dt];

%% Design the partial gain matrix
PartialControllabilityMatrixDIP % Obtain the partial gain matrix
    sttK1

%% Simulation of double inverted pendulum control
for t = tSpan
    %% Control method
    acc = 0; % No control of cart acceleration
    % Partial control by first inverted pendulum angular
    acceleration
    acc1 = -sttK1'*stt;

    %% Double inverted pendulum dynamics
    stt = DynamicsDIP(SimConfig, stt, acc, acc1);
    sttC = num2cell(stt); [y1, dy1, y2, dy2, x, dx] = sttC{:};
```

```
    if (abs(y1)>=pi/2 && abs(y2)>=pi/2)
        fprintf('Control failure!\n'); break;
    end
    k = k+1; sttAll(:,k) = stt;
    %% Double inverted pendulum visualization
    if (rem(k,20) == 0)
        DisplayDIP(x, y1, y2, L1, L2); pause(20*dt);
    end
end
```

After trying above Matlab code, readers will find that part of the double inverted pendulum state namely both inverted pendulum angles θ_1 and θ_2 and their angular velocities can indeed be controlled to the equilibrium state by the control input part of first inverted pendulum angular acceleration a_1, if given a stable initial condition of the uncontrollable state part namely the cart position and velocity.

6.5.2 Control Input Decomposition and Iterative Design

With knowledge on partial controllability of control systems, we can move forward to *multiple-input-multiple-output* control problems. Given a control system that adopts linear state-space modelling described by (6.5), with its state denoted as $\boldsymbol{x}$ (suppose that the state dimension is n) and its control input to the target process denoted as $\boldsymbol{u}$, for the moment, suppose the control input $\boldsymbol{u}$ is double-input denoted as $\boldsymbol{u} \equiv [u_1 \;\; u_2]^{\mathrm{T}}$. We focus on presenting extension from single-input-multiple-output control to double-input-multiple-output control, after which extension from single-input-multiple-output control to generic multiple-input-multiple-output control will become natural.

Decompose the control input $\boldsymbol{u}$ of the state differential equation described in (6.5) as

$$\frac{\mathrm{d}}{\mathrm{d}t}\boldsymbol{x} = \boldsymbol{A}\boldsymbol{x} + \boldsymbol{B}\boldsymbol{u} = \boldsymbol{A}\boldsymbol{x} + \boldsymbol{B}_1 u_1 + \boldsymbol{B}_2 u_2, \tag{6.70}$$

where the n-by-n square matrix $\boldsymbol{A}$ is the state transition matrix and the n-by-1 matrices $\boldsymbol{B}_1$ and $\boldsymbol{B}_2$ are the control input matrices of u_1 and u_2 respectively.

For a generic matrix $\boldsymbol{M}$, denote its **span** namely the span of its column vectors as $\langle \boldsymbol{M} \rangle$. Partial controllability (or controllability) of control input u_1, u_2, and $\boldsymbol{u}$ are characterized by spans of their controllability matrices $\langle \boldsymbol{C}_{\boldsymbol{A},\boldsymbol{B}_1} \rangle$, $\langle \boldsymbol{C}_{\boldsymbol{A},\boldsymbol{B}_2} \rangle$, and $\langle \boldsymbol{C}_{\boldsymbol{A},\boldsymbol{B}} \rangle$ respectively, where

$$\begin{aligned}
\langle \boldsymbol{C}_{\boldsymbol{A},\boldsymbol{B}_1} \rangle &= \langle [\boldsymbol{B}_1 \;\; \boldsymbol{A}\boldsymbol{B}_1 \;\; \cdots \;\; \boldsymbol{A}^{n-1}\boldsymbol{B}_1] \rangle, \\
\langle \boldsymbol{C}_{\boldsymbol{A},\boldsymbol{B}_2} \rangle &= \langle [\boldsymbol{B}_2 \;\; \boldsymbol{A}\boldsymbol{B}_2 \;\; \cdots \;\; \boldsymbol{A}^{n-1}\boldsymbol{B}_2] \rangle, \\
\langle \boldsymbol{C}_{\boldsymbol{A},\boldsymbol{B}} \rangle &= \langle [\boldsymbol{B} \;\; \boldsymbol{A}\boldsymbol{B} \;\; \cdots \;\; \boldsymbol{A}^{n-1}\boldsymbol{B}] \rangle = \langle \boldsymbol{C}_{\boldsymbol{A},\boldsymbol{B}_1}, \boldsymbol{C}_{\boldsymbol{A},\boldsymbol{B}_2} \rangle.
\end{aligned}$$

Suppose that neither u_1 nor u_2 is redundant control input part, which implies $m \equiv$ rank of $\boldsymbol{C}_{\boldsymbol{A},\boldsymbol{B}_1} <$ rank of $\boldsymbol{C}_{\boldsymbol{A},\boldsymbol{B}} \leqslant n$ (the reason is left to readers for reflection).

Following the method presented in Sect. 6.5.1, we can compute the state transform matrix $\bar{\boldsymbol{P}}_{\boldsymbol{A},\boldsymbol{B}_1}$ via (6.59) by replacing $\boldsymbol{B}$ with $\boldsymbol{B}_1$ and take state transform on (6.70) as

$$\bar{\boldsymbol{x}} = \bar{\boldsymbol{P}}_{\boldsymbol{A},\boldsymbol{B}_1}^{-1}\boldsymbol{x} \iff \boldsymbol{x} = \bar{\boldsymbol{P}}_{\boldsymbol{A},\boldsymbol{B}_1}\bar{\boldsymbol{x}}, \tag{6.71}$$

via which we can obtain

$$\frac{\mathrm{d}}{\mathrm{d}t}\bar{\boldsymbol{x}} = \begin{bmatrix} \bar{\boldsymbol{S}}_{\boldsymbol{A},\boldsymbol{B}_1} & \bar{\boldsymbol{A}}_{12} \\ \boldsymbol{0} & \bar{\boldsymbol{A}}_{22} \end{bmatrix}\bar{\boldsymbol{x}} + \begin{bmatrix} \bar{\boldsymbol{e}}_m \\ \boldsymbol{0} \end{bmatrix} u_1 + \bar{\boldsymbol{P}}_{\boldsymbol{A},\boldsymbol{B}_1}^{-1}\boldsymbol{B}_2 u_2, \tag{6.72}$$

where $\bar{\boldsymbol{S}}_{\boldsymbol{A},\boldsymbol{B}_1}$ is of the form

$$\bar{\boldsymbol{S}}_{\boldsymbol{A},\boldsymbol{B}_1} = \begin{bmatrix} & 1 & & \\ & & \ddots & \\ & & & 1 \\ -\bar{a}_0 & -\bar{a}_1 & \cdots & -\bar{a}_{m-1} \end{bmatrix}.$$

Decompose the transformed state $\bar{\boldsymbol{x}}$ into two blocks namely the first m-element block $\bar{\boldsymbol{x}}_c$ and the second $(n-m)$-element block $\bar{\boldsymbol{x}}_s$, which are partially controllable and uncontrollable by u_1 respectively.

As presented in Sect. 6.5.1, via the partial controllability canonical form described in (6.72) for $\bar{\boldsymbol{x}}_c$ given u_1, we can conveniently design a full-state feedback control law of u_1 as

$$u_1 = -\bar{\boldsymbol{K}}_c^{\mathrm{T}}\bar{\boldsymbol{x}}_c = -\begin{bmatrix} \bar{\boldsymbol{K}}_c \\ \boldsymbol{0} \end{bmatrix}^{\mathrm{T}} \bar{\boldsymbol{x}} \tag{6.73}$$

with transformed gain matrix $\bar{\boldsymbol{K}}_c$ such that $\bar{\boldsymbol{S}}_{\boldsymbol{A},\boldsymbol{B}_1} - \bar{\boldsymbol{e}}_m\bar{\boldsymbol{K}}_c^{\mathrm{T}}$ has eigenvalues all with negative real part.

On the other hand, consider dynamics of $\bar{\boldsymbol{x}}_s$ which is modelled by

$$\frac{\mathrm{d}}{\mathrm{d}t}\bar{\boldsymbol{x}}_s = \bar{\boldsymbol{A}}_{22}\bar{\boldsymbol{x}}_s + [\boldsymbol{0} \;\; \boldsymbol{I}_{n-m}]\bar{\boldsymbol{P}}_{\boldsymbol{A},\boldsymbol{B}_1}^{-1}\boldsymbol{B}_2 u_2 \equiv \bar{\boldsymbol{A}}_{22}\bar{\boldsymbol{x}}_s + \bar{\boldsymbol{B}}_{2s}u_2.$$

If the state $\boldsymbol{x}$ of the control system is controllable by control input $\boldsymbol{u}$, then the state part $\bar{\boldsymbol{x}}_s$ which is uncontrollable by u_1 is definitely controllable by u_2. This implies that we can design a full-state feedback control law of u_2 as

$$u_2 = -\bar{\boldsymbol{K}}_s^{\mathrm{T}}\bar{\boldsymbol{x}}_s = -\begin{bmatrix} \boldsymbol{0} \\ \bar{\boldsymbol{K}}_s \end{bmatrix}^{\mathrm{T}} \bar{\boldsymbol{x}} \tag{6.74}$$

with transformed gain matrix $\bar{\boldsymbol{K}}_s$ such that $\bar{A}_{22} - \bar{\boldsymbol{B}}_{2s}\bar{\boldsymbol{K}}_s^{\mathrm{T}}$ has eigenvalues all with negative real part.

Substitute (6.73) and (6.74) into (6.72) and obtain

$$\frac{\mathrm{d}}{\mathrm{d}t}\bar{\boldsymbol{x}} = \begin{bmatrix} \bar{\boldsymbol{S}}_{\boldsymbol{A},\boldsymbol{B}_1} - \bar{\boldsymbol{e}}_m\bar{\boldsymbol{K}}_c^{\mathrm{T}} & * \\ \boldsymbol{0} & \bar{A}_{22} - \bar{\boldsymbol{B}}_{2s}\bar{\boldsymbol{K}}_s^{\mathrm{T}} \end{bmatrix}\bar{\boldsymbol{x}},$$

where eigenvalues of the state transition matrix consist of eigenvalues of $\bar{\boldsymbol{S}}_{\boldsymbol{A},\boldsymbol{B}_1} - \bar{\boldsymbol{e}}_m\bar{\boldsymbol{K}}_c^{\mathrm{T}}$ and eigenvalues of $\bar{A}_{22} - \bar{\boldsymbol{B}}_{2s}\bar{\boldsymbol{K}}_s^{\mathrm{T}}$ and hence are all with negative real part. So the transformed state $\bar{\boldsymbol{x}}$ converges to $\boldsymbol{0}$ when the double-input $\boldsymbol{u}$ adopts full-state feedback control laws described in (6.73) and (6.74).

This further implies if u_1 adopts the full-state feedback control law described in (6.73) with transformed gain matrix $\bar{\boldsymbol{K}}_c$ designed according to the partial controllability canonical form described in (6.72), or in other words, if u_1 adopts the full-state feedback control law as

$$u_1 = -\begin{bmatrix} \bar{\boldsymbol{K}}_c \\ \boldsymbol{0} \end{bmatrix}^{\mathrm{T}} \bar{\boldsymbol{x}} = -\begin{bmatrix} \bar{\boldsymbol{K}}_c \\ \boldsymbol{0} \end{bmatrix}^{\mathrm{T}} \bar{\boldsymbol{P}}_{A,\boldsymbol{B}_1}^{-1}\boldsymbol{x} \equiv -\boldsymbol{K}_c^{\mathrm{T}}\boldsymbol{x} \tag{6.75}$$

with gain matrix $\boldsymbol{K}_c$ for full-state feedback of the original state $\boldsymbol{x}$, then (6.70) will be reduced to a state differential equation as

$$\frac{\mathrm{d}}{\mathrm{d}t}\boldsymbol{x} = (\boldsymbol{A} - \boldsymbol{B}_1\boldsymbol{K}_c^{\mathrm{T}})\boldsymbol{x} + \boldsymbol{B}_2u_2 \tag{6.76}$$

with single-input u_2, which maintains controllability of the original control system by double-input $\boldsymbol{u}$. We reduce the original double-input-multiple-output control system to a single-input-multiple-output control system that *maintains controllability of the original control system.*

We can naturally extend above methodology to generic multiple-input-multiple-output control. Now suppose that the generic control input $\boldsymbol{u}$ is of dimension r. Decompose the control input $\boldsymbol{u}$ of the state differential equation described in (6.5) as

$$\frac{\mathrm{d}}{\mathrm{d}t}\boldsymbol{x} = \boldsymbol{A}\boldsymbol{x} + \boldsymbol{B}\boldsymbol{u} = \boldsymbol{A}\boldsymbol{x} + \boldsymbol{B}_1u_1 + \boldsymbol{B}_{2:r}\boldsymbol{u}_{2:r}, \tag{6.77}$$

where u_1 represents the first dimension of control input $\boldsymbol{u}$ and $\boldsymbol{u}_{2:r}$ represents the remaining $r-1$ dimensions of $\boldsymbol{u}$.

Compute the state transform matrix $\bar{\boldsymbol{P}}_{\boldsymbol{A},\boldsymbol{B}_1}$ and take the state transform described in (6.71) on (6.77) to obtain

$$\frac{\mathrm{d}}{\mathrm{d}t}\bar{\boldsymbol{x}} = \begin{bmatrix} \bar{\boldsymbol{S}}_{\boldsymbol{A},\boldsymbol{B}_1} & * \\ \boldsymbol{0} & * \end{bmatrix}\bar{\boldsymbol{x}} + \begin{bmatrix} \bar{\boldsymbol{e}}_m \\ \boldsymbol{0} \end{bmatrix}u_1 + \bar{\boldsymbol{P}}_{A,\boldsymbol{B}_1}^{-1}\boldsymbol{B}_{2:r}\boldsymbol{u}_{2:r}, \tag{6.78}$$

where $\bar{S}_{A,B_1}$ is of the same form as that described in (6.72).

Design the transformed gain matrix $\bar{K}_c$ of full-state feedback control law of u_1 according to the partial controllability canonical form described in (6.78) such that $\bar{S}_{A,B_1} - \bar{e}_m \bar{K}_c^{\mathrm{T}}$ has eigenvalues all with negative real part, then recover via (6.75) the gain matrix K_c for full-state feedback of the original state x and reduce (6.77) to a state differential equation as

$$\frac{\mathrm{d}}{\mathrm{d}t}x = (A - B_1 K_c^{\mathrm{T}})x + B_{2:r}u_{2:r}, \tag{6.79}$$

which *has control input* $u_{2:r}$ *of one dimension less than the original control input* u and *maintains controllability of the original control system*. We can further decompose the control input $u_{2:r}$ and *apply above methodology iteratively*, via which we can design gain matrices of full-state feedback control laws of $u_1, u_2, \ldots, u_r$ gradually and finally obtain the gain matrix of full-state feedback control law of the holistic control input u.

Variant of double inverted pendulum control

We apply the methodology of *control input decomposition and iterative design* to handle the variant of double inverted pendulum control which serves as a concrete example of *multiple-input-multiple-output* control. Suppose that both inverted pendulum angles θ_1 and θ_2 are close to zero, and hence the linear state differential equation described in (6.51) is fairly adopted:

$$\frac{\mathrm{d}}{\mathrm{d}t}x = \begin{bmatrix} 0 & 1 & 0 & 0\,0\,0 \\ \left(1+\frac{m_2}{m_1}\right)\frac{g}{L_1} & 0 & -\frac{m_2}{m_1}\frac{g}{L_1} & 0\,0\,0 \\ 0 & 0 & 0 & 1\,0\,0 \\ -\left(1+\frac{m_2}{m_1}\right)\frac{g}{L_2} & 0 & \left(1+\frac{m_2}{m_1}\right)\frac{g}{L_2} & 0\,0\,0 \\ 0 & 0 & 0 & 0\,0\,1 \\ 0 & 0 & 0 & 0\,0\,0 \end{bmatrix} x + \begin{bmatrix} 0 & 0 \\ -\frac{1}{L_1} & 1 \\ 0 & 0 \\ 0 & -\frac{L_1}{L_2} \\ 0 & 0 \\ 1 & 0 \end{bmatrix} \begin{bmatrix} a \\ a_1 \end{bmatrix}$$

$$\equiv Ax + Bu \equiv Ax + B_0 a + B_1 a_1.$$

The control input u is the multiple-input of cart acceleration a and first inverted pendulum angular acceleration a_1. As in Sect. 6.4, we take the same set of double inverted pendulum parameters: Let $m_1 = 1$, $m_2 = 1$, $L_1 = 1$, $L_2 = 1$, and $g = 10$. Then the state transition matrix A and the control input matrix B are

$$A = \begin{bmatrix} 0 & 1 & 0 & 0\,0\,0 \\ 20 & 0 & -10 & 0\,0\,0 \\ 0 & 0 & 0 & 1\,0\,0 \\ -20 & 0 & 20 & 0\,0\,0 \\ 0 & 0 & 0 & 0\,0\,1 \\ 0 & 0 & 0 & 0\,0\,0 \end{bmatrix}, \quad B = \begin{bmatrix} 0 & 0 \\ -1 & 1 \\ 0 & 0 \\ 0 & -1 \\ 0 & 0 \\ 1 & 0 \end{bmatrix} \equiv \begin{bmatrix} B_0 & B_1 \end{bmatrix}.$$

Consider the control input part a_1 and let it adopt the full-state feedback control law described in (6.75) as

$$a_1 = -\begin{bmatrix} \bar{\boldsymbol{K}}_1 \\ \boldsymbol{0} \end{bmatrix}^{\mathrm{T}} \bar{x} = -\begin{bmatrix} \bar{\boldsymbol{K}}_1 \\ \boldsymbol{0} \end{bmatrix}^{\mathrm{T}} \bar{\boldsymbol{P}}_{A,B_1}^{-1} \boldsymbol{x} \equiv -\boldsymbol{K}_1^{\mathrm{T}} \boldsymbol{x},$$

where $\bar{\boldsymbol{P}}_{A,B_1}$, $\bar{\boldsymbol{K}}_1$, and $\boldsymbol{K}_1$ are already obtained in Sect. 6.5.1 (given the same set of expected eigenvalues), respectively as

$$\bar{\boldsymbol{P}}_{A,B_1} = \begin{bmatrix} -10 & 0 & 1 & 0 & 0 & 0 \\ 0 & -10 & 0 & 1 & 0 & 0 \\ 0 & 0 & -1 & 0 & 0 & 0 \\ 0 & 0 & 0 & -1 & 0 & 0 \\ 0 & 0 & 0 & 0 & 1 & 0 \\ 0 & 0 & 0 & 0 & 0 & 1 \end{bmatrix}, \quad \begin{bmatrix} \bar{\boldsymbol{K}}_1 \\ \boldsymbol{0} \end{bmatrix} = \begin{bmatrix} 56 \\ 256 \\ 136 \\ 16 \\ 0 \\ 0 \end{bmatrix}, \quad \boldsymbol{K}_1 = \begin{bmatrix} -5.6 \\ -25.6 \\ -141.6 \\ -41.6 \\ 0 \\ 0 \end{bmatrix}.$$

Substitute $a_1 = -\boldsymbol{K}_1^{\mathrm{T}} \boldsymbol{x}$ into (6.51) and obtain

$$\frac{\mathrm{d}}{\mathrm{d}t}\boldsymbol{x} = \boldsymbol{A}\boldsymbol{x} + \boldsymbol{B}_0 a + \boldsymbol{B}_1 a_1 = (\boldsymbol{A} - \boldsymbol{B}_1 \boldsymbol{K}_1^{\mathrm{T}})\boldsymbol{x} + \boldsymbol{B}_0 a$$
$$= \begin{bmatrix} 0 & 1 & 0 & 0 & 0 & 0 \\ 25.6 & 25.6 & 131.6 & 41.6 & 0 & 0 \\ 0 & 0 & 0 & 1 & 0 & 0 \\ -25.6 & -25.6 & -121.6 & -41.6 & 0 & 0 \\ 0 & 0 & 0 & 0 & 0 & 1 \\ 0 & 0 & 0 & 0 & 0 & 0 \end{bmatrix} \boldsymbol{x} + \begin{bmatrix} 0 \\ -1 \\ 0 \\ 0 \\ 0 \\ 1 \end{bmatrix} a.$$

Still set all expected eigenvalues to −4 as in previous sections—There is no special reason for setting the expected eigenvalues in this way but only for demonstration convenience—Then compute the gain matrix of full-state feedback control law of a with the state transition matrix $\boldsymbol{A} - \boldsymbol{B}_1 \boldsymbol{K}_1^{\mathrm{T}}$ and the control input matrix $\boldsymbol{B}_0$ and finally obtain the gain matrix of full-state feedback control law of the holistic control input $\boldsymbol{u}$ as

$$\begin{bmatrix} a \\ a_1 \end{bmatrix} = \begin{bmatrix} 0 & 0 & 0 & 0 & 16 & 8 \\ -5.6 & -25.6 & -141.6 & -41.6 & 0 & 0 \end{bmatrix} \boldsymbol{x}.$$

Matlab simulation code for complete demonstration of the variant of double inverted pendulum control by multiple-input of cart acceleration a and first inverted pendulum angular acceleration a_1 is given as follows.

DoubleInvertedPendulumMIMO.m

```
%% Double inverted pendulum parameters
m1 = 1; m2 = 1; L1 = 1; L2 = 1; g = 10;
%% Simulation preliminary configuration
```

```
dt = 0.001; % Numerical computation step
tSpan = 0:dt:3; % Simulation time span
x = 0.2; dx = 0; % Cart position and its velocity
% Inverted pendulum angle theta-1 and its angular velocity
y1 = 0.2; dy1 = 0;
% Inverted pendulum angle theta-2 and its angular velocity
y2 = 0; dy2 = 0;
sttSIMO = [y1; dy1; y2; dy2; x; dx]; % Double inverted pendulum
    state
sttMIMO = [y1; dy1; y2; dy2; x; dx]; % Double inverted pendulum
    state
SimConfig = [m1, m2, L1, L2, g, dt];

%% Design the gain matrix iteratively
A = [0, 1, 0, 0, 0, 0; ...
    (m1+m2)*g/(m1*L1), 0, -m2*g/(m1*L1), 0, 0, 0; ...
    0, 0, 0, 1, 0, 0; ...
    -(m1+m2)*g/(m1*L2), 0, (m1+m2)*g/(m1*L2), 0, 0, 0; ...
    0, 0, 0, 0, 0, 1; ...
    0, 0, 0, 0, 0, 0];
B = [0, 0; -1/L1, 1; 0, 0; 0, -L1/L2; 0, 0; 1, 0];
lambdaE = [-4;-4;-4;-4;-4;-4]; % Expected eigenvalues
% For single-input-multiple-output control
sttK = DesignGainMatrixSub(A, B(:,1), lambdaE);
fprintf('Gain matrix K: '); sttK'
% For multiple-input-multiple-output control
sttK2 = DesignGainMatrixSub(A, B(:,2), lambdaE);
sttK1 = DesignGainMatrixSub(A-B(:,2)*sttK2', B(:,1), lambdaE);
fprintf('Gain matrix K: '); [sttK1'; sttK2']

k = 0; figure(1), set(gcf, 'Position', [100, 100, 1200, 600]);
%% Simulation of double inverted pendulum control
for t = tSpan
    %% Multiple-input-multiple-output control method
    acc = -sttK1'*sttMIMO; acc1 = -sttK2'*sttMIMO;
    %% Double inverted pendulum dynamics
    sttMIMO = DynamicsDIP(SimConfig, sttMIMO, acc, acc1);
    sttC = num2cell(sttMIMO);
    [y1M, dy1M, y2M, dy2M, xM, dxM] = sttC{:};
    if (abs(y1M)>=pi/2 && abs(y2M)>=pi/2)
        fprintf('Control failure!\n'); break;
    end

    %% Single-input-multiple-output control method
    acc = -sttK'*sttSIMO;
    %% Double inverted pendulum dynamics
    sttSIMO = DynamicsDIP(SimConfig, sttSIMO, acc);
    sttC = num2cell(sttSIMO);
    [y1S, dy1S, y2S, dy2S, xS, dxS] = sttC{:};
    if (abs(y1S)>=pi/2 && abs(y2S)>=pi/2)
        fprintf('Control failure!\n'); break;
    end
    k = k+1;
```

```
    %% Double inverted pendulum visualization
    if (rem(k,20) == 0)
        subplot(1,2,1); DisplayDIP(xS, y1S, y2S, L1, L2); title(
    'DIP SIMO Control');
        subplot(1,2,2); DisplayDIP(xM, y1M, y2M, L1, L2); title(
    'DIP MIMO Control');
        pause(20*dt);
    end
end
```

The code **DesignGainMatrixSub.m** that encapsulates the function of gain matrix design for both controllable and partially controllable cases is given as follows.

DesignGainMatrixSub.m

```
% A: State transition matrix
% B: Control input matrix (for single-input)
% lambdaE: Eigenvalues of expected characteristic polynomial
function Kc = DesignGainMatrixSub(A, B, lambdaE)
    n = size(A,1); % State dimension
    %% Compute the partial controllability matrix
    CM = zeros(n); CM(:,1) = B;
    for k = 2:n
        CM(:,k) = A*CM(:,k-1);
    end
    m = rank(CM); PCM = zeros(n,m); PCM(:,1) = B;
    for k = 2:m
        PCM(:,k) = A*PCM(:,k-1);
    end
    if (m<n)
        PCMorg = PCM;
        %% Gaussian pivot elimination
        r = 1; rL = []; k = 1; % Linearly independent rows
        while (k<=size(PCM,2))
            pI = -1; pM = 0; % Pivot
            for i = k:size(PCM,2)
                if (abs(PCM(r,i)) > pM)
                    pM = abs(PCM(r,i)); pI = i;
                end
            end
            if (pM<10*eps) r = r+1; continue; end
            if (k ~= pI) PCM(r:end,[k, pI]) = PCM(r:end,[pI, k])
    ; end
            ck = (k+1):size(PCM,2);
            PCM(r:end,ck) = PCM(r:end,ck)
                            - PCM(r:end,k)*PCM(r,ck)/PCM(r,k);
            rL = [rL, r]; r = r+1; k = k+1;
        end
        PCM = PCMorg;
        %% Generate the complemented controllability matrix
        rC = ones(1,n); rC(rL) = 0; rC = find(rC == 1);
        % Complement matrix
        CompM = zeros(n,n-m); CompM(rC,:) = eye(n-m);
        % Complemented controllability matrix
```

```
        CompCM = [PCM, CompM];
    else
        CompCM = PCM; CompM = zeros(n,0);
    end

    %% Compute the state transform matrix
    aA = PCM\(A*PCM(:,m)); aA = [-aA', 1];
    M = zeros(m, m);
    for k = 1:m % Auxiliary matrix M
        M(k,1:(m+1-k)) = aA((k+1):(m+1));
    end
    P = [PCM*M, CompM]; % State transform matrix

    %% Design the transformed gain matrix
    aE = GetCharacteristicPolynomialCoefficients(lambdaE(1:m));
    transformedK = zeros(n,1);
    transformedK(1:m) = transpose(aE(1:m) - aA(1:m));
    %% Recover the gain matrix
    Kc = inv(transpose(P))*transformedK;

    %% Compute the characteristic polynomial coefficients from
    eigenvalues
    function a = GetCharacteristicPolynomialCoefficients(lambda)
        lambdaN = length(lambda);
        a = zeros(1,lambdaN+2);
        a(2) = 1;
        for k = 1:lambdaN
            aOld = a;
            for i = 2:(k+1)
                a(i) = aOld(i-1) + aOld(i)*(-lambda(k));
            end
            a(k+2) = 1;
        end
        % Characteristic polynomial coefficients a0, a1, ..., an
        a = a(2:end);
    end
end
```

The Matlab simulation code demonstrates comparison between a multiple-input-multiple-output version and a single-input-multiple-output version of double inverted pendulum control. Readers can try the code, observe differences between the two versions, and reflect on why such differences exist.

The variant of the double inverted pendulum control system has redundant control input part, namely the control input part of first inverted pendulum angular acceleration a_1. Although the redundant control input part plays no essential role in determining controllability of the double inverted pendulum, it can help *optimize* double inverted pendulum control in terms of certain control cost. Clarification of this optimization point is postponed to Sect. 7.2 in Chap. 7.

Cooperative longitudinal control of vehicle platooning
We also apply the methodology of control input decomposition and iterative design to handle cooperative longitudinal control of vehicle platooning which serves as another concrete example of multiple-input-multiple-output control, as illustrated in Fig. 6.4.

Denote the control system's state as $\boldsymbol{x} \equiv [\Delta x - \Delta x_{\mathrm{E}} \;\; v_1 - v_{\mathrm{E}} \;\; v_2 - v_{\mathrm{E}}]^{\mathrm{T}}$ which represents the difference between the absolute state[4] and certain expected absolute state, namely the absolute state error. Dynamics of the state $\boldsymbol{x}$ can be modelled by the state differential equation described in (6.52):

$$
\begin{aligned}
\frac{\mathrm{d}}{\mathrm{d}t}\boldsymbol{x} &= \begin{bmatrix} 0 & 1 & -1 \\ 0 & 0 & 0 \\ 0 & 0 & 0 \end{bmatrix} \boldsymbol{x} + \begin{bmatrix} 0 & 0 \\ 1 & 0 \\ 0 & 1 \end{bmatrix} \begin{bmatrix} a_1 \\ a_2 \end{bmatrix} \\
&\equiv \boldsymbol{A}\boldsymbol{x} + \boldsymbol{B}\boldsymbol{u} \equiv \boldsymbol{A}\boldsymbol{x} + \boldsymbol{B}_1 a_1 + \boldsymbol{B}_2 a_2
\end{aligned}
$$

The control input is the multiple-input of first vehicle acceleration a_1 and second vehicle acceleration a_2. Unlike the variant of the double inverted pendulum control system, the cooperative longitudinal control system has no redundant control input part. Both a_1 and a_2 need to be used to achieve the control objective of converging the state $\boldsymbol{x}$ namely the absolute state error to $\mathbf{0}$.

Decompose the control input and consider the control input part a_1. Compute the controllability matrix $\boldsymbol{C}_{\boldsymbol{A},\boldsymbol{B}_1}$ via (6.24) as

$$
\boldsymbol{C}_{\boldsymbol{A},\boldsymbol{B}_1} = [\boldsymbol{B}_1 \;\; \boldsymbol{A}\boldsymbol{B}_1 \;\; \boldsymbol{A}^2\boldsymbol{B}_1] = \begin{bmatrix} 0 & 1 & 0 \\ 1 & 0 & 0 \\ 0 & 0 & 0 \end{bmatrix},
$$

which is of rank $m = 2$. Then compute the partial controllability matrix $\bar{\boldsymbol{C}}_{\boldsymbol{A},\boldsymbol{B}_1}$ via (6.54) as

$$
\bar{\boldsymbol{C}}_{\boldsymbol{A},\boldsymbol{B}_1} = [\boldsymbol{B}_1 \;\; \boldsymbol{A}\boldsymbol{B}_1] = \begin{bmatrix} 0 & 1 \\ 1 & 0 \\ 0 & 0 \end{bmatrix},
$$

whose first two rows are linearly independent, so we can generate the complemented controllability matrix $\bar{\boldsymbol{C}}^{*}_{\boldsymbol{A},\boldsymbol{B}_1}$ via (6.55) as

$$
\bar{\boldsymbol{C}}^{*}_{\boldsymbol{A},\boldsymbol{B}_1} \equiv [\bar{\boldsymbol{C}}_{\boldsymbol{A},\boldsymbol{B}_1} \;\; \bar{\boldsymbol{C}}^{+}_{\boldsymbol{A},\boldsymbol{B}_1}] = \begin{bmatrix} 0 & 1 & 0 \\ 1 & 0 & 0 \\ 0 & 0 & 1 \end{bmatrix}.
$$

Compute the coefficients $\bar{a}_0, \bar{a}_1$ via (6.57) as

---

[4] The absolute state consists of the inter-vehicle position $\Delta x = x_1 - x_2$, the first vehicle velocity $v_1 \equiv \frac{\mathrm{d}x_1}{\mathrm{d}t}$, and the second vehicle velocity $v_2 \equiv \frac{\mathrm{d}x_2}{\mathrm{d}t}$.

$$\begin{bmatrix} -\bar{a}_0 \\ -\bar{a}_1 \end{bmatrix} = (\bar{\boldsymbol{C}}^{\mathrm{T}}_{\boldsymbol{A},\boldsymbol{B}_1} \bar{\boldsymbol{C}}_{\boldsymbol{A},\boldsymbol{B}_1})^{-1} \bar{\boldsymbol{C}}^{\mathrm{T}}_{\boldsymbol{A},\boldsymbol{B}_1} \boldsymbol{A}^m \boldsymbol{B}_1 = \begin{bmatrix} 0 \\ 0 \end{bmatrix}$$

such that $\boldsymbol{A}^2\boldsymbol{B}_1 + \bar{a}_1\boldsymbol{A}\boldsymbol{B}_1 + \bar{a}_0\boldsymbol{B}_1 = \boldsymbol{0}$. Generate the auxiliary matrix $\bar{\boldsymbol{M}}_{\boldsymbol{A},\boldsymbol{B}_1}$ via (6.58) as

$$\bar{\boldsymbol{M}}_{\boldsymbol{A},\boldsymbol{B}_1} = \begin{bmatrix} & 1 \\ 1 & \end{bmatrix}$$

and compute the state transform matrix $\bar{\boldsymbol{P}}_{\boldsymbol{A},\boldsymbol{B}_1}$ via (6.59) as

$$\bar{\boldsymbol{P}}_{\boldsymbol{A},\boldsymbol{B}_1} = \begin{bmatrix} \bar{\boldsymbol{C}}_{\boldsymbol{A},\boldsymbol{B}_1} \bar{\boldsymbol{M}}_{\boldsymbol{A},\boldsymbol{B}_1} & \bar{\boldsymbol{C}}^{+}_{\boldsymbol{A},\boldsymbol{B}_1} \end{bmatrix} = \begin{bmatrix} 1 & 0 & 0 \\ 0 & 1 & 0 \\ 0 & 0 & 1 \end{bmatrix}.$$

Note that the state transform $\bar{\boldsymbol{x}} = \bar{\boldsymbol{P}}^{-1}_{\boldsymbol{A},\boldsymbol{B}_1}\boldsymbol{x} = \boldsymbol{x}$ with the state transform matrix $\bar{\boldsymbol{P}}_{\boldsymbol{A},\boldsymbol{B}_1} = \boldsymbol{I}$ enables (6.52) to transform into the partial controllability canonical form for a_1 as

$$\frac{\mathrm{d}}{\mathrm{d}t}\bar{\boldsymbol{x}} = \begin{bmatrix} 0 & 1 & \\ 0 & 0 & * \\ \boldsymbol{0} & & * \end{bmatrix} \bar{\boldsymbol{x}} + \begin{bmatrix} 0 \\ 1 \\ \boldsymbol{0} \end{bmatrix} a_1 + \boldsymbol{B}_2 a_2,$$

which is exactly the original state differential equation form described in (6.52) and it seems that we have done nothing after above tour of procedures. In fact, the "useless" tour of procedures is only to help readers review control theory presented in Sect. 6.5.1.

Set all expected eigenvalues to -1 and hence the expected closed-loop characteristic polynomial is

$$\bar{C}_{\mathrm{E}}(s) = (s+1)^m = s^2 + 2s + 1.$$

Compute the transformed gain matrix via (6.43) as

$$\begin{cases} \bar{k}_1 = \bar{a}_0^{\mathrm{E}} - \bar{a}_0 = 1 - 0 = 1 \\ \bar{k}_2 = \bar{a}_1^{\mathrm{E}} - \bar{a}_1 = 2 - 0 = 2 \end{cases} \quad \bar{\boldsymbol{K}}_1 \equiv \begin{bmatrix} \bar{k}_1 \\ \bar{k}_2 \end{bmatrix} = \begin{bmatrix} 1 \\ 2 \end{bmatrix}.$$

Recover the gain matrix $\boldsymbol{K}_1$ of full-state feedback control law of a_1 from the transformed gain matrix $\bar{\boldsymbol{K}}_1$ as

$$\boldsymbol{K}_1 = \bar{\boldsymbol{P}}^{-\mathrm{T}}_{\boldsymbol{A},\boldsymbol{B}_1} \begin{bmatrix} \bar{\boldsymbol{K}}_1 \\ \boldsymbol{0} \end{bmatrix} = \begin{bmatrix} 1 \\ 2 \\ 0 \end{bmatrix}.$$

Substitute $a_1 = -\boldsymbol{K}_1^{\mathrm{T}}\boldsymbol{x}$ into (6.52) and obtain

$$\frac{\mathrm{d}}{\mathrm{d}t}\boldsymbol{x} = \boldsymbol{A}\boldsymbol{x} + \boldsymbol{B}_1 a_1 + \boldsymbol{B}_2 a_2 = (\boldsymbol{A} - \boldsymbol{B}_1\boldsymbol{K}_1^{\mathrm{T}})\boldsymbol{x} + \boldsymbol{B}_2 a_2$$
$$= \begin{bmatrix} 0 & 1 & -1 \\ -1 & -2 & 0 \\ 0 & 0 & 0 \end{bmatrix}\boldsymbol{x} + \begin{bmatrix} 0 \\ 0 \\ 1 \end{bmatrix} a_2.$$

Still set all expected eigenvalues to -1. Then compute the gain matrix of full-state feedback control law of a_2 with the state transition matrix $\boldsymbol{A} - \boldsymbol{B}_1\boldsymbol{K}_1^{\mathrm{T}}$ and the control input matrix $\boldsymbol{B}_2$ and finally obtain the gain matrix of full-state feedback control law of the holistic control input $\boldsymbol{u}$ as

$$\begin{bmatrix} a_1 \\ a_2 \end{bmatrix} = \begin{bmatrix} 1 & 2 & 0 \\ 0 & 0 & 1 \end{bmatrix}\boldsymbol{x}.$$

Matlab simulation code for complete demonstration of two-vehicle cooperative longitudinal control by multiple-input of first vehicle acceleration a_1 and second vehicle acceleration a_2 is given as follows.

CooperativeLongitudinalControlMIMO.m

```
%% Cooperative longitudinal control model for two-vehicle
    platooning
A = [0, 1, -1; 0, 0, 0; 0, 0, 0];
B = [0, 0; 1, 0; 0, 1];
%% Simulation preliminary configuration
dt = 0.001; % Numerical computation step
tSpan = 0:dt:10; % Simulation time span
x = 2; % Inter-vehicle position error
v1 = 1; % First vehicle velocity error
v2 = -1; % Second vehicle velocity error
stt = [x; v1; v2]; % Cooperative longitudinal state
% Record states in simulation
sttAll = zeros(length(stt), length(tSpan)); k = 0;

%% Design the gain matrix iteratively
lambdaE = [-1;-1;-1]; % Expected eigenvalues
sttK1 = DesignGainMatrixSub(A, B(:,1), lambdaE);
sttK2 = DesignGainMatrixSub(A-B(:,1)*sttK1', B(:,2), lambdaE);
fprintf('Gain matrix K: '); [sttK1'; sttK2']

k = 0; figure(1), set(gcf, 'Position', [100, 100, 600, 700]);
%% Simulation of cooperative longitudinal control
for t = tSpan
    %% Multiple-input-multiple-output control method
    acc1 = -sttK1'*stt;
    acc2 = -sttK2'*stt;

    %% Cooperative longitudinal dynamics
```

```
    x = x + (v1 - v2)*dt;
    v1 = v1 + acc1*dt;
    v2 = v2 + acc2*dt;
    stt = [x; v1; v2];
    k = k+1; sttAll(:,k) = stt;
end
subplot(3,1,1), plot(tSpan, sttAll(1,:), 'LineWidth', 2);
ylabel('Inter-Vehicle Position Error'); grid on;
subplot(3,1,2), plot(tSpan, sttAll(2,:), 'LineWidth', 2);
ylabel('1st vehicle velocity error'); grid on;
subplot(3,1,3), plot(tSpan, sttAll(3,:), 'LineWidth', 2);
xlabel('Time'); ylabel('2nd vehicle velocity error'); grid on;
```

As demonstrated above, the two concrete examples of multiple-input-multiple-output control both have double-input namely two-dimensional control input. Although the presented methodology of *control input decomposition and iterative design* is not limited to double-input-multiple-output control, readers may still be interested in a concrete example of multiple-input-multiple-output control with higher-dimensional control input. For this purpose, we can simply extend the already demonstrated cooperative longitudinal control of two-vehicle platooning to that of more vehicle platooning, for example, cooperative longitudinal control of four-vehicle platooning.

The absolute state of the four-vehicle cooperative longitudinal control system, denoted as $\boldsymbol{x}_{C4}$, consists of the first and second vehicle inter-vehicle position $\Delta x_1 = x_1 - x_2$, the second and third vehicle inter-vehicle position $\Delta x_2 = x_2 - x_3$, the third and fourth vehicle inter-vehicle position $\Delta x_3 = x_3 - x_4$, the first vehicle velocity $v_1 \equiv \frac{dx_1}{dt}$, the second vehicle velocity $v_2 \equiv \frac{dx_2}{dt}$, the third vehicle velocity $v_3 \equiv \frac{dx_3}{dt}$, and the fourth vehicle velocity $v_4 \equiv \frac{dx_4}{dt}$. Denote the control system's state as $\boldsymbol{x} \equiv \boldsymbol{x}_{C4} - \boldsymbol{x}_{C4}^{E}$ which represents the error between the absolute state $\boldsymbol{x}_{C4}$ and certain expected absolute state $\boldsymbol{x}_{C4}^{E}$ (note that the four vehicles share a common expected vehicle velocity). Dynamics of the state $\boldsymbol{x}$ can be modelled by a state differential equation as

$$
\begin{aligned}
\frac{d}{dt}\boldsymbol{x} &= \begin{bmatrix} 0 & 0 & 0 & 1 & -1 & 0 & 0 \\ 0 & 0 & 0 & 0 & 1 & -1 & 0 \\ 0 & 0 & 0 & 0 & 0 & 1 & -1 \\ 0 & 0 & 0 & 0 & 0 & 0 & 0 \\ 0 & 0 & 0 & 0 & 0 & 0 & 0 \\ 0 & 0 & 0 & 0 & 0 & 0 & 0 \\ 0 & 0 & 0 & 0 & 0 & 0 & 0 \end{bmatrix} \boldsymbol{x} + \begin{bmatrix} 0 & 0 & 0 & 0 \\ 0 & 0 & 0 & 0 \\ 0 & 0 & 0 & 0 \\ 1 & 0 & 0 & 0 \\ 0 & 1 & 0 & 0 \\ 0 & 0 & 1 & 0 \\ 0 & 0 & 0 & 1 \end{bmatrix} \begin{bmatrix} a_1 \\ a_2 \\ a_3 \\ a_4 \end{bmatrix} \\
&\equiv \boldsymbol{A}\boldsymbol{x} + \boldsymbol{B}\boldsymbol{u} \equiv \boldsymbol{A}\boldsymbol{x} + \boldsymbol{B}_1 a_1 + \boldsymbol{B}_2 a_2 + \boldsymbol{B}_3 a_3 + \boldsymbol{B}_4 a_4.
\end{aligned} \tag{6.80}
$$

The control input is the multiple-input of first vehicle acceleration a_1, second vehicle acceleration a_2, third vehicle acceleration a_3, and fourth vehicle acceleration a_4. The four-vehicle cooperative longitudinal control system aims at controlling the multiple-input of a_1, a_2, a_3, and a_4 simultaneously to converge the state $\boldsymbol{x}$ namely the absolute state error to $\boldsymbol{0}$.

Decompose the control input, consider the control input part a_1, and set all eigenvalues of the expected closed-loop characteristic polynomial to -1. Compute the state transform matrix $\bar{\boldsymbol{P}}_{\boldsymbol{A},\boldsymbol{B}_1}$ via (6.59), design the transformed gain matrix $\bar{\boldsymbol{K}}_1$ via (6.43), and recover the gain matrix $\boldsymbol{K}_1$ of full-state feedback control law of a_1, respectively as

$$\bar{\boldsymbol{P}}_{\boldsymbol{A},\boldsymbol{B}_1} = \begin{bmatrix} 1 & 0 & 0 & 0 & 0 & 0 & 0 \\ 0 & 0 & 1 & 0 & 0 & 0 & 0 \\ 0 & 0 & 0 & 1 & 0 & 0 & 0 \\ 0 & 1 & 0 & 0 & 0 & 0 & 0 \\ 0 & 0 & 0 & 0 & 1 & 0 & 0 \\ 0 & 0 & 0 & 0 & 0 & 1 & 0 \\ 0 & 0 & 0 & 0 & 0 & 0 & 1 \end{bmatrix}, \quad \begin{bmatrix} \bar{\boldsymbol{K}}_1 \\ \boldsymbol{0} \end{bmatrix} = \begin{bmatrix} 1 \\ 2 \\ 0 \\ 0 \\ 0 \\ 0 \\ 0 \end{bmatrix}, \quad \boldsymbol{K}_1 = \bar{\boldsymbol{P}}_{\boldsymbol{A},\boldsymbol{B}_1}^{-\mathrm{T}} \begin{bmatrix} \bar{\boldsymbol{K}}_1 \\ \boldsymbol{0} \end{bmatrix} = \begin{bmatrix} 1 \\ 0 \\ 0 \\ 2 \\ 0 \\ 0 \\ 0 \end{bmatrix}.$$

Substitute $a_1 = -\boldsymbol{K}_1^{\mathrm{T}}\boldsymbol{x}$ into (6.80) and obtain

$$\begin{aligned} \frac{\mathrm{d}}{\mathrm{d}t}\boldsymbol{x} &= \boldsymbol{A}\boldsymbol{x} + \boldsymbol{B}_1 a_1 + \boldsymbol{B}_2 a_2 + \boldsymbol{B}_3 a_3 + \boldsymbol{B}_4 a_4 \\ &= (\boldsymbol{A} - \boldsymbol{B}_1\boldsymbol{K}_1^{\mathrm{T}})\boldsymbol{x} + \boldsymbol{B}_2 a_2 + \boldsymbol{B}_3 a_3 + \boldsymbol{B}_4 a_4 \\ &\equiv \boldsymbol{A}_2\boldsymbol{x} + \boldsymbol{B}_2 a_2 + \boldsymbol{B}_3 a_3 + \boldsymbol{B}_4 a_4. \end{aligned}$$

Always set expected eigenvalues to -1 throughout all following gain matrix design process for this concrete example of cooperative longitudinal control. Then compute the gain matrix of full-state feedback control law of a_2 with the state transition matrix $\boldsymbol{A}_2 \equiv \boldsymbol{A} - \boldsymbol{B}_1\boldsymbol{K}_1^{\mathrm{T}}$ and the control input matrix $\boldsymbol{B}_2$ as

$$\boldsymbol{K}_2 = [0 \ \ 1 \ \ 0 \ \ 0 \ \ 2 \ \ 0 \ \ 0]^{\mathrm{T}}.$$

Further substitute $a_2 = -\boldsymbol{K}_2^{\mathrm{T}}\boldsymbol{x}$ into (6.80) and obtain

$$\frac{\mathrm{d}}{\mathrm{d}t}\boldsymbol{x} = (\boldsymbol{A}_2 - \boldsymbol{B}_2\boldsymbol{K}_2^{\mathrm{T}})\boldsymbol{x} + \boldsymbol{B}_3 a_3 + \boldsymbol{B}_4 a_4 \equiv \boldsymbol{A}_3\boldsymbol{x} + \boldsymbol{B}_3 a_3 + \boldsymbol{B}_4 a_4.$$

Compute the gain matrix of full-state feedback control law of a_3 with $\boldsymbol{A}_3 \equiv \boldsymbol{A}_2 - \boldsymbol{B}_2\boldsymbol{K}_2^{\mathrm{T}}$ and $\boldsymbol{B}_3$ as

$$\boldsymbol{K}_3 = [0 \ \ 0 \ \ 1 \ \ 0 \ \ 0 \ \ 2 \ \ 0]^{\mathrm{T}}.$$

Finally substitute $a_3 = -\boldsymbol{K}_3^{\mathrm{T}}\boldsymbol{x}$ into (6.80) and obtain

$$\frac{\mathrm{d}}{\mathrm{d}t}\boldsymbol{x} = (\boldsymbol{A}_3 - \boldsymbol{B}_3\boldsymbol{K}_3^{\mathrm{T}})\boldsymbol{x} + \boldsymbol{B}_4 a_4.$$

Compute the gain matrix of full-state feedback control law of a_4 with $\boldsymbol{A}_3 - \boldsymbol{B}_3\boldsymbol{K}_3^{\mathrm{T}}$ and $\boldsymbol{B}_4$ as

$$\boldsymbol{K}_4 = [0\ 0\ 0\ 0\ 0\ 0\ 1]^{\mathrm{T}}$$

and finally obtain the gain matrix of full-state feedback control law of the holistic control input $\boldsymbol{u}$ as

$$\begin{bmatrix} a_1 \\ a_2 \\ a_3 \\ a_4 \end{bmatrix} = \begin{bmatrix} 1\,0\,0\,2\,0\,0\,0 \\ 0\,1\,0\,0\,2\,0\,0 \\ 0\,0\,1\,0\,0\,2\,0 \\ 0\,0\,0\,0\,0\,0\,1 \end{bmatrix} \boldsymbol{x}.$$

Matlab simulation code for complete demonstration of four-vehicle cooperative longitudinal control by multiple-input of four vehicle accelerations a_1, a_2, a_3, and a_4 is given as follows.

CooperativeLongitudinalControlMIMO2.m

```
%% Cooperative longitudinal control model for four-vehicle
    platooning
A = [0, 0, 0, 1, -1, 0, 0; ...
     0, 0, 0, 0, 1, -1, 0; ...
     0, 0, 0, 0, 0, 1, -1; ...
     0, 0, 0, 0, 0, 0, 0; ...
     0, 0, 0, 0, 0, 0, 0; ...
     0, 0, 0, 0, 0, 0, 0; ...
     0, 0, 0, 0, 0, 0, 0];
B = [0, 0, 0, 0; 0, 0, 0, 0; 0, 0, 0, 0; ...
     1, 0, 0, 0; 0, 1, 0, 0; 0, 0, 1, 0; 0, 0, 0, 1];
%% Simulation preliminary configuration
dt = 0.001; % Numerical computation step
tSpan = 0:dt:15; % Simulation time span
x1 = 1; % First and second vehicle inter-vehicle position error
x2 = -1; % Second and third vehicle inter-vehicle position error
x3 = 1; % Third and fourth vehicle inter-vehicle position error
v1 = 1; % First vehicle velocity error
v2 = -1; % Second vehicle velocity error
v3 = 1; % Third vehicle velocity error
v4 = -1; % Fourth vehicle velocity error
stt = [x1; x2; x3; v1; v2; v3; v4]; % Cooperative longitudinal
    state
% Record states in simulation
sttAll = zeros(length(stt), length(tSpan)); k = 0;

%% Design the gain matrix iteratively
lambdaE = -ones(1,7); % Expected eigenvalues
sttK1 = DesignGainMatrixSub(A, B(:,1), lambdaE);
A2 = A-B(:,1)*sttK1';
sttK2 = DesignGainMatrixSub(A2, B(:,2), lambdaE);
A3 = A2-B(:,2)*sttK2';
```

```
sttK3 = DesignGainMatrixSub(A3, B(:,3), lambdaE);
A4 = A3-B(:,3)*sttK3';
sttK4 = DesignGainMatrixSub(A4, B(:,4), lambdaE);
fprintf('Gain matrix K: '); [sttK1'; sttK2'; sttK3'; sttK4']

k = 0; figure(1), set(gcf, 'Position', [100, 0, 1100, 800]);
%% Simulation of cooperative longitudinal control
for t = tSpan
    %% Multiple-input-multiple-output control method
    acc1 = -sttK1'*stt;
    acc2 = -sttK2'*stt;
    acc3 = -sttK3'*stt;
    acc4 = -sttK4'*stt;

    %% Cooperative longitudinal dynamics
    x1 = x1 + (v1 - v2)*dt;
    x2 = x2 + (v2 - v3)*dt;
    x3 = x3 + (v3 - v4)*dt;
    v1 = v1 + acc1*dt;
    v2 = v2 + acc2*dt;
    v3 = v3 + acc3*dt;
    v4 = v4 + acc4*dt;
    stt = [x1; x2; x3; v1; v2; v3; v4];
    k = k+1; sttAll(:,k) = stt;
end
subplot(4,2,3), plot(tSpan, sttAll(1,:), 'LineWidth', 2);
ylabel('\Delta x_1 Error'); grid on;
subplot(4,2,5), plot(tSpan, sttAll(2,:), 'LineWidth', 2);
ylabel('\Delta x_2 Error'); grid on;
subplot(4,2,7), plot(tSpan, sttAll(3,:), 'LineWidth', 2);
xlabel('Time'); ylabel('\Delta x_3 Error'); grid on;
subplot(4,2,2), plot(tSpan, sttAll(4,:), 'LineWidth', 2);
ylabel('v_1 Error'); grid on;
subplot(4,2,4), plot(tSpan, sttAll(5,:), 'LineWidth', 2);
ylabel('v_2 Error'); grid on;
subplot(4,2,6), plot(tSpan, sttAll(6,:), 'LineWidth', 2);
ylabel('v_3 Error'); grid on;
subplot(4,2,8), plot(tSpan, sttAll(7,:), 'LineWidth', 2);
xlabel('Time'); ylabel('v_4 Error'); grid on;
```

Readers may try the Matlab simulation code and will observe that the designed full-state feedback control law of multiple-input of four vehicle accelerations a_1, a_2, a_3, and a_4 can indeed achieve the cooperative longitudinal control objective, namely to converge the absolute state error to $\mathbf{0}$.

After demonstration of three concrete examples of multiple-input-multiple-output control, we believe readers have better understanding of the methodology of control input decomposition and iterative design and have better knowledge of how to apply the methodology to handle multiple-input-multiple-output control.

We have finished presentation of fundamentals of modern control theory, namely state-space analysis especially linear state-space analysis and full-state feedback control for both single-input-multiple-output and multiple-input-multiple-output control problems. In the remaining part of Chap. 6, we will briefly present a topic of control theory that does not involve control itself but is by no means trivial to control systems.

6.6 State Estimation and Observability

By so far, we have assumed by default that the control system's state is known. In practical applications, a question arises naturally: *how do we know the state*? Readers may think this is a trivial issue, because it seems that we can simply install suitable sensors to measure the complete state directly namely to measure all elements of the state directly.

However, above question is not a trivial issue at all in practical applications. Reasons are two-folds: First, even suppose that we do have suitable sensors to measure the complete state directly, can we have some way to know the state better than raw sensor measurements of the state? Second, we may not have suitable sensors to measure the complete state directly but only have sensors to measure the state partially in practical applications. For example, in GPS (global positioning system such as Bei Dou) based vehicle navigation, the GPS sensor only measures the vehicle position which is part of the vehicle state, whereas the vehicle orientation is the vehicle state part that is not directly measurable. For another example, in landmark ranging based robot localization, the landmark ranging module only measures the distance between the robot and certain registered landmark at each control period and apparently has no ability to measure the complete robot state directly. Then can we have some way to completely know the state with raw sensor measurements that only reflect the state partially?

These two aspects of practical need, namely to know the state better and to know the state completely, involve systematic knowledge of **state estimation** or **recursive state estimation** (Li 2022) especially in the context of control systems which are normally intended for real-time applications.

Given a control system that adopts linear state-space modelling described by (6.5), with its state denoted as $\boldsymbol{x}$ and its control input to the target process denoted as $\boldsymbol{u}$:

$$\frac{\mathrm{d}}{\mathrm{d}t}\boldsymbol{x} = \boldsymbol{A}\boldsymbol{x} + \boldsymbol{B}\boldsymbol{u}.$$

Suppose that its **linearized measurement model** is

$$z = \boldsymbol{H}\boldsymbol{x}, \tag{6.81}$$

where z denotes raw sensor measurement of the state $\boldsymbol{x}$ and $\boldsymbol{H}$ denotes the **measurement matrix**. State estimation aims at inferring or revealing the complete state with measurements available in finite time.

A control system's target process is **observable** if its state $\boldsymbol{x}$ can be inferred with its available measurements, be the state inferred directly or indirectly. To better understand **observability**, it is worth clarifying the difference between *unobservability* and *stochastic uncertainty*. In practical applications, the estimate of any entity suffers from stochastic uncertainty that exists universally, and in this sense we cannot have absolute knowledge of any entity. However, this does not mean that any entity is unobservable.

Readers may resort to an example to understand the difference between unobservability and stochastic uncertainty. Imagine two vehicles that are both located at the origin initially: the first vehicle is equipped with an speedometer, whereas the second vehicle is not equipped with any motion sensor. For the first vehicle, although the equipped speedometer can never be absolutely accurate, we at least have a way to infer the vehicle speed within certain error tolerance associated with the odometer. Despite existence of stochastic uncertainty, we still treat the speed of the first vehicle as observable. Besides, we can also treat the position of the first vehicle as observable because we can infer the position indirectly by integral of the speed, though the estimated position also suffers from certain stochastic uncertainty inevitably. In contrast, for the second vehicle, both its speed and position are unobservable, because we do not have any way to infer its speed and position, neither directly nor indirectly.

From this example we can see that whether a control system's target process is observable depends on whether there is a way to infer its state, regardless of the fact that the estimated state suffers from stochastic uncertainty. Therefore, to analyse observability of the target process, we can neglect stochastic uncertainty (though concrete state estimation methods normally need to take stochastic uncertainty into account).

When the measurement matrix $\boldsymbol{H}$ is of full rank in terms of row vectors or in other words the rank of $\boldsymbol{H}$ equals the state dimension n, then we can take measurement transform as

$$\bar{z} = \boldsymbol{H}^{-1}z \quad \text{or} \quad \bar{z} = (\boldsymbol{H}^{\mathrm{T}}\boldsymbol{H})^{-1}\boldsymbol{H}^{\mathrm{T}}z$$

and transform the measurement model into

$$\bar{z} = \boldsymbol{x},$$

which implies that the sensors can measure the complete state directly and hence the target process is apparently observable.

On the other hand, when the measurement matrix $\boldsymbol{H}$ is rank-deficient, we need some systematic method to theoretically determine observability of the target process. Consider the solution of (6.5), which is described in (6.20):

$$\boldsymbol{x} = \mathrm{e}^{At}\boldsymbol{x}_0 + \int_0^t \mathrm{e}^{A(t-\tau)}\boldsymbol{B}\boldsymbol{u}(\tau)\mathrm{d}\tau.$$

The solution described in (6.20) implies that once the initial state $\boldsymbol{x}_0$ is known, we can infer the state at any time via (6.20). Transform (6.20) into

$$\boldsymbol{x}_0 = \mathrm{e}^{-\boldsymbol{A}t}\left(\boldsymbol{x} - \int_0^t \mathrm{e}^{\boldsymbol{A}(t-\tau)}\boldsymbol{B}\boldsymbol{u}(\tau)\mathrm{d}\tau\right),$$

which further implies that we can also recover the initial state from the state at any time. Therefore, observability of the target process is equivalent to observability of the initial state $\boldsymbol{x}_0$. In other words, whether a control system's target process is observable depends on whether we can infer its initial state $\boldsymbol{x}_0$ with measurements available in finite time.

Substitute (6.20) into (6.81) and obtain

$$\begin{aligned} z = \boldsymbol{H}\boldsymbol{x} &= \boldsymbol{H}\mathrm{e}^{\boldsymbol{A}t}\boldsymbol{x}_0 + \boldsymbol{H}\int_0^t \mathrm{e}^{\boldsymbol{A}(t-\tau)}\boldsymbol{B}\boldsymbol{u}(\tau)\mathrm{d}\tau \\ \Longleftrightarrow \boldsymbol{H}\mathrm{e}^{\boldsymbol{A}t}\boldsymbol{x}_0 &= z - \boldsymbol{H}\int_0^t \mathrm{e}^{\boldsymbol{A}(t-\tau)}\boldsymbol{B}\boldsymbol{u}(\tau)\mathrm{d}\tau \equiv \bar{z}. \end{aligned} \tag{6.82}$$

For each time t, (6.82) provides a linear equation in terms of the initial state $\boldsymbol{x}_0$. Given t in a time interval $[0, T]$, an infinite number of linear equations in terms of the initial state $\boldsymbol{x}_0$ can be established. Whether we can infer the initial state $\boldsymbol{x}_0$ with measurements available in finite time depends on whether we can always solve the initial state $\boldsymbol{x}_0$ according to the linear equations established by (6.82) for certain finite time interval $[0, T]$, which further depends on whether $\langle \boldsymbol{H}\mathrm{e}^{\boldsymbol{A}t}\rangle_R$ namely the span of $\boldsymbol{H}\mathrm{e}^{\boldsymbol{A}t}$ in terms of row vectors for certain finite time interval $[0, T]$ is of full rank or in other words whether the rank of $\langle \boldsymbol{H}\mathrm{e}^{\boldsymbol{A}t}\rangle_R$ for certain finite time interval $[0, T]$ equals the state dimension n.

$$\boldsymbol{H}\mathrm{e}^{\boldsymbol{A}t} = \boldsymbol{H}\sum_{k=0}^{\infty}\boldsymbol{A}^k\frac{t^k}{k!} = \boldsymbol{U}^{\mathrm{T}}\boldsymbol{O}_{\boldsymbol{H},\boldsymbol{A}},$$

where

$$\boldsymbol{U} = \begin{bmatrix} \boldsymbol{I} \\ t\boldsymbol{I} \\ \frac{t^2}{2!}\boldsymbol{I} \\ \frac{t^3}{3!}\boldsymbol{I} \\ \vdots \end{bmatrix}, \quad \boldsymbol{O}_{\boldsymbol{H},\boldsymbol{A}} = \begin{bmatrix} \boldsymbol{H} \\ \boldsymbol{H}\boldsymbol{A} \\ \boldsymbol{H}\boldsymbol{A}^2 \\ \boldsymbol{H}\boldsymbol{A}^3 \\ \vdots \end{bmatrix}.$$

We have

$$\langle \boldsymbol{H}\mathrm{e}^{\boldsymbol{A}t}\rangle_R = \langle \boldsymbol{U}^{\mathrm{T}}\rangle_R \boldsymbol{O}_{H,A}.$$

Since $1, t, \frac{t^2}{2!}, \frac{t^3}{3!}, \cdots$ are linearly independent, $\langle \boldsymbol{U}^{\mathrm{T}}\rangle_R$ for any real-value interval is of full rank[5] and

$$\text{rank of } \langle \boldsymbol{H}\mathrm{e}^{\boldsymbol{A}t}\rangle_R = \text{rank of } \boldsymbol{O}_{H,A}.$$

Therefore, the necessary and sufficient condition for the target process to be observable is that $\boldsymbol{O}_{H,A}$ is of full rank in terms of row vectors or in other words the rank of $\boldsymbol{O}_{H,A}$ equals the state dimension n. The infinite matrix $\boldsymbol{O}_{H,A}$ is called the **observability matrix** of the control system.

The *Hamilton-Cayley theorem* described by (6.23) implies each element after $\boldsymbol{H}\boldsymbol{A}^{n-1}$ in the observability matrix $\boldsymbol{O}_{H,A}$ namely each $\boldsymbol{H}\boldsymbol{A}^k$ for $k \geqslant n$ can be transformed into a linear combination of $\boldsymbol{H}, \boldsymbol{H}\boldsymbol{A}, \boldsymbol{H}\boldsymbol{A}^2, \cdots, \boldsymbol{H}\boldsymbol{A}^{n-1}$. So we have

$$\text{rank of} \begin{bmatrix} \boldsymbol{H} \\ \boldsymbol{H}\boldsymbol{A} \\ \boldsymbol{H}\boldsymbol{A}^2 \\ \boldsymbol{H}\boldsymbol{A}^3 \\ \vdots \end{bmatrix} = \text{rank of} \begin{bmatrix} \boldsymbol{H} \\ \boldsymbol{H}\boldsymbol{A} \\ \boldsymbol{H}\boldsymbol{A}^2 \\ \vdots \\ \boldsymbol{H}\boldsymbol{A}^{n-1} \end{bmatrix}$$

and we can define the **observability matrix** simply as

$$\boldsymbol{O}_{H,A} = \begin{bmatrix} \boldsymbol{H} \\ \boldsymbol{H}\boldsymbol{A} \\ \boldsymbol{H}\boldsymbol{A}^2 \\ \vdots \\ \boldsymbol{H}\boldsymbol{A}^{n-1} \end{bmatrix}. \tag{6.83}$$

If the observability matrix $\boldsymbol{O}_{H,A}$ defined in (6.83) is rank-deficient in terms of row vectors, then the target process is unobservable. If the observability matrix $\boldsymbol{O}_{H,A}$ is of full rank in terms of row vectors, then the target process is observable.

Control system observability criterion: *For a linear control system, if its observability matrix is rank-deficient in terms of row vectors, then its target process is unobservable. If its observability matrix is of full rank in terms of row vectors, then its target process is observable.*

We apply the observability criterion to determine observability of the double inverted pendulum that adopts linear state-space modelling described by (6.3)

[5] This rank is an abstract "number" that equals the total "number" of natural numbers.

or (6.51). Let $m_1 = 1, m_2 = 1, L_1 = 1, L_2 = 1$, and $g = 10$. Then the state transition matrix $\boldsymbol{A}$ is

$$\boldsymbol{A} = \begin{bmatrix} 0 & 1 & 0 & 0 & 0 & 0 \\ 20 & 0 & -10 & 0 & 0 & 0 \\ 0 & 0 & 0 & 1 & 0 & 0 \\ -20 & 0 & 20 & 0 & 0 & 0 \\ 0 & 0 & 0 & 0 & 0 & 1 \\ 0 & 0 & 0 & 0 & 0 & 0 \end{bmatrix}.$$

Suppose that we can measure the first inverted pendulum angle θ_1 and the cart position x directly. The measurement model is

$$z = \begin{bmatrix} 1 & 0 & 0 & 0 & 0 & 0 \\ 0 & 0 & 0 & 0 & 1 & 0 \end{bmatrix} \boldsymbol{x} \equiv \boldsymbol{H}\boldsymbol{x}.$$

Compute the observability matrix via (6.83). Matlab code for such purpose is given as follows.

ObservabilityMatrixDIP.m

```
%% Double inverted pendulum parameters
m1 = 1; m2 = 1; L1 = 1; L2 = 1; g = 10;
A = [0, 1, 0, 0, 0, 0; ...
    (m1+m2)*g/(m1*L1), 0, -m2*g/(m1*L1), 0, 0, 0; ...
    0, 0, 0, 1, 0, 0; ...
    -(m1+m2)*g/(m1*L2), 0, (m1+m2)*g/(m1*L2), 0, 0, 0; ...
    0, 0, 0, 0, 0, 1; ...
    0, 0, 0, 0, 0, 0];
H = [1, 0, 0, 0, 0, 0; 0, 0, 0, 0, 1, 0];
n = size(A,1); % State dimension
nz = size(H,1); % Measurement dimension

OM = zeros(n*nz, n); OM(1:nz,:) = H;
for k = 2:n
    OM((k*nz-nz+1):(k*nz),:) = OM((k*nz-2*nz+1):(k*nz-nz),:)*A;
end % OM = [H; H*A; H*A^2; H*A^3; H*A^4; H*A^5];
fprintf('Observability matrix: '); OM

%% Check if the observability matrix is of full rank
if (rank(OM) == n)
    fprintf('The double inverted pendulum is observable\n');
else
    fprintf('The double inverted pendulum is unobservable\n');
end
```

After trying above Matlab code, readers will find that the double inverted pendulum is indeed observable given measurements of first inverted pendulum angle θ_1 and cart position x. In other words, the complete double inverted pendulum state $\boldsymbol{x}$ can be inferred or revealed with measurements only of first inverted pendulum angle θ_1 and cart position x.

As already mentioned, state estimation especially recursive state estimation is by no means trivial to control systems in practical applications, yet state estimation does not involve control itself and is beyond the scope of this book. So we refrain from presenting more knowledge on state estimation such as representative estimation methods. Readers may refer to (Li 2022) for relevant fundamentals.

References

J. Ackermann. *Sampled-data control systems: analysis and synthesis, robust system design*. Springer-Verlag Berlin, 1985.

G.H. Golub, C.F. Van Loan. *Matrix computations*. Johns Hopkins University Press, 1996.

R.A. Horn, C.R. Johnson. *Matrix analysis*. Cambridge University Press, 2012.

H. Li. *Fundamentals and applications of recursive estimation theory*. Shanghai Jiao Tong University Press, 2022.

A. Quarteroni, R. Sacco, F. Saleri. *Numerical mathematics*. Springer, 2007.

Chapter 7
Advanced Control

Following fundamentals of state-space analysis and control presented in Chap. 6, we briefly present a number of advanced topics of modern control theory in Chap. 7.

7.1 Sliding Mode Control

We start with a representative control problem which readers are now familiar with, namely the double inverted pendulum control problem. Readers may think this problem has already been handled in Sect. 6.4 in Chap. 6. If the initial deviation of the cart position x is not large, the double inverted pendulum control problem has been indeed handled by so far. However, what if the initial deviation of the cart position x is large?

We suggest that readers try the Matlab simulation code **DoubleInvertedPendulum3.m** presented in Sect. 6.4 in Chap. 6 and see what would happen if the initial cart position x is set to a large value, for example, as follows.

DoubleInvertedPendulum3.m

```
%% Double inverted pendulum parameters
m1 = 1; m2 = 1; L1 = 1; L2 = 1; g = 10;
%% Simulation preliminary configuration
dt = 0.001; % Numerical computation step
tSpan = 0:dt:3; % Simulation time span
x = 20; dx = 0; % Cart position and its velocity
% Inverted pendulum angle theta-1 and its angular velocity
y1 = 0.2; dy1 = 0;
% Inverted pendulum angle theta-2 and its angular velocity
y2 = 0; dy2 = 0;
stt = [y1; dy1; y2; dy2; x; dx]; % Double inverted pendulum
    state
% Record states in simulation
```


H. Li, *Control Theory for Practical Applications*,
https://doi.org/10.1007/978-981-97-5008-5_7

```
sttAll = zeros(length(stt), length(tSpan)); k = 0;
% Expected equilibrium status
xExpected = 0; y1Expected = 0; y2Expected = 0;
SimConfig = [m1, m2, L1, L2, g, dt];

%% Design the gain matrix
A = [0, 1, 0, 0, 0, 0; ...
    (m1+m2)*g/(m1*L1), 0, -m2*g/(m1*L1), 0, 0, 0; ...
    0, 0, 0, 1, 0, 0; ...
    -(m1+m2)*g/(m1*L2), 0, (m1+m2)*g/(m1*L2), 0, 0, 0; ...
    0, 0, 0, 0, 0, 1; ...
    0, 0, 0, 0, 0, 0];
B = [0; -1/L1; 0; 0; 0; 1];
lambdaE = [-4;-4;-4;-4;-4;-4]; % Expected eigenvalues
sttK = DesignGainMatrixSub(A, B(:,1), lambdaE);
fprintf('Gain matrix K: '); sttK'

%% Simulation of double inverted pendulum control
for t = tSpan
    %% Control method
    acc = -sttK'*stt; % Full-state feedback control of the cart
    acceleration

    %% Double inverted pendulum dynamics
    stt = DynamicsDIP(SimConfig, stt, acc);
    sttC = num2cell(stt); [y1, dy1, y2, dy2, x, dx] = sttC{:};
    if (abs(y1)>=pi/2 && abs(y2)>=pi/2)
        fprintf('Control failure!\n'); break;
    end
    k = k+1; sttAll(:,k) = stt;
    %% Double inverted pendulum visualization
    if (rem(k,20) == 0)
        DisplayDIP2(x, y1, y2, L1, L2); pause(20*dt);
    end
end
```

The visualization code **DisplayDIP2.m** which is a modified version of **DisplayDIP.m** is given as follows. It better visualizes dynamics of the double inverted pendulum in a wide range of cart position.

DisplayDIP2.m

```
%% Double inverted pendulum visualization
function DisplayDIP2(x, y1, y2, L1, L2)
    Mx = x; My = 0;
    m1x = Mx + L1*sin(y1); m1y = My + L1*cos(y1);
    m2x = m1x + L2*sin(y2); m2y = m1y + L2*cos(y2);
    figure(1); cla; hold on;
    plot(Mx, My, 'or', 'MarkerSize', 16, 'LineWidth', 2);
    plot(m1x, m1y, 'ob', 'MarkerSize', 8, 'LineWidth', 2);
    plot(m2x, m2y, 'ob', 'MarkerSize', 8, 'LineWidth', 2);
    line([Mx,m1x], [My,m1y], 'Color', 'b', 'LineWidth', 2);
    line([m1x,m2x], [m1y,m2y], 'Color', 'b', 'LineWidth', 2);
```

```
    axis equal; grid on; ylim([0, 3]);
    if (abs(x)<2) xlim([-2,2]); else xlim([x-2, x+2]); end; hold
     off;
end
```

After trials with the Matlab simulation code, readers will find that the adopted full-state feedback control method, which worked previously when the initial deviation of the cart position x was not large, no longer works when the initial deviation of the cart position x is large. Why does such control phenomenon take place?

Reasons are two-folds: First, the state differential equation described in (6.1) that models double inverted pendulum dynamics is nonlinear and hence its linearly-approximated version described in (6.3) based on which the adopted full-state feedback control method is designed is not always valid for configuration space of the double inverted pendulum state. Second, the adopted full-state feedback control method only focuses on converging the final state to the expected state as soon as possible, without considering intermediate state evolution during the control process. When the initial deviation of the cart position is large, the adopted full-state feedback control method tends to generate drastic control input of cart acceleration, without considering how the double inverted pendulum state will evolve accordingly. Consequently, drastic control input of cart acceleration causes the double inverted pendulum state to evolve into configuration space for which the linearly-approximated model version described in (6.3) is no longer valid.

It may seem to be a frustrating fact to readers: after a long journey in Chap. 6, we have finally obtained a complete realization of the full-state feedback control method for double inverted pendulum control, but the obtained full-state feedback control method works only for a small portion of cart position in the state space (namely if the initial deviation of the cart position x is close to zero), whereas it does not work for much larger portion of cart position in the state space.

In fact, for the double inverted pendulum control problem given arbitrary initial cart position, the full-state feedback control method still works. We only need to improve the original full-state feedback control method moderately by incorporating the basic spirit of **sliding mode control** (Utkin 1992), which is to *force the state to evolve onto certain designed sliding mode manifold or surface and then evolve towards the expected state in sliding mode on the manifold.*[1] Figuratively speaking, this is like to drive the state to slide along certain sliding path on the designed manifold until the state arrives at the destination, as illustrated in Fig. 7.1.

For above problem, we can intuitively set the sliding mode manifold as

$$[\boldsymbol{I}_4 \;\; \boldsymbol{0}]\boldsymbol{x} = \boldsymbol{0}$$

[1] Originally, sliding mode control theory aims at control systems with *discontinuous dynamics*—Existence of discontinuous dynamics may be due to a variety of factors, for example, imperfections of switching devices (time delay, dead zones, hysteresis loops), model order mismatch, state feedback delay, piecewise linear approximation, etc.—Yet the basic spirit of sliding mode control is applicable to control systems with or without discontinuous dynamics.

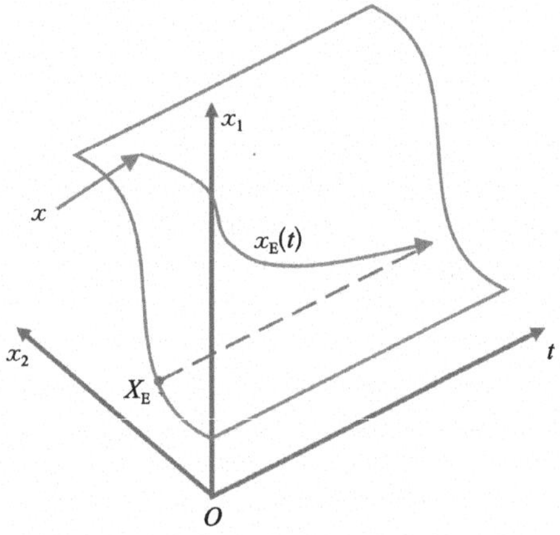

Fig. 7.1 Sliding mode control

namely

$$\left[\theta_1 \ \frac{d\theta_1}{dt} \ \theta_2 \ \frac{d\theta_2}{dt}\right]^T = [0 \ 0 \ 0 \ 0]^T$$

and design a sliding mode as[2]

$$\boldsymbol{x}_E = [0 \ 0 \ 0 \ 0 \ \text{sign}(x_0) \max\{|x_0| - s_v t, 0\} \ 0]^T, \tag{7.1}$$

where s_v is a constant parameter that tunes the sliding velocity. For example, set $s_v = 8$. Then the full-state feedback control law of cart acceleration a is

$$a = -\boldsymbol{K}^T(\boldsymbol{x} - \boldsymbol{x}_E),$$

where $\boldsymbol{x}_E$ is no longer a constant but a time-variant function described in (7.1).

Matlab simulation code for demonstrating sliding mode control of the double inverted pendulum is given as follows.

DoubleInvertedPendulum4.m

```
%% Double inverted pendulum parameters
m1 = 1; m2 = 1; L1 = 1; L2 = 1; g = 10;
%% Simulation preliminary configuration
dt = 0.001; % Numerical computation step
tSpan = 0:dt:8; % Simulation time span
x = 20; dx = 0; % Cart position and its velocity
% Inverted pendulum angle theta-1 and its angular velocity
y1 = 0.2; dy1 = 0;
```

[2] For more complicated control systems especially control systems with discontinuous dynamics, we usually cannot resort to intuition to determine sliding mode manifolds. Instead, we need systematic analysis and computation to determine sliding mode manifolds.

```
% Inverted pendulum angle theta-2 and its angular velocity
y2 = 0; dy2 = 0;
stt = [y1; dy1; y2; dy2; x; dx]; % Double inverted pendulum
    state
% Record states in simulation
sttAll = zeros(length(stt), length(tSpan)); k = 0;
% Expected equilibrium status
xExpected = 0; y1Expected = 0; y2Expected = 0;
SimConfig = [m1, m2, L1, L2, g, dt];

%% Design the gain matrix
A = [0, 1, 0, 0, 0, 0; ...
    (m1+m2)*g/(m1*L1), 0, -m2*g/(m1*L1), 0, 0, 0; ...
    0, 0, 0, 1, 0, 0; ...
    -(m1+m2)*g/(m1*L2), 0, (m1+m2)*g/(m1*L2), 0, 0, 0; ...
    0, 0, 0, 0, 0, 1; ...
    0, 0, 0, 0, 0, 0];
B = [0; -1/L1; 0; 0; 0; 1];
lambdaE = [-4;-4;-4;-4;-4;-4]; % Expected eigenvalues
sttK = DesignGainMatrixSub(A, B, lambdaE);
fprintf('Gain matrix K: '); sttK'

sttE = [0; 0; 0; 0; x; 0]; % Sliding mode initialization
%% Simulation of double inverted pendulum control
for t = tSpan
    %% Control method
    if (sttE(5)>0) sttE(5) = max(sttE(5) - 8*dt, 0); % Sliding
    mode design
    else sttE(5) = min(sttE(5) + 8*dt, 0); end
    acc = -sttK'*(stt-sttE); % Full-state feedback control

    %% Double inverted pendulum dynamics
    stt = DynamicsDIP(SimConfig, stt, acc);
    sttC = num2cell(stt); [y1, dy1, y2, dy2, x, dx] = sttC{:};
    if (abs(y1)>=pi/2 && abs(y2)>=pi/2)
        fprintf('Control failure!\n'); break;
    end
    k = k+1; sttAll(:,k) = stt;
    %% Double inverted pendulum visualization
    if (rem(k,20) == 0)
        DisplayDIP2(x, y1, y2, L1, L2); pause(20*dt);
    end
end
```

The performance of double inverted pendulum sliding mode control is demonstrated in Fig. 7.2.

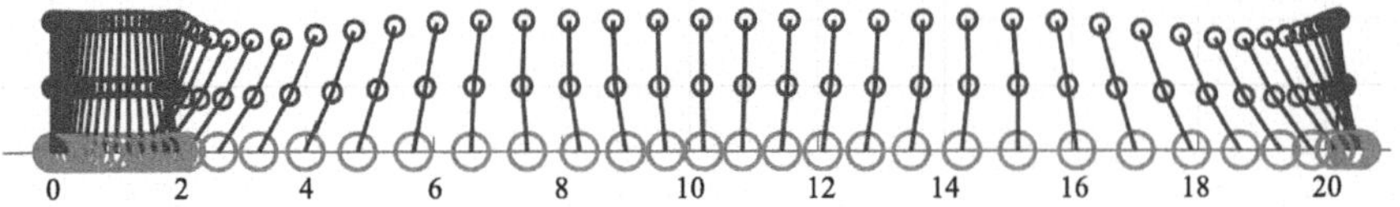

Fig. 7.2 Double inverted pendulum sliding mode control

7.2 Optimal Control

7.2.1 Minimizing Control Cost

We have demonstrated that the full-state feedback control method adopted in Chap. 6 does not work if the initial deviation of the cart position is no longer close to zero. We have explained why the adopted full-state feedback control method incurs the control failure: It only focuses on converging the final state to the expected state as soon as possible, *without considering intermediate state evolution during the control process*. Consequently, it may generate drastic control input and cause the state to evolve into configuration space where the essential modelling assumption is violated.

How to take *intermediate state evolution during the control process* into account? One methodology is to regulate intermediate state evolution indirectly by incorporating the spirit of sliding mode control, as already demonstrated in Sect. 7.1. Another methodology is to regulate intermediate state evolution directly by minimizing certain **control cost** of the control process, which may include state related cost as well as control input related cost. This second methodology is called the **optimal control**.

Given a control system that adopts generic state-space modelling described by (6.2), with its state denoted as $\boldsymbol{x}$ and its control input to the target process denoted as $\boldsymbol{u}$:

$$\frac{\mathrm{d}}{\mathrm{d}t}\boldsymbol{x} = f(\boldsymbol{x}, \boldsymbol{u}).$$

Throughout this book, we have been denoting any time-variant variable at generic time t simply by the variable notation itself without subscript time index, parenthesis time index, or other kind of explicit time index. For example, the state notation $\boldsymbol{x}$ refers to the time-variant state at generic time t, namely $\boldsymbol{x}_t$ or $\boldsymbol{x}(t)$ if we want to highlight the time index t. Omitting explicit time index is simply for expression conciseness yet without causing confusion.

On the other hand, when we intend to denote any time-variant variable during a generic time interval from t_1 to t_2, we will always add the subscript time index "$_{t_1:t_2}$" or parenthesis time index "$(t_1 : t_2)$" explicitly. For example, $\boldsymbol{x}_{t_1:t_2}$ or $\boldsymbol{x}(t_1 : t_2)$ denotes the time-variant state from t_1 to t_2. Note that $\boldsymbol{x}$ can be treated as a function in terms of time t, so $\boldsymbol{x}_{t_1:t_2}$ or $\boldsymbol{x}(t_1 : t_2)$ can be regarded as a segment of the function $\boldsymbol{x}$ defined on the time interval from t_1 to t_2. We also denote the subscript time index

"$_{0:\infty}$" and the parenthesis time index "$(0:\infty)$" simply as the subscript time index "$_{:}$" and the parenthesis time index "$(:)$" respectively, so $\boldsymbol{x}_{:}$ and $\boldsymbol{x}(:)$ actually denote $\boldsymbol{x}_{0:\infty}$ and $\boldsymbol{x}(0:\infty)$ respectively.

Suppose that the control system's target process is observable and hence its initial state $\boldsymbol{x}_0$ can be inferred. Once $\boldsymbol{u}_{0:t}$ is given, $\boldsymbol{x}$ namely $\boldsymbol{x}_t$ or $\boldsymbol{x}(t)$ can be derived with $\boldsymbol{u}_{0:t}$ via the system model described in (6.2). So we can *treat* $\boldsymbol{x}_{:}$ *as a functional in terms of* $\boldsymbol{u}_{:}$.

Further suppose that we have designed certain **control cost functional** in terms of the state and the control input during the control process, denoted as the functional $c(\boldsymbol{x}_{:}, \boldsymbol{u}_{:})$ in terms of the state function $\boldsymbol{x}_{:}$ and the control input function $\boldsymbol{u}_{:}$. For example, the control cost functional $c(\boldsymbol{x}_{:}, \boldsymbol{u}_{:})$ can be of a representative formalism as

$$c(\boldsymbol{x}_{:}, \boldsymbol{u}_{:}) = \int_0^{\infty} \|\boldsymbol{x}_{\mathrm{E}} - \boldsymbol{x}\|_2^2 \mathrm{d}t + \int_0^{\infty} \|\boldsymbol{u}\|_2^2 \mathrm{d}t, \tag{7.2}$$

where $\|.\|_2$ denotes the L_2-norm.

In practical applications, the practice of incorporating state related cost into the control cost functional such as in (7.2) is natural, because reducing state related cost is directly consistent with the control objective. The practice of incorporating control input related cost into the control cost functional also has reasons, usually two reasons: First, control input related cost does matter as cost in literal sense, namely as economic cost, because large control input usually incurs more consumption of energy. Second, control input related cost may also matter as cost in abstract sense, because large and even drastic control input is after all undesirable for many practical applications.

Since we can treat $\boldsymbol{x}_{:}$ as a functional in terms of $\boldsymbol{u}_{:}$, we can also *treat the control cost functional* $c(\boldsymbol{x}_{:}, \boldsymbol{u}_{:})$ *as a functional implicitly in terms of* $\boldsymbol{u}_{:}$. Then the optimal control law of $\boldsymbol{u}_{:}$ is obtained by minimizing the control cost functional $c(\boldsymbol{x}_{:}, \boldsymbol{u}_{:})$:

$$\boldsymbol{u}_{:} = \arg\min_{\boldsymbol{u}_{:}} c(\boldsymbol{x}_{:}, \boldsymbol{u}_{:}), \tag{7.3}$$

which formalizes the strategy of optimal control.

7.2.2 Linear Quadratic Regulator

It is normally difficult to solve (7.3) analytically and even numerically. On the other hand, if linear state-space modelling described by (6.5) i.e.

$$\frac{\mathrm{d}}{\mathrm{d}t}\boldsymbol{x} = \boldsymbol{A}\boldsymbol{x} + \boldsymbol{B}\boldsymbol{u}$$

can be fairly adopted for the control system and if the control cost functional $c(\boldsymbol{x}_{:}, \boldsymbol{u}_{:})$ adopts a quadratic form as

$$c(\boldsymbol{x}_{:}, \boldsymbol{u}_{:}) = \int_{0}^{\infty} \left(\boldsymbol{x}^{\mathrm{T}}\boldsymbol{Q}\boldsymbol{x} + \boldsymbol{u}^{\mathrm{T}}\boldsymbol{R}\boldsymbol{u}\right) \mathrm{d}t \tag{7.4}$$

with normally *positive definite* cost matrices $\boldsymbol{Q}$ and $\boldsymbol{R}$, then the optimal control law of $\boldsymbol{u}_{:}$ can be obtained analytically.

The instantiation of the optimal control strategy described by (7.3), which adopts linear state-space modelling described by (6.5) and the quadratic control cost functional described in (7.4), is called the **linear quadratic regulator** (Andersonand-Moore 1990):

$$\boldsymbol{u}_{:} = \arg\min_{\boldsymbol{u}_{:}} \int_{0}^{\infty} \left(\boldsymbol{x}^{\mathrm{T}}\boldsymbol{Q}\boldsymbol{x} + \boldsymbol{u}^{\mathrm{T}}\boldsymbol{R}\boldsymbol{u}\right) \mathrm{d}t. \tag{7.5}$$

To solve (7.5) analytically, we can resort to **calculus of variations**, yet we postpone presentation of such analysis and the solution of (7.5) to Sect. 7.2.5.

For the moment, we consider a simplified version of the linear quadratic regulator. More specifically, instead of considering generic control input $\boldsymbol{u}_{:}$, we focus on the family of control input functions that are generated according to the full-state feedback control strategy. Then (7.5) is reduced to the following optimization problem:

$$\boldsymbol{K} = \arg\min_{\boldsymbol{K}} \int_{0}^{\infty} \left(\boldsymbol{x}^{\mathrm{T}}\boldsymbol{Q}\boldsymbol{x} + \boldsymbol{u}^{\mathrm{T}}\boldsymbol{R}\boldsymbol{u}\right) \mathrm{d}t\,|_{\boldsymbol{u}=-\boldsymbol{K}^{\mathrm{T}}\boldsymbol{x}}, \tag{7.6}$$

which aims at obtaining the *optimal gain matrix* $\boldsymbol{K}$ of full-state feedback control. By default, we only consider the following set of gain matrices:

$$\boldsymbol{K}_{\Omega} = \{\boldsymbol{K} | \boldsymbol{A} - \boldsymbol{B}\boldsymbol{K}^{\mathrm{T}} \text{ is stable.}\}, \tag{7.7}$$

where the property that a matrix is stable means that the matrix has eigenvalues all with negative real part. In other words, we only consider gain matrices that can stabilize the control system. Suppose that the target process is controllable and hence $\boldsymbol{K}_{\Omega}$ is non-empty.

In the simplified version of the linear quadratic regulator, both the state function $\boldsymbol{x}_{:}$ and the control input function $\boldsymbol{u}_{:}$ can be expressed in terms of the gain matrix $\boldsymbol{K}$. To understand this, substitute the full-state feedback control law $\boldsymbol{u} = -\boldsymbol{K}^{\mathrm{T}}\boldsymbol{x}$ into the linear state differential equation described in (6.5) and obtain

$$\frac{\mathrm{d}}{\mathrm{d}t}\boldsymbol{x} = \boldsymbol{A}\boldsymbol{x} + \boldsymbol{B}\boldsymbol{u} = (\boldsymbol{A} - \boldsymbol{B}\boldsymbol{K}^{\mathrm{T}})\boldsymbol{x},$$

from which we can derive

$$\boldsymbol{x} = \mathrm{e}^{(\boldsymbol{A}-\boldsymbol{B}\boldsymbol{K}^{\mathrm{T}})t}\boldsymbol{x}_0,$$
$$\boldsymbol{u} = -\boldsymbol{K}^{\mathrm{T}}\mathrm{e}^{(\boldsymbol{A}-\boldsymbol{B}\boldsymbol{K}^{\mathrm{T}})t}\boldsymbol{x}_0.$$

We abuse the control cost functional notation to simply denote $c(\boldsymbol{K}) \equiv c(\boldsymbol{x}_{:}, \boldsymbol{u}_{:})$, which is computed as

$$c(\boldsymbol{K}) = \int_0^{\infty} \left(\boldsymbol{x}^{\mathrm{T}}\boldsymbol{Q}\boldsymbol{x} + \boldsymbol{u}^{\mathrm{T}}\boldsymbol{R}\boldsymbol{u}\right) \mathrm{d}t = \boldsymbol{x}_0^{\mathrm{T}}\boldsymbol{P}(\boldsymbol{K})\boldsymbol{x}_0, \tag{7.8}$$

where

$$\boldsymbol{P}(\boldsymbol{K}) \equiv \int_0^{\infty} \mathrm{e}^{(\boldsymbol{A}-\boldsymbol{B}\boldsymbol{K}^{\mathrm{T}})^{\mathrm{T}}t}(\boldsymbol{Q} + \boldsymbol{K}\boldsymbol{R}\boldsymbol{K}^{\mathrm{T}})\mathrm{e}^{(\boldsymbol{A}-\boldsymbol{B}\boldsymbol{K}^{\mathrm{T}})t}\mathrm{d}t.$$

The positive definite matrix $\boldsymbol{P}(\boldsymbol{K})$ determines the control cost of full-state feedback control associated with the gain matrix $\boldsymbol{K}$. It exists (or is finite) if and only if $\boldsymbol{K} \in \boldsymbol{K}_{\Omega}$.[3]

7.2.3 Lyapunov Equation

Readers may wonder how to obtain the positive definite matrix $\boldsymbol{P}(\boldsymbol{K})$ described in (7.8), as it seems difficult to compute the integral of $\boldsymbol{P}(\boldsymbol{K})$ directly. In fact, to obtain $\boldsymbol{P}(\boldsymbol{K})$, we do not need to compute such integral directly, but can resort to a mathematical skill based on Lyapunov equation solving. Before presentation of how to obtain the positive definite matrix $\boldsymbol{P}(\boldsymbol{K})$, relevant preliminary knowledge is necessary.

The generic formalism of **Lyapunov equation** is

$$\boldsymbol{P}\boldsymbol{X} + \boldsymbol{X}^{\mathrm{T}}\boldsymbol{P} = \boldsymbol{Y}, \tag{7.9}$$

where $\boldsymbol{P}$ is the unknown square matrix to be solved with known square matrices $\boldsymbol{X}$ and $\boldsymbol{Y}$. By default, $\boldsymbol{X}$ is a real-value matrix and $\boldsymbol{Y}$ is a real-value symmetric matrix. Instead of generally discussing the Lyapunov equation described in (7.9), we focus on special cases of the Lyapunov equation that are closely related to control problems.

Special case 1: The matrix $\boldsymbol{X}$ *is stable.* Decompose $\boldsymbol{X}$ via *Schur decomposition* (Horn and Johnson 2012) as

$$\boldsymbol{X} = \boldsymbol{U}\boldsymbol{\Sigma}\boldsymbol{U}^*$$

such that $\boldsymbol{\Sigma}$ is an upper-triangular matrix and $\boldsymbol{U}$ is a unitary matrix satisfying $\boldsymbol{U}\boldsymbol{U}^* = \boldsymbol{I}$. Note that $\boldsymbol{X}$ is a real-value matrix and hence $\boldsymbol{X}^{\mathrm{T}} = \boldsymbol{X}^*$, transform (7.9) into an

[3] Note that the integrand $\mathrm{e}^{(\boldsymbol{A}-\boldsymbol{B}\boldsymbol{K}^{\mathrm{T}})^{\mathrm{T}}t}(\boldsymbol{Q} + \boldsymbol{K}\boldsymbol{R}\boldsymbol{K}^{\mathrm{T}})\mathrm{e}^{(\boldsymbol{A}-\boldsymbol{B}\boldsymbol{K}^{\mathrm{T}})t} > 0$ always holds, namely the integrand is always positive definite. So *finiteness* of $\boldsymbol{P}(\boldsymbol{K})$ implies that the integrand converges to 0 as $t \to \infty$, which further implies $\mathrm{e}^{(\boldsymbol{A}-\boldsymbol{B}\boldsymbol{K}^{\mathrm{T}})t} \to 0$ as $t \to \infty$ and hence $\boldsymbol{K} \in \boldsymbol{K}_{\Omega}$. On the other hand, if $\boldsymbol{K} \in \boldsymbol{K}_{\Omega}$, we can choose a matrix norm and know that the norm of $\mathrm{e}^{(\boldsymbol{A}-\boldsymbol{B}\boldsymbol{K}^{\mathrm{T}})t}$ is bounded within certain exponentially-decaying limit. Since the integral of an exponentially-decaying function definitely exists, $\boldsymbol{P}(\boldsymbol{K})$ is bounded and hence exists.

equivalent Lyapunov equation as

$$\begin{aligned} \boldsymbol{P}\boldsymbol{X} + \boldsymbol{X}^{\mathrm{T}}\boldsymbol{P} = \boldsymbol{Y} &\iff \boldsymbol{P}\boldsymbol{U}\boldsymbol{\Sigma}\boldsymbol{U}^{*} + \boldsymbol{U}\boldsymbol{\Sigma}^{*}\boldsymbol{U}^{*}\boldsymbol{P} = \boldsymbol{Y} \\ &\iff \boldsymbol{U}^{*}\boldsymbol{P}\boldsymbol{U}\boldsymbol{\Sigma} + \boldsymbol{\Sigma}^{*}\boldsymbol{U}^{*}\boldsymbol{P}\boldsymbol{U} = \boldsymbol{U}^{*}\boldsymbol{Y}\boldsymbol{U} \end{aligned}$$

namely

$$\bar{\boldsymbol{P}}\boldsymbol{\Sigma} + \boldsymbol{\Sigma}^{*}\bar{\boldsymbol{P}} = \bar{\boldsymbol{Y}}, \tag{7.10}$$

where $\bar{\boldsymbol{P}} = \boldsymbol{U}^{*}\boldsymbol{P}\boldsymbol{U}$ and $\bar{\boldsymbol{Y}} = \boldsymbol{U}^{*}\boldsymbol{Y}\boldsymbol{U}$.

Perform *matrix vectorization*[4] on both sides of (7.10) and obtain

$$\mathrm{vec}(\bar{\boldsymbol{P}}\boldsymbol{\Sigma} + \boldsymbol{\Sigma}^{*}\bar{\boldsymbol{P}}) = \mathrm{vec}(\bar{\boldsymbol{Y}}) \iff (\boldsymbol{\Sigma}^{\mathrm{T}} \otimes \boldsymbol{I} + \boldsymbol{I} \otimes \boldsymbol{\Sigma}^{*})\mathrm{vec}(\bar{\boldsymbol{P}}) = \mathrm{vec}(\bar{\boldsymbol{Y}}), \tag{7.11}$$

where $\otimes$ denotes the *Kronecker product* (Horn and Johnson 1991).[5] Let $\lambda_1, \ldots, \lambda_n$ denote the eigenvalues of $\boldsymbol{X}$, which are all with negative real part as $\boldsymbol{X}$ is stable. Since

$$\boldsymbol{\Sigma}^{\mathrm{T}} \otimes \boldsymbol{I} + \boldsymbol{I} \otimes \boldsymbol{\Sigma}^{*} = \begin{bmatrix} \lambda_1\boldsymbol{I} + \boldsymbol{\Sigma}^{*} & & \\ & \ddots & \\ * & & \lambda_n\boldsymbol{I} + \boldsymbol{\Sigma}^{*} \end{bmatrix}$$

and each of its diagonal blocks is of the form

$$\lambda_i\boldsymbol{I} + \boldsymbol{\Sigma}^{*} = \begin{bmatrix} \lambda_i + \bar{\lambda}_1 & & \\ & \ddots & \\ * & & \lambda_i + \bar{\lambda}_n \end{bmatrix}$$

whose diagonal elements are all with negative real part, $\boldsymbol{\Sigma}^{\mathrm{T}} \otimes \boldsymbol{I} + \boldsymbol{I} \otimes \boldsymbol{\Sigma}^{*}$ is a lower-triangular matrix with non-zero diagonal elements. Therefore, the equation described in (7.11) has a unique solution of $\mathrm{vec}(\bar{\boldsymbol{P}})$, which implies the original Lyapunov equation described in (7.9) has a unique solution of $\boldsymbol{P}$.

> **Lyapunov criterion I**: *If the matrix* $\boldsymbol{X}$ *is stable, the Lyapunov equation described in* (7.9) *has a unique solution of* $\boldsymbol{P}$.

---

[4] Matrix vectorization: $\mathrm{vec}([\boldsymbol{b}_1\ \boldsymbol{b}_2\ \cdots\ \boldsymbol{b}_n]) \equiv \begin{bmatrix} \boldsymbol{b}_1 \\ \boldsymbol{b}_2 \\ \vdots \\ \boldsymbol{b}_n \end{bmatrix}$ where $\boldsymbol{b}_1, \boldsymbol{b}_2, \ldots, \boldsymbol{b}_n$ are vectors.

[5] Kronecker product: $\boldsymbol{A} \otimes \boldsymbol{B} \equiv \begin{bmatrix} a_{11}\boldsymbol{B} & \cdots & a_{1n}\boldsymbol{B} \\ \vdots & & \vdots \\ a_{m1}\boldsymbol{B} & \cdots & a_{mn}\boldsymbol{B} \end{bmatrix}$. We have $\mathrm{vec}(\boldsymbol{A}\boldsymbol{B}\boldsymbol{C}) = (\boldsymbol{C}^{\mathrm{T}} \otimes \boldsymbol{A})\mathrm{vec}(\boldsymbol{B})$.

It is worth noting that $\boldsymbol{\Sigma}^{\mathrm{T}} \otimes \boldsymbol{I} + \boldsymbol{I} \otimes \boldsymbol{\Sigma}^*$ is a lower-triangular matrix, so the linear equation group described in (7.11) can be solved efficiently. For a Lyapunov equation of large scale, we can take advantage of *Schur decomposition* to transform it into an equivalent Lyapunov equation of the form described in (7.11) that can be solved efficiently. On the other hand, for a Lyapunov equation of moderate scale, we may perform matrix vectorization directly on the original Lyapunov equation described in (7.9) as

$$\mathrm{vec}(\boldsymbol{P}\boldsymbol{X} + \boldsymbol{X}^{\mathrm{T}}\boldsymbol{P}) = \mathrm{vec}(\boldsymbol{Y}) \iff (\boldsymbol{X}^{\mathrm{T}} \otimes \boldsymbol{I} + \boldsymbol{I} \otimes \boldsymbol{X}^{\mathrm{T}})\mathrm{vec}(\boldsymbol{P}) = \mathrm{vec}(\boldsymbol{Y})$$

and solve $\boldsymbol{P}$. Matlab code for demonstration of Lyapunov equation solving is given as follows.

SolveLyapunovEquation.m

```
%% Lyapunov equation: P X + X' P = Y
function P = SolveLyapunovEquation(X, Y)
    n = size(X,1); vecY = reshape(Y,[],1); % [Y(:,1); Y(:,2);
    ...; Y(:,n)]
    M = zeros(n*n,n*n); k = 0;
    for c = 1:n
        for r = 1:n
            k = k+1;
            M(k,(c*n-n+1):(c*n)) = M(k,(c*n-n+1):(c*n)) + X(:,r)
    ';
            M(k,r:n:end) = M(k,r:n:end) + X(:,c)';
        end
    end
    vecP = M\vecY; % [P(:,1); P(:,2); ...; P(:,n)]
    P = reshape(vecP,n,n);
end
```

Special case 2: The matrix $\boldsymbol{X}$ *is stable and the matrix* $\boldsymbol{Y}$ *is negative definite.* According to the *Lyapunov criterion I*, the Lyapunov equation described in (7.9) has a unique solution of $\boldsymbol{P}$. Since $-\boldsymbol{Y}$ is positive definite namely $-\boldsymbol{Y} > 0$, we further have

$$\begin{aligned}\boldsymbol{P} &= \mathrm{e}^{\boldsymbol{X}^{\mathrm{T}}t}\boldsymbol{P}\mathrm{e}^{\boldsymbol{X}t}|_{t=0} - \mathrm{e}^{\boldsymbol{X}^{\mathrm{T}}t}\boldsymbol{P}\mathrm{e}^{\boldsymbol{X}t}|_{t\to\infty} = -\int_0^{\infty} \frac{\mathrm{d}}{\mathrm{d}t}[\mathrm{e}^{\boldsymbol{X}^{\mathrm{T}}t}\boldsymbol{P}\mathrm{e}^{\boldsymbol{X}t}]\mathrm{d}t \\ &= -\int_0^{\infty} \mathrm{e}^{\boldsymbol{X}^{\mathrm{T}}t}(\boldsymbol{X}^{\mathrm{T}}\boldsymbol{P} + \boldsymbol{P}\boldsymbol{X})\mathrm{e}^{\boldsymbol{X}t}\mathrm{d}t = \int_0^{\infty} \mathrm{e}^{\boldsymbol{X}^{\mathrm{T}}t}(-\boldsymbol{Y})\mathrm{e}^{\boldsymbol{X}t}\mathrm{d}t > 0,\end{aligned}$$

which implies the Lyapunov equation described in (7.9) has a unique solution of $\boldsymbol{P}$ that is positive definite.

Lyapunov criterion II: *If the matrix* $\boldsymbol{X}$ *is stable and the matrix* $\boldsymbol{Y}$ *is negative definite, the Lyapunov equation described in* (7.9) *has a unique solution of* $\boldsymbol{P}$ *that is positive definite.*

Special case 3: The matrix $\boldsymbol{Y}$ *is negative definite and the solution* $\boldsymbol{P}$ *is positive definite.* In the spirit of the Lyapunov strategy (Liapounoff 1907), construct a dynamic system with state z that follows

$$\frac{\mathrm{d}}{\mathrm{d}t}z = \boldsymbol{X}z.$$

Define a positive definite function V as

$$V \equiv V(z) = z^{\mathrm{T}}\boldsymbol{P}z.$$

We have

$$\frac{\mathrm{d}}{\mathrm{d}t}V = \frac{\mathrm{d}z^{\mathrm{T}}}{\mathrm{d}t}\boldsymbol{P}z + z^{\mathrm{T}}\boldsymbol{P}\frac{\mathrm{d}z}{\mathrm{d}t} = z^{\mathrm{T}}(\boldsymbol{X}^{\mathrm{T}}\boldsymbol{P} + \boldsymbol{P}\boldsymbol{X})z = z^{\mathrm{T}}\boldsymbol{Y}z < 0,$$

which implies that the positive definite function V decreases monotonically and definitely converges to a limit with $\frac{\mathrm{d}}{\mathrm{d}t}V = 0$. Note that

$$\frac{\mathrm{d}}{\mathrm{d}t}V = z^{\mathrm{T}}\boldsymbol{Y}z = 0 \iff z = 0$$

and all above derivation holds regardless of the initial state z_0, so the constructed dynamic system always converges to the stable state $z = 0$ and hence the matrix $\boldsymbol{X}$ is stable. In other words, for the Lyapunov equation described in (7.9), if the matrix $\boldsymbol{Y}$ is negative definite and the solution $\boldsymbol{P}$ is positive definite, the matrix $\boldsymbol{X}$ is stable.

Lyapunov criterion III: *For the Lyapunov equation described in* (7.9), *if the matrix* $\boldsymbol{Y}$ *is negative definite and the solution* $\boldsymbol{P}$ *is positive definite, the matrix* $\boldsymbol{X}$ *is stable.*

Now we can return to computation of the positive definite matrix $\boldsymbol{P}(\boldsymbol{K})$ described in (7.8). As $\boldsymbol{K} \in \boldsymbol{K}_{\Omega}$ namely $\boldsymbol{A} - \boldsymbol{B}\boldsymbol{K}^{\mathrm{T}}$ is stable, the Lyapunov equation

$$\boldsymbol{P}(\boldsymbol{A} - \boldsymbol{B}\boldsymbol{K}^{\mathrm{T}}) + (\boldsymbol{A} - \boldsymbol{B}\boldsymbol{K}^{\mathrm{T}})^{\mathrm{T}}\boldsymbol{P} = -\boldsymbol{Q} - \boldsymbol{K}\boldsymbol{R}\boldsymbol{K}^{\mathrm{T}} \tag{7.12}$$

has a unique solution of $\boldsymbol{P}$. Besides, the right side of (7.12) is negative definite, so the Lyapunov equation described in (7.12) has a unique solution of $\boldsymbol{P}$ that is positive definite. We have

$$
\begin{aligned}
P(K) &\equiv \int_0^{\infty} e^{(A-BK^{T})^{T}t}(Q+KRK^{T})e^{(A-BK^{T})t}\,dt \\
&= \int_0^{\infty} e^{(A-BK^{T})^{T}t}(-P(A-BK^{T})-(A-BK^{T})^{T}P)e^{(A-BK^{T})t}\,dt \\
&= -\int_0^{\infty} \frac{d}{dt}\left[e^{(A-BK^{T})^{T}t}Pe^{(A-BK^{T})t}\right]dt \\
&= e^{(A-BK^{T})^{T}t}Pe^{(A-BK^{T})t}\Big|_{\infty}^{0} = P.
\end{aligned}
$$

In other words, the positive definite matrix $P(K)$ is the unique (positive definite) solution of the Lyapunov equation described in (7.12), namely

$$
P(K)(A-BK^{T})+(A-BK^{T})^{T}P(K)=-Q-KRK^{T}. \tag{7.13}
$$

7.2.4 Riccati Equation

In the light of (7.13), the optimization problem described in (7.6) becomes

$$
K=\arg\min_{K} x_0^{T}Px_0|_{P(A-BK^{T})+(A-BK^{T})^{T}P=-Q-KRK^{T}}. \tag{7.14}
$$

The optimal gain matrix K is apparently not at infinity. Otherwise, an infinite optimal control cost will be incurred, which contradicts the fact that an arbitrary finite $K \in K_{\Omega}$ incurs only a finite control cost.

For the optimal gain matrix K which is finite, consider an infinitesimal variation ΔK in the Lyapunov equation described in (7.12) and obtain

$$
\begin{aligned}
&(P+\Delta P)[A-B(K+\Delta K)^{T}]+[A-B(K+\Delta K)^{T}]^{T}(P+\Delta P) \\
&= -Q-(K+\Delta K)R(K+\Delta K)^{T} \\
\Longleftrightarrow\ &\Delta P(A-BK^{T})+(A-BK^{T})^{T}\Delta P \\
&= \Delta K(B^{T}P-RK^{T})+(PB-KR)\Delta K^{T},
\end{aligned}
$$

which is a Lyapunov equation in terms of ΔP. Recall the *Lyapunov criterion I*, since $A-BK^{T}$ is stable, above Lyapunov equation always has a unique solution of ΔP that varies in proportion to ΔK. As the infinitesimal variation ΔK can be arbitrary, *for optimality of the gain matrix* K, we must have

$$
\Delta P \equiv 0 \Longleftrightarrow \Delta K(B^{T}P-RK^{T})+(PB-KR)\Delta K^{T} \equiv 0
$$

no matter for what ΔK, which can hold only when

$$
B^{T}P-RK^{T}=(PB-KR)^{T}=0 \Longleftrightarrow K=PBR^{-1} \text{ or } K^{T}=R^{-1}B^{T}P. \tag{7.15}
$$

Substitute (7.15) into the Lyapunov equation described in (7.12) and obtain

$$\boldsymbol{P}(\boldsymbol{A}-\boldsymbol{B}\boldsymbol{R}^{-1}\boldsymbol{B}^{\mathrm{T}}\boldsymbol{P})+(\boldsymbol{A}-\boldsymbol{B}\boldsymbol{R}^{-1}\boldsymbol{B}^{\mathrm{T}}\boldsymbol{P})^{\mathrm{T}}\boldsymbol{P}+\boldsymbol{Q}+\boldsymbol{P}\boldsymbol{B}\boldsymbol{R}^{-1}\boldsymbol{R}\boldsymbol{R}^{-1}\boldsymbol{B}^{\mathrm{T}}\boldsymbol{P}=\boldsymbol{0}$$

namely

$$\boldsymbol{P}\boldsymbol{A}+\boldsymbol{A}^{\mathrm{T}}\boldsymbol{P}-\boldsymbol{P}\boldsymbol{B}\boldsymbol{R}^{-1}\boldsymbol{B}^{\mathrm{T}}\boldsymbol{P}+\boldsymbol{Q}=\boldsymbol{0}, \tag{7.16}$$

which is called the **matrix Riccati equation** or simply the **Riccati equation**. Therefore, to obtain the optimal gain matrix $\boldsymbol{K}$, we can first solve the Riccati equation described in (7.16) to obtain the (unique) positive definite solution of $\boldsymbol{P}$ and then follow (7.15).

If the *Hamiltonian matrix* $\boldsymbol{H}$ is diagonalizable with half of its eigenvalues having negative real part (corresponding to the diagonal block $\boldsymbol{\Sigma}_-$) as

$$\boldsymbol{H}\equiv\begin{bmatrix}\boldsymbol{A} & -\boldsymbol{B}\boldsymbol{R}^{-1}\boldsymbol{B}^{\mathrm{T}}\\ -\boldsymbol{Q} & -\boldsymbol{A}^{\mathrm{T}}\end{bmatrix}=\begin{bmatrix}\boldsymbol{U}_{11} & \boldsymbol{U}_{12}\\ \boldsymbol{U}_{21} & \boldsymbol{U}_{22}\end{bmatrix}\begin{bmatrix}\boldsymbol{\Sigma}_- & \\ & \boldsymbol{\Sigma}_+\end{bmatrix}\begin{bmatrix}\boldsymbol{U}_{11} & \boldsymbol{U}_{12}\\ \boldsymbol{U}_{21} & \boldsymbol{U}_{22}\end{bmatrix}^{-1}$$

and if the block $\boldsymbol{U}_{11}$ is invertible, then the Riccati equation described in (7.16) is solved as $\boldsymbol{P}=\boldsymbol{U}_{21}\boldsymbol{U}_{11}^{-1}$ which is the positive definite solution.

The Riccati equation described in (7.16) may also be solved in iterative way as follows, which is computationally more desirable and general.

Riccati equation iterative solving (Kleinman 1968)
Initialization:
Find $\boldsymbol{K}_0$ such that $\boldsymbol{A}-\boldsymbol{B}\boldsymbol{K}_0^{\mathrm{T}}$ is stable.
Iteration:
Compute $\boldsymbol{A}_i=\boldsymbol{A}-\boldsymbol{B}\boldsymbol{K}_i^{\mathrm{T}}$ and $\boldsymbol{Q}_i=-\boldsymbol{K}_i\boldsymbol{R}\boldsymbol{K}_i^{\mathrm{T}}-\boldsymbol{Q}$.
Solve the *Lyapunov equation* $\boldsymbol{P}_i\boldsymbol{A}_i+\boldsymbol{A}_i^{\mathrm{T}}\boldsymbol{P}_i=\boldsymbol{Q}_i$.
Update $\boldsymbol{K}_{i+1}^{\mathrm{T}}=\boldsymbol{R}^{-1}\boldsymbol{B}^{\mathrm{T}}\boldsymbol{P}_i \iff \boldsymbol{K}_{i+1}=\boldsymbol{P}_i\boldsymbol{B}\boldsymbol{R}^{-1}$.

Suppose that the control system's target process is controllable and hence there is $\boldsymbol{K}_0$ such that $\boldsymbol{A}-\boldsymbol{B}\boldsymbol{K}_0^{\mathrm{T}}$ is stable. Resort to methods presented in Chap. 6 to obtain such $\boldsymbol{K}_0$. In each iteration of the Riccati equation iterative solving algorithm, apply the method presented in Sect. 7.4 to solve the following Lyapunov equation:

$$\boldsymbol{P}_i(\boldsymbol{A}-\boldsymbol{B}\boldsymbol{K}_i^{\mathrm{T}})+(\boldsymbol{A}-\boldsymbol{B}\boldsymbol{K}_i^{\mathrm{T}})^{\mathrm{T}}\boldsymbol{P}_i=-\boldsymbol{Q}-\boldsymbol{K}_i\boldsymbol{R}\boldsymbol{K}_i^{\mathrm{T}}. \tag{7.17}$$

Three points need to be clarified for the Riccati equation iterative solving algorithm.

First, the matrices $\boldsymbol{A}_i=\boldsymbol{A}-\boldsymbol{B}\boldsymbol{K}_i^{\mathrm{T}}$ encountered during iteration of the Riccati equation iterative solving algorithm are stable. This can be proved via *mathematical induction*: Initially, the matrix $\boldsymbol{A}_0=\boldsymbol{A}-\boldsymbol{B}\boldsymbol{K}_0^{\mathrm{T}}$ is stable. Suppose that the matrix $\boldsymbol{A}_i=\boldsymbol{A}-\boldsymbol{B}\boldsymbol{K}_i^{\mathrm{T}}$ is stable and consider $\boldsymbol{A}_{i+1}=\boldsymbol{A}-\boldsymbol{B}\boldsymbol{K}_{i+1}^{\mathrm{T}}$.

Since $A_i = A - BK_i^{\mathrm{T}}$ is stable and $Q_i = -K_iRK_i^{\mathrm{T}} - Q < 0$, the matrix P_i, which is the solution of the Lyapunov equation $P_iA_i + A_i^{\mathrm{T}}P_i = Q_i$, is positive definite (according to the *Lyapunov criterion II*). Further note that

$$\begin{aligned}
&P_iA_{i+1} + A_{i+1}^{\mathrm{T}}P_i \\
&= P_i[A_i + B(K_i - K_{i+1})^{\mathrm{T}}] + [A_i + B(K_i - K_{i+1})^{\mathrm{T}}]^{\mathrm{T}}P_i \\
&= P_iA_i + A_i^{\mathrm{T}}P_i + P_iBR^{-1}R(K_i - K_{i+1})^{\mathrm{T}} + (K_i - K_{i+1})RR^{-1}B^{\mathrm{T}}P_i \\
&= -Q - K_iRK_i^{\mathrm{T}} + K_{i+1}R(K_i - K_{i+1})^{\mathrm{T}} + (K_i - K_{i+1})RK_{i+1}^{\mathrm{T}} \\
&= -Q - (K_i - K_{i+1})R(K_i - K_{i+1})^{\mathrm{T}} - K_{i+1}RK_{i+1}^{\mathrm{T}} < 0,
\end{aligned}$$

which implies that $A_{i+1} = A - BK_{i+1}^{\mathrm{T}}$ is stable (according to the *Lyapunov criterion III*). Therefore, all matrices $A_i = A - BK_i^{\mathrm{T}}$ $(i = 0, 1, 2, \ldots)$ are stable.

Second, the matrices P_i encountered during iteration of the Riccati equation iterative solving algorithm are positive definite and converge to the positive definite solution of the Riccati equation described in (7.16). Their positive definiteness is already clarified above during clarification of the first point, so we focus on their convergence. Denote $\Delta K_i = K_i - K_{i+1}$, then we have

$$\begin{aligned}
&(P_i - P_{i+1})A_{i+1} + A_{i+1}^{\mathrm{T}}(P_i - P_{i+1}) \\
&= (P_iA_{i+1} + A_{i+1}^{\mathrm{T}}P_i) - (P_{i+1}A_{i+1} + A_{i+1}^{\mathrm{T}}P_{i+1}) \\
&= -Q - \Delta K_iR\Delta K_i^{\mathrm{T}} - K_{i+1}RK_{i+1}^{\mathrm{T}} - (-Q - K_{i+1}RK_{i+1}^{\mathrm{T}}) \\
&= -\Delta K_iR\Delta K_i^{\mathrm{T}} \leqslant 0
\end{aligned}$$

and

$$\begin{aligned}
P_i - P_{i+1} &= \mathrm{e}^{A_{i+1}^{\mathrm{T}}t}(P_i - P_{i+1})\mathrm{e}^{A_{i+1}t}|_{t=0} - \mathrm{e}^{A_{i+1}^{\mathrm{T}}t}(P_i - P_{i+1})\mathrm{e}^{A_{i+1}t}|_{t\to\infty} \\
&= -\int_0^{\infty}\frac{\mathrm{d}}{\mathrm{d}t}[\mathrm{e}^{A_{i+1}^{\mathrm{T}}t}(P_i - P_{i+1})\mathrm{e}^{A_{i+1}t}]\mathrm{d}t \\
&= -\int_0^{\infty}\mathrm{e}^{A_{i+1}^{\mathrm{T}}t}[(P_i - P_{i+1})A_{i+1} + A_{i+1}^{\mathrm{T}}(P_i - P_{i+1})]\mathrm{e}^{A_{i+1}t}\mathrm{d}t \\
&= \int_0^{\infty}\mathrm{e}^{A_{i+1}^{\mathrm{T}}t}\Delta K_iR\Delta K_i^{\mathrm{T}}\mathrm{e}^{A_{i+1}t}\mathrm{d}t \geqslant 0,
\end{aligned}$$

which implies that the positive definite matrices P_i $(i = 0, 1, 2, \ldots)$ decrease monotonically in terms of positive definiteness. Therefore, they definitely converge to a limit P with a convergent $K = PBR^{-1}$ as well. Consider limits on both sides of (7.17) and obtain

$$\begin{aligned}
&P(A - BK^{\mathrm{T}}) + (A - BK^{\mathrm{T}})^{\mathrm{T}}P = -Q - KRK^{\mathrm{T}} \\
&\Longleftrightarrow PA + A^{\mathrm{T}}P - PBR^{-1}B^{\mathrm{T}}P + Q = 0,
\end{aligned}$$

which is of the Riccati equation form described in (7.16). In other words, the convergent $\boldsymbol{P}$ is right the positive definite solution of the Riccati equation described in (7.16).

Third, the convergent $\boldsymbol{P}$ is the unique positive definite solution of the Riccati equation described in (7.16). Suppose that there is another positive definite solution $\bar{\boldsymbol{P}}$. Denote $\bar{\boldsymbol{K}} = \bar{\boldsymbol{P}}\boldsymbol{B}\boldsymbol{R}^{-1}$, $\Delta\boldsymbol{K} = \boldsymbol{K} - \bar{\boldsymbol{K}}$, $\boldsymbol{A}_{\boldsymbol{P}} = \boldsymbol{A} - \boldsymbol{B}\boldsymbol{K}^{\mathrm{T}}$, and $\boldsymbol{A}_{\bar{\boldsymbol{P}}} = \boldsymbol{A} - \boldsymbol{B}\bar{\boldsymbol{K}}^{\mathrm{T}}$. We have

$$\bar{\boldsymbol{P}}\boldsymbol{A}_{\bar{\boldsymbol{P}}} + \boldsymbol{A}_{\bar{\boldsymbol{P}}}^{\mathrm{T}}\bar{\boldsymbol{P}} = \bar{\boldsymbol{P}}(\boldsymbol{A} - \boldsymbol{B}\bar{\boldsymbol{K}}^{\mathrm{T}}) + (\boldsymbol{A} - \boldsymbol{B}\bar{\boldsymbol{K}}^{\mathrm{T}})^{\mathrm{T}}\bar{\boldsymbol{P}} = -\boldsymbol{Q} - \bar{\boldsymbol{K}}\boldsymbol{R}\bar{\boldsymbol{K}}^{\mathrm{T}},$$

which implies that $\boldsymbol{A}_{\bar{\boldsymbol{P}}}$ is stable (according to the *Lyapunov criterion III*). We further have

$$\begin{aligned}
&(\boldsymbol{P} - \bar{\boldsymbol{P}})\boldsymbol{A}_{\bar{\boldsymbol{P}}} + \boldsymbol{A}_{\bar{\boldsymbol{P}}}^{\mathrm{T}}(\boldsymbol{P} - \bar{\boldsymbol{P}}) \\
&= (\boldsymbol{P}\boldsymbol{A}_{\boldsymbol{P}} + \boldsymbol{A}_{\boldsymbol{P}}^{\mathrm{T}}\boldsymbol{P}) + \boldsymbol{P}\boldsymbol{B}\Delta\boldsymbol{K}^{\mathrm{T}} + \Delta\boldsymbol{K}\boldsymbol{B}^{\mathrm{T}}\boldsymbol{P} - (\bar{\boldsymbol{P}}\boldsymbol{A}_{\bar{\boldsymbol{P}}} + \boldsymbol{A}_{\bar{\boldsymbol{P}}}^{\mathrm{T}}\bar{\boldsymbol{P}}) \\
&= (-\boldsymbol{Q} - \boldsymbol{K}\boldsymbol{R}\boldsymbol{K}^{\mathrm{T}}) + \boldsymbol{K}\boldsymbol{R}\Delta\boldsymbol{K}^{\mathrm{T}} + \Delta\boldsymbol{K}\boldsymbol{R}\boldsymbol{K}^{\mathrm{T}} - (-\boldsymbol{Q} - \bar{\boldsymbol{K}}\boldsymbol{R}\bar{\boldsymbol{K}}^{\mathrm{T}}) \\
&= \Delta\boldsymbol{K}\boldsymbol{R}\Delta\boldsymbol{K}^{\mathrm{T}} \geqslant 0
\end{aligned}$$

and similarly by symmetry of above derivation we have

$$\begin{aligned}
&(\boldsymbol{P} - \bar{\boldsymbol{P}})\boldsymbol{A}_{\boldsymbol{P}} + \boldsymbol{A}_{\boldsymbol{P}}^{\mathrm{T}}(\boldsymbol{P} - \bar{\boldsymbol{P}}) = -[(\bar{\boldsymbol{P}} - \boldsymbol{P})\boldsymbol{A}_{\boldsymbol{P}} + \boldsymbol{A}_{\boldsymbol{P}}^{\mathrm{T}}(\bar{\boldsymbol{P}} - \boldsymbol{P})] \\
&= -(-\Delta\boldsymbol{K})\boldsymbol{R}(-\Delta\boldsymbol{K})^{\mathrm{T}} = -\Delta\boldsymbol{K}\boldsymbol{R}\Delta\boldsymbol{K}^{\mathrm{T}} \leqslant 0.
\end{aligned}$$

On one hand

$$\begin{aligned}
\boldsymbol{P} - \bar{\boldsymbol{P}} &= -\int_0^\infty \frac{\mathrm{d}}{\mathrm{d}t}\left[\mathrm{e}^{\boldsymbol{A}_{\bar{\boldsymbol{P}}}^{\mathrm{T}} t}(\boldsymbol{P} - \bar{\boldsymbol{P}})\mathrm{e}^{\boldsymbol{A}_{\bar{\boldsymbol{P}}} t}\right]\mathrm{d}t \\
&= -\int_0^\infty \mathrm{e}^{\boldsymbol{A}_{\bar{\boldsymbol{P}}}^{\mathrm{T}} t}\left[(\boldsymbol{P} - \bar{\boldsymbol{P}})\boldsymbol{A}_{\bar{\boldsymbol{P}}} + \boldsymbol{A}_{\bar{\boldsymbol{P}}}^{\mathrm{T}}(\boldsymbol{P} - \bar{\boldsymbol{P}})\right]\mathrm{e}^{\boldsymbol{A}_{\bar{\boldsymbol{P}}} t}\mathrm{d}t \\
&= -\int_0^\infty \mathrm{e}^{\boldsymbol{A}_{\bar{\boldsymbol{P}}}^{\mathrm{T}} t}\Delta\boldsymbol{K}\boldsymbol{R}\Delta\boldsymbol{K}^{\mathrm{T}}\mathrm{e}^{\boldsymbol{A}_{\bar{\boldsymbol{P}}} t}\mathrm{d}t \leqslant 0,
\end{aligned}$$

and on the other hand

$$\begin{aligned}
\boldsymbol{P} - \bar{\boldsymbol{P}} &= -\int_0^\infty \frac{\mathrm{d}}{\mathrm{d}t}\left[\mathrm{e}^{\boldsymbol{A}_{\boldsymbol{P}}^{\mathrm{T}} t}(\boldsymbol{P} - \bar{\boldsymbol{P}})\mathrm{e}^{\boldsymbol{A}_{\boldsymbol{P}} t}\right]\mathrm{d}t \\
&= -\int_0^\infty \mathrm{e}^{\boldsymbol{A}_{\boldsymbol{P}}^{\mathrm{T}} t}\left[(\boldsymbol{P} - \bar{\boldsymbol{P}})\boldsymbol{A}_{\boldsymbol{P}} + \boldsymbol{A}_{\boldsymbol{P}}^{\mathrm{T}}(\boldsymbol{P} - \bar{\boldsymbol{P}})\right]\mathrm{e}^{\boldsymbol{A}_{\boldsymbol{P}} t}\mathrm{d}t \\
&= \int_0^\infty \mathrm{e}^{\boldsymbol{A}_{\boldsymbol{P}}^{\mathrm{T}} t}\Delta\boldsymbol{K}\boldsymbol{R}\Delta\boldsymbol{K}^{\mathrm{T}}\mathrm{e}^{\boldsymbol{A}_{\boldsymbol{P}} t}\mathrm{d}t \geqslant 0.
\end{aligned}$$

Therefore, $\boldsymbol{P} = \bar{\boldsymbol{P}}$ and the uniqueness of the positive definite solution of the Riccati equation described in (7.16) is proved.

Matlab code for demonstration of Riccati equation iterative solving is given as follows.

SolveRiccatiEquationLQR.m

```
%% Riccati equation of LQR: P A + A' P - P B inv(R) B' P + Q = 0
%% Solve the symmetric matrix P and obtain Kc = (inv(R)*B'*P)'
function [P, Kc] = SolveRiccatiEquationLQR(A, B, Q, R, Kinit)
    Kc = Kinit; errK = 10000;
    while (errK>0.00001)
        P = SolveLyapunovEquation(A-B*Kc', -Kc*R*Kc'-Q);
        Kold = Kc; Kc = (inv(R)*B'*P)'; % Transpose
        errK = trace((Kc-Kold)'*(Kc-Kold));
    end
end
```

We have finally solved the simplified version of the linear quadratic regulator, namely have determined the optimal gain matrix of full-state feedback control. Then we move forward to solve the original version of the linear quadratic regulator.

7.2.5 Solution of Linear Quadratic Regulator

By so far, we have determined the optimal gain matrix of full-state feedback control, yet this does not mean that we have found the optimal solution of the linear quadratic regulator described by (7.5). There might be certain optimal control solution better than the optimal full-state feedback control method. We resort to **calculus of variations** to solve (7.5) analytically.

Since we can treat the control cost functional $c(\boldsymbol{x}_:, \boldsymbol{u}_:)$ as a functional implicitly in terms of $\boldsymbol{u}_:$, we abuse the control cost functional notation to simply denote $c(\boldsymbol{u}_:) \equiv c(\boldsymbol{x}_:, \boldsymbol{u}_:)$. The optimal control input function $\boldsymbol{u}_:$ is apparently not at infinity. Otherwise, an infinite optimal control cost will be incurred, which contradicts the fact that the optimal full-state feedback control method incurs only a finite control cost.

Consider infinitesimal variation $\Delta\boldsymbol{u}_:$ on the optimal control input function $\boldsymbol{u}_:$ which is bounded. Note the solution described in (6.20):

$$\boldsymbol{x} = \mathrm{e}^{\boldsymbol{A}t}\boldsymbol{x}_0 + \int_0^t \mathrm{e}^{\boldsymbol{A}(t-\tau)}\boldsymbol{B}\boldsymbol{u}(\tau)\mathrm{d}\tau.$$

The state variation caused by the variation $\Delta\boldsymbol{u}_:$ is

$$\Delta\boldsymbol{x} = \int_0^t \mathrm{e}^{\boldsymbol{A}(t-\tau)}\boldsymbol{B}\Delta\boldsymbol{u}(\tau)\mathrm{d}\tau.$$

We have

$$
\begin{aligned}
c(\boldsymbol{u}_{:} + \Delta \boldsymbol{u}_{:}) &= \int_0^{\infty} \left[(\boldsymbol{x} + \Delta \boldsymbol{x})^{\mathrm{T}} \boldsymbol{Q} (\boldsymbol{x} + \Delta \boldsymbol{x}) + (\boldsymbol{u} + \Delta \boldsymbol{u})^{\mathrm{T}} \boldsymbol{R} (\boldsymbol{u} + \Delta \boldsymbol{u})\right] \mathrm{d}t \\
&= c(\boldsymbol{u}_{:}) + 2 \int_0^{\infty} \left(\boldsymbol{x}^{\mathrm{T}} \boldsymbol{Q} \Delta \boldsymbol{x} + \boldsymbol{u}^{\mathrm{T}} \boldsymbol{R} \Delta \boldsymbol{u}\right) \mathrm{d}t
\end{aligned}
$$

and

$$
\begin{aligned}
\Delta c(\boldsymbol{u}_{:}) \equiv c(\boldsymbol{u}_{:} + \Delta \boldsymbol{u}_{:}) - c(\boldsymbol{u}_{:}) &= 2 \int_0^{\infty} \left(\boldsymbol{x}^{\mathrm{T}} \boldsymbol{Q} \Delta \boldsymbol{x} + \boldsymbol{u}^{\mathrm{T}} \boldsymbol{R} \Delta \boldsymbol{u}\right) \mathrm{d}t \\
&= 2 \int_0^{\infty} \left(\boldsymbol{x}^{\mathrm{T}} \boldsymbol{Q} \int_0^{t} \mathrm{e}^{A(t-\tau)} \boldsymbol{B} \Delta \boldsymbol{u}(\tau) \mathrm{d}\tau + \boldsymbol{u}^{\mathrm{T}} \boldsymbol{R} \Delta \boldsymbol{u}\right) \mathrm{d}t \\
&= 2 \int_0^{\infty} \int_0^{t} \boldsymbol{x}^{\mathrm{T}} \boldsymbol{Q} \mathrm{e}^{A(t-\tau)} \boldsymbol{B} \Delta \boldsymbol{u}(\tau) \mathrm{d}\tau \mathrm{d}t + 2 \int_0^{\infty} \boldsymbol{u}^{\mathrm{T}} \boldsymbol{R} \Delta \boldsymbol{u} \mathrm{d}t.
\end{aligned}
$$

Use the following integral transform

$$
\int_0^{\infty} \int_0^{t} F(\tau, t) \mathrm{d}\tau \mathrm{d}t = \int_0^{\infty} \int_{\tau}^{\infty} F(\tau, t) \mathrm{d}t \mathrm{d}\tau = \int_0^{\infty} \int_{t}^{\infty} F(t, \tau) \mathrm{d}\tau \mathrm{d}t
$$

in above equation and obtain

$$
\begin{aligned}
\Delta c(\boldsymbol{u}_{:}) &= 2 \int_0^{\infty} \int_t^{\infty} \boldsymbol{x}(\tau)^{\mathrm{T}} \boldsymbol{Q} \mathrm{e}^{A(\tau-t)} \boldsymbol{B} \Delta \boldsymbol{u} \mathrm{d}\tau \mathrm{d}t + 2 \int_0^{\infty} \boldsymbol{u}^{\mathrm{T}} \boldsymbol{R} \Delta \boldsymbol{u} \mathrm{d}t \\
&= 2 \int_0^{\infty} \left(\int_t^{\infty} \boldsymbol{x}(\tau)^{\mathrm{T}} \boldsymbol{Q} \mathrm{e}^{A(\tau-t)} \boldsymbol{B} \mathrm{d}\tau + \boldsymbol{u}^{\mathrm{T}} \boldsymbol{R}\right) \Delta \boldsymbol{u} \mathrm{d}t.
\end{aligned}
$$

Since the infinitesimal variation $\Delta \boldsymbol{u}_{:}$ can be arbitrary, *for optimality of the control input function* $\boldsymbol{u}_{:}$, we must have

$$
\begin{aligned}
&\int_t^{\infty} \boldsymbol{x}(\tau)^{\mathrm{T}} \boldsymbol{Q} \mathrm{e}^{A(\tau-t)} \boldsymbol{B} \mathrm{d}\tau + \boldsymbol{u}^{\mathrm{T}} \boldsymbol{R} = 0 \\
\Longleftrightarrow\ &\boldsymbol{u} = -\boldsymbol{R}^{-1} \boldsymbol{B}^{\mathrm{T}} \int_t^{\infty} \mathrm{e}^{A^{\mathrm{T}}(\tau-t)} \boldsymbol{Q} \boldsymbol{x}(\tau) \mathrm{d}\tau.
\end{aligned}
\tag{7.18}
$$

Define the function transform on part of the right-hand side of (7.18) as

$$
\boldsymbol{y} \equiv \boldsymbol{P}^{-1} \int_t^{\infty} \mathrm{e}^{A^{\mathrm{T}}(\tau-t)} \boldsymbol{Q} \boldsymbol{x}(\tau) \mathrm{d}\tau, \tag{7.19}
$$

where $\boldsymbol{P}$ denotes the positive definite solution of the Riccati equation described in (7.16). As we suppose that the target process is controllable, $\boldsymbol{P}$ can be obtained via the Riccati equation iterative solving algorithm and is the unique positive definite solution of the Riccati equation described in (7.16). Denote $\boldsymbol{K} = \boldsymbol{P}\boldsymbol{B}\boldsymbol{R}^{-1}$ namely the optimal gain matrix of full-state feedback control, then (7.18) becomes a compact formalism as

$$u = -K^{\mathrm{T}} y. \tag{7.20}$$

Substitute (7.20) into the linear state differential equation described in (6.5) and associate (7.19) to establish a dual state differential equation group as

$$\begin{cases} \frac{\mathrm{d}}{\mathrm{d}t} x &= Ax - BK^{\mathrm{T}} y \\ P\frac{\mathrm{d}}{\mathrm{d}t} y &= -A^{\mathrm{T}} P y - Qx \end{cases} \tag{7.21}$$

We have

$$\begin{aligned} \frac{\mathrm{d}}{\mathrm{d}t}(x - y) &= P^{-1}[(PA + Q)x - (PBK^{\mathrm{T}} - A^{\mathrm{T}} P)y] \\ &= P^{-1}(PA + Q)(x - y) \end{aligned}$$

namely

$$\frac{\mathrm{d}}{\mathrm{d}t} z = Mz, \tag{7.22}$$

where $z \equiv x - y$ and $M \equiv P^{-1}(PA + Q)$. The matrix $-M$ satisfies

$$P(-M) + (-M)^{\mathrm{T}} P = -(PA + A^{\mathrm{T}} P + 2Q) = -(PBR^{-1} B^{\mathrm{T}} P + Q) < 0.$$

Also note that the matrix P is positive definite, so the matrix $-M$ is stable (according to the *Lyapunov criterion III*) and hence M has eigenvalues all with positive real part.

Solve (7.22) and obtain

$$z = \mathrm{e}^{Mt} z_0. \tag{7.23}$$

Substitute (7.23) into (7.20) and obtain

$$u = -K^{\mathrm{T}} y = -K^{\mathrm{T}}(x - z).$$

For optimality of the control input function u, which incurs a finite control cost, we must have $x \to 0$ and $u \to 0$ as $t \to \infty$, which implies $K^{\mathrm{T}} z = K^{\mathrm{T}} \mathrm{e}^{Mt} z_0 \to 0$ as $t \to \infty$. Each non-zero element of $K^{\mathrm{T}} \mathrm{e}^{Mt} z_0$ must be a linear combination of linearly independent function terms, in the form of $\sum_i c_i \mathrm{e}^{\lambda_i t} t^{k_i}$ where all λ_i belong to the set of eigenvalues of M. Since all eigenvalues of M are with positive real part, we definitely have $\sum_i c_i \mathrm{e}^{\lambda_i t} t^{k_i} \not\to 0$ as $t \to \infty$, which contradicts the condition that $K^{\mathrm{T}} \mathrm{e}^{Mt} z_0 \to 0$ as $t \to \infty$. To avoid such contradiction, we must have $K^{\mathrm{T}} z = K^{\mathrm{T}} \mathrm{e}^{Mt} z_0 \equiv 0$ and hence have

$$u = -K^{\mathrm{T}}(x - z) = -K^{\mathrm{T}}x. \tag{7.24}$$

The derived result given in (7.24) conveys an important and interesting fact: For the linear quadratic regulator, the optimal control input function $u_{\cdot}$ and the optimal state function $x_{\cdot}$ caused by the optimal $u_{\cdot}$ mutually satisfy a relationship the same to that of optimal full-state feedback control. In other words, for the linear quadratic regulator, the optimal control method is right the optimal full-state feedback control method.

Linear quadratic regulator solution: *For the linear quadratic regulator, the optimal control method is right the optimal full-state feedback control method.*

Double inverted pendulum control

We apply the optimal control method of linear quadratic regulator to perform double inverted pendulum control. First, we consider the original single-input double inverted pendulum control system that adopts linear state-space modelling described by (6.3) and obtain the optimal gain matrix of full-state feedback control. We take the same set of double inverted pendulum parameters as in Sect. 6.4 in Chap. 6. Let $m_1 = 1$, $m_2 = 1$, $L_1 = 1$, $L_2 = 1$, and $g = 10$. Then the state transition matrix A and the control input matrix B are

$$A = \begin{bmatrix} 0 & 1 & 0 & 0 & 0 & 0 \\ 20 & 0 & -10 & 0 & 0 & 0 \\ 0 & 0 & 0 & 1 & 0 & 0 \\ -20 & 0 & 20 & 0 & 0 & 0 \\ 0 & 0 & 0 & 0 & 0 & 1 \\ 0 & 0 & 0 & 0 & 0 & 0 \end{bmatrix}, \quad B = \begin{bmatrix} 0 \\ -1 \\ 0 \\ 0 \\ 0 \\ 1 \end{bmatrix}.$$

Set the cost matrices Q and R as

$$Q = \begin{bmatrix} 1 & & & & & \\ & 1 & & & & \\ & & 1 & & & \\ & & & 1 & & \\ & & & & 1 & \\ & & & & & 1 \end{bmatrix}, \quad R = 0.6.$$

Use the Riccati equation iterative solving algorithm to obtain the unique positive definite solution P of the Riccati equation described in (7.16) and compute the optimal gain matrix $K = PBR^{-1}$ as

$$\boldsymbol{P} = \begin{bmatrix} 1\,194.99 & 68.15 & -1\,635.62 & -375.96 & -11.72 & -35.09 \\ 68.15 & 13.89 & -35.09 & -4.46 & 2.58 & 4.81 \\ -1\,635.62 & -35.09 & 2\,583.58 & 614.77 & 35.08 & 88.10 \\ -375.96 & -4.46 & 614.77 & 147.54 & 9.24 & 22.72 \\ -11.72 & 2.58 & 35.08 & 9.24 & 2.77 & 3.35 \\ -35.09 & 4.81 & 88.10 & 22.72 & 3.35 & 6.96 \end{bmatrix},$$

$$\boldsymbol{K}^{\mathrm{T}} = [-172.07 \;\; -15.19 \;\; 205.32 \;\; 45.29 \;\; 1.29 \;\; 3.58]. \tag{7.25}$$

Matlab simulation code for demonstrating linear quadratic regulator control of the original double inverted pendulum control system is given as follows.

DoubleInvertedPendulum5.m

```
%% Double inverted pendulum parameters
m1 = 1; m2 = 1; L1 = 1; L2 = 1; g = 10;
%% Simulation preliminary configuration
dt = 0.001; % Numerical computation step
tSpan = 0:dt:8; % Simulation time span
x = 20; dx = 0; % Cart position and its velocity
% Inverted pendulum angle theta-1 and its angular velocity
y1 = 0.2; dy1 = 0;
% Inverted pendulum angle theta-2 and its angular velocity
y2 = 0; dy2 = 0;
stt = [y1; dy1; y2; dy2; x; dx]; % Double inverted pendulum
    state
% Record states in simulation
sttAll = zeros(length(stt), length(tSpan)); k = 0;
% Expected equilibrium status
xExpected = 0; y1Expected = 0; y2Expected = 0;
SimConfig = [m1, m2, L1, L2, g, dt];

%% Design the optimal LQR gain matrix (linear quadratic
    regulator)
A = [0, 1, 0, 0, 0, 0; ...
    (m1+m2)*g/(m1*L1), 0, -m2*g/(m1*L1), 0, 0, 0; ...
    0, 0, 0, 1, 0, 0; ...
    -(m1+m2)*g/(m1*L2), 0, (m1+m2)*g/(m1*L2), 0, 0, 0; ...
    0, 0, 0, 0, 0, 1; ...
    0, 0, 0, 0, 0, 0];
B = [0; -1/L1; 0; 0; 0; 1];
lambdaE = [-4;-4;-4;-4;-4;-4]; % Expected eigenvalues
sttK = DesignGainMatrixSub(A, B, lambdaE);
fprintf('Initial gain matrix K: '); sttK'
Q = eye(6); R = 0.6; % LQR cost matrices
[P, sttK] = SolveRiccatiEquationLQR(A, B, Q, R, sttK);
fprintf('Optimal gain matrix K: '); sttK'

%% Simulation of double inverted pendulum control
for t = tSpan
    %% Control method
    acc = -sttK'*stt; % Full-state feedback control of LQR
```

```
    %% Double inverted pendulum dynamics
    stt = DynamicsDIP(SimConfig, stt, acc);
    sttC = num2cell(stt); [y1, dy1, y2, dy2, x, dx] = sttC{:};
    if (abs(y1)>=pi/2 && abs(y2)>=pi/2)
        fprintf('Control failure!\n'); break;
    end
    k = k+1; sttAll(:,k) = stt;
    %% Double inverted pendulum visualization
    if (rem(k,20) == 0)
        DisplayDIP2(x, y1, y2, L1, L2); pause(20*dt);
    end
end
```

After trials with the Matlab simulation code, readers will find that the optimal control method of linear quadratic regulator indeed works regardless of whether the initial deviation of the cart position is small or large. This attributes to the merit of the optimal control methodology in regulating intermediate state evolution directly via minimization of certain control cost in terms of state related cost as well as control input related cost.

Second, we consider the multiple-input variant of the double inverted pendulum control system that adopts linear state-space modelling described by (6.51) and obtain the optimal gain matrix of full-state feedback control. We take the same set of double inverted pendulum parameters as in Sect. 6.4 in Chap. 6. Let $m_1 = 1, m_2 = 1, L_1 = 1$, $L_2 = 1$, and $g = 10$. Then the state transition matrix $\boldsymbol{A}$ and the control input matrix $\boldsymbol{B}$ are

$$\boldsymbol{A} = \begin{bmatrix} 0 & 1 & 0 & 0 & 0 & 0 \\ 20 & 0 & -10 & 0 & 0 & 0 \\ 0 & 0 & 0 & 1 & 0 & 0 \\ -20 & 0 & 20 & 0 & 0 & 0 \\ 0 & 0 & 0 & 0 & 0 & 1 \\ 0 & 0 & 0 & 0 & 0 & 0 \end{bmatrix}, \quad \boldsymbol{B} = \begin{bmatrix} 0 & 0 \\ -1 & 1 \\ 0 & 0 \\ 0 & -1 \\ 0 & 0 \\ 1 & 0 \end{bmatrix}.$$

Set the cost matrices $\boldsymbol{Q}$ and $\boldsymbol{R}$ as

$$\boldsymbol{Q} = \begin{bmatrix} 1 & & & & & \\ & 1 & & & & \\ & & 1 & & & \\ & & & 1 & & \\ & & & & 1 & \\ & & & & & 1 \end{bmatrix}, \quad \boldsymbol{R} = \begin{bmatrix} 0.6 & \\ & 0.6 \end{bmatrix}.$$

Apply the control input decomposition and iterative design method presented in Sect. 6.5 in Chap. 6 to find an initial gain matrix $\boldsymbol{K}_0$. Then use the Riccati equation iterative solving algorithm to obtain the unique positive definite solution $\boldsymbol{P}$ of the Riccati equation described in (7.16) and compute the optimal gain matrix $\boldsymbol{K} = \boldsymbol{P}\boldsymbol{B}\boldsymbol{R}^{-1}$ as

$$P = \begin{bmatrix} 74.50 & 17.01 & -25.28 & -1.10 & 3.46 & 6.66 \\ 17.01 & 10.65 & 33.00 & 11.07 & 2.97 & 6.08 \\ -25.28 & 33.00 & 236.93 & 66.56 & 11.47 & 24.32 \\ -1.10 & 11.07 & 66.56 & 19.36 & 3.63 & 7.64 \\ 3.46 & 2.97 & 11.47 & 3.63 & 2.47 & 2.57 \\ 6.66 & 6.08 & 24.32 & 7.64 & 2.57 & 4.96 \end{bmatrix}, \tag{7.26}$$

$$K^{\mathrm{T}} = \begin{bmatrix} -17.26 & -7.61 & -14.46 & -5.71 & -0.66 & -1.87 \\ 30.18 & -0.70 & -55.95 & -13.82 & -1.11 & -2.60 \end{bmatrix}.$$

Matlab simulation code for demonstrating linear quadratic regulator control of the variant of the double inverted pendulum control system is given as follows.

DoubleInvertedPendulumMIMO2.m

```
%% Double inverted pendulum parameters
m1 = 1; m2 = 1; L1 = 1; L2 = 1; g = 10;
%% Simulation preliminary configuration
dt = 0.001; % Numerical computation step
tSpan = 0:dt:8; % Simulation time span
x = 20; dx = 0; % Cart position and its velocity
% Inverted pendulum angle theta-1 and its angular velocity
y1 = 0.2; dy1 = 0;
% Inverted pendulum angle theta-2 and its angular velocity
y2 = 0; dy2 = 0;
stt = [y1; dy1; y2; dy2; x; dx]; % Double inverted pendulum
    state
% Record states in simulation
sttAll = zeros(length(stt), length(tSpan)); k = 0;
% Expected equilibrium status
xExpected = 0; y1Expected = 0; y2Expected = 0;
SimConfig = [m1, m2, L1, L2, g, dt];

%% Design the optimal LQR gain matrix (linear quadratic
    regulator)
A = [0, 1, 0, 0, 0, 0; ...
    (m1+m2)*g/(m1*L1), 0, -m2*g/(m1*L1), 0, 0, 0; ...
    0, 0, 0, 1, 0, 0; ...
    -(m1+m2)*g/(m1*L2), 0, (m1+m2)*g/(m1*L2), 0, 0, 0; ...
    0, 0, 0, 0, 0, 1; ...
    0, 0, 0, 0, 0, 0];
B = [0, 0; -1/L1, 1; 0, 0; 0, -L1/L2; 0, 0; 1, 0];
lambdaE = [-4;-4;-4;-4;-4;-4]; % Expected eigenvalues
sttK2 = DesignGainMatrixSub(A, B(:,2), lambdaE);
sttK1 = DesignGainMatrixSub(A-B(:,2)*sttK2', B(:,1), lambdaE);
sttK = [sttK1, sttK2]; fprintf('Initial gain matrix K: '); sttK'
Q = eye(6); R = 0.6*eye(2); % LQR cost matrices
[P, sttK] = SolveRiccatiEquationLQR(A, B, Q, R, sttK);
fprintf('Optimal gain matrix K: '); sttK'

%% Simulation of double inverted pendulum control
for t = tSpan
    %% Multiple-input-multiple-output control method
```

```
    accU = -sttK'*stt; % Full-state feedback control of LQR

    %% Double inverted pendulum dynamics
    stt = DynamicsDIP(SimConfig, stt, accU(1), accU(2));
    sttC = num2cell(stt); [y1, dy1, y2, dy2, x, dx] = sttC{:};
    if (abs(y1)>=pi/2 && abs(y2)>=pi/2)
        fprintf('Control failure!\n'); break;
    end
    k = k+1; sttAll(:,k) = stt;
    %% Double inverted pendulum visualization
    if (rem(k,20) == 0)
        DisplayDIP2(x, y1, y2, L1, L2); pause(20*dt);
    end
end
```

After trials with the Matlab simulation code, readers will find that the optimal control method of linear quadratic regulator also works for the variant of the double inverted pendulum control system. The control input part of first inverted pendulum angular acceleration a_1 is redundant in the sense that it plays no essential role in determining controllability of the double inverted pendulum, yet it enhances control flexibility and helps optimize double inverted pendulum control by reducing the control cost in comparison with the original single-input double inverted pendulum control system.

More specifically, for the variant of the double inverted pendulum control system, if we do not use the redundant control input part of first inverted pendulum angular acceleration a_1, we can remove its associated cost weight from the cost matrix $\boldsymbol{R}$ which will then be reduced to the same cost matrix $\boldsymbol{R}$ used for the original double inverted pendulum control system. As (7.8) conveys, the positive definite matrix $\boldsymbol{P}$ determines the control cost. So $\boldsymbol{P}_{\mathrm{S}}$ which denotes the positive definite solution $\boldsymbol{P}$ described in (7.25) determines the optimal control cost of the original double inverted pendulum control system, whereas $\boldsymbol{P}_{\mathrm{M}}$ which denotes the positive definite solution $\boldsymbol{P}$ described in (7.26) determines the optimal control cost of the variant of the double inverted pendulum control system. As $\boldsymbol{P}_{\mathrm{M}} < \boldsymbol{P}_{\mathrm{S}}$ namely $\boldsymbol{P}_{\mathrm{M}} - \boldsymbol{P}_{\mathrm{S}}$ is negative definite, we can see that the redundant control input part of first inverted pendulum angular acceleration a_1 indeed contributes to reduction of the control cost.

This reflects another merit of the optimal control methodology: For multiple-input-multiple-output control, the optimal control methodology can take "best" advantage of multiple-input (including cooperation among various control input parts) by minimizing the control cost.

Cooperative longitudinal control of vehicle platooning

We also apply the optimal control method of linear quadratic regulator to perform cooperative longitudinal control of vehicle platooning which serves as another concrete example, as illustrated in Fig. 6.4. Suppose that there are four vehicles in platooning. Dynamics of the four-vehicle cooperative longitudinal control system's state $\boldsymbol{x}$ is modelled by the state differential equation described in (6.80):

$$\frac{\mathrm{d}}{\mathrm{d}t}\boldsymbol{x} = \begin{bmatrix} 0&0&0&1&-1&0&0 \\ 0&0&0&0&1&-1&0 \\ 0&0&0&0&0&1&-1 \\ 0&0&0&0&0&0&0 \\ 0&0&0&0&0&0&0 \\ 0&0&0&0&0&0&0 \\ 0&0&0&0&0&0&0 \end{bmatrix}\boldsymbol{x} + \begin{bmatrix} 0&0&0&0 \\ 0&0&0&0 \\ 0&0&0&0 \\ 1&0&0&0 \\ 0&1&0&0 \\ 0&0&1&0 \\ 0&0&0&1 \end{bmatrix}\begin{bmatrix} a_1 \\ a_2 \\ a_3 \\ a_4 \end{bmatrix} \equiv \boldsymbol{A}\boldsymbol{x} + \boldsymbol{B}\boldsymbol{u},$$

where the state $\boldsymbol{x}$ represents the error between the absolute state of the four vehicles and certain expected absolute state (note that the four vehicles share a common expected vehicle velocity). The control input is the multiple-input of first vehicle acceleration a_1, second vehicle acceleration a_2, third vehicle acceleration a_3, and fourth vehicle acceleration a_4.

In the spirit of optimal control, the four-vehicle cooperative longitudinal control system aims at controlling the multiple-input of a_1, a_2, a_3, and a_4 simultaneously to converge the state $\boldsymbol{x}$ namely the absolute state error to $\boldsymbol{0}$ at the minimum control cost. Set the cost matrices $\boldsymbol{Q}$ and $\boldsymbol{R}$ as

$$\boldsymbol{Q} = \begin{bmatrix} 1 &&&&&& \\ & 1 &&&&& \\ && 1 &&&& \\ &&& 1 &&& \\ &&&& 1 && \\ &&&&& 1 & \\ &&&&&& 1 \end{bmatrix}, \quad \boldsymbol{R} = \begin{bmatrix} 0.6 &&& \\ & 0.6 && \\ && 0.6 & \\ &&& 0.6 \end{bmatrix}.$$

Apply the control input decomposition and iterative design method presented in Sect. 6.5 in Chap. 6 to find an initial gain matrix $\boldsymbol{K}_0$. Solve the following Lyapunov equation

$$\boldsymbol{P}_0(\boldsymbol{A} - \boldsymbol{B}\boldsymbol{K}_0^{\mathrm{T}}) + (\boldsymbol{A} - \boldsymbol{B}\boldsymbol{K}_0^{\mathrm{T}})^{\mathrm{T}}\boldsymbol{P}_0 = -\boldsymbol{Q} - \boldsymbol{K}_0\boldsymbol{R}\boldsymbol{K}_0^{\mathrm{T}}$$

to obtain the unique positive definite solution of $\boldsymbol{P}_0$ which determines the control cost associated with the gain matrix $\boldsymbol{K}_0$:

$$\boldsymbol{P}_0 = \begin{bmatrix} 1.65 & 0.62 & 0.21 & 0.80 & -0.41 & -0.16 & -0.12 \\ 0.62 & 2.48 & 1.18 & 0.41 & 0.80 & -0.57 & -0.78 \\ 0.21 & 1.18 & 3.12 & 0.16 & 0.57 & 0.80 & -2.33 \\ 0.80 & 0.41 & 0.16 & 1.25 & -0.20 & -0.10 & -0.09 \\ -0.41 & 0.80 & 0.57 & -0.20 & 1.46 & -0.16 & -0.41 \\ -0.16 & -0.57 & 0.80 & -0.10 & -0.16 & 1.53 & -0.79 \\ -0.12 & -0.78 & -2.33 & -0.09 & -0.41 & -0.79 & 3.13 \end{bmatrix}.$$

Use the Riccati equation iterative solving algorithm to obtain the unique positive definite solution $\boldsymbol{P}$ of the Riccati equation described in (7.16) and compute the optimal gain matrix $\boldsymbol{K} = \boldsymbol{PBR}^{-1}$ as

$$\boldsymbol{P} = \begin{bmatrix} 1.38 & 0.31 & 0.12 & 0.63 & -0.42 & -0.13 & -0.08 \\ 0.31 & 1.50 & 0.31 & 0.21 & 0.51 & -0.51 & -0.21 \\ 0.12 & 0.31 & 1.38 & 0.08 & 0.13 & 0.42 & -0.63 \\ 0.63 & 0.21 & 0.08 & 1.14 & -0.22 & -0.09 & -0.06 \\ -0.42 & 0.51 & 0.13 & -0.22 & 1.27 & -0.19 & -0.09 \\ -0.13 & -0.51 & 0.42 & -0.09 & -0.19 & 1.27 & -0.22 \\ -0.08 & -0.21 & -0.63 & -0.06 & -0.09 & -0.22 & 1.14 \end{bmatrix},$$

$$\boldsymbol{K}^{\mathrm{T}} = \begin{bmatrix} 1.05 & 0.35 & 0.14 & 1.90 & -0.36 & -0.14 & -0.10 \\ -0.70 & 0.84 & 0.21 & -0.36 & 2.12 & -0.32 & -0.14 \\ -0.21 & -0.84 & 0.70 & -0.14 & -0.32 & 2.12 & -0.36 \\ -0.14 & -0.35 & -1.05 & -0.10 & -0.14 & -0.36 & 1.90 \end{bmatrix}.$$

Note that $\boldsymbol{P} < \boldsymbol{P}_0$ namely $\boldsymbol{P} - \boldsymbol{P}_0$ is negative definite, which implies that the optimal gain matrix $\boldsymbol{K}$ does incur less control cost than the initial gain matrix $\boldsymbol{K}_0$. Readers can try with various gain matrices to check the optimality of the optimal gain matrix $\boldsymbol{K}$.

Matlab simulation code for demonstrating linear quadratic regulator control of the four-vehicle cooperative longitudinal control system is given as follows.

CooperativeLongitudinalControlMIMO3.m

```
%% Cooperative longitudinal control model for four-vehicle
   platooning
A = [0, 0, 0, 1, -1, 0, 0; ...
     0, 0, 0, 0, 1, -1, 0; ...
     0, 0, 0, 0, 0, 1, -1; ...
     0, 0, 0, 0, 0, 0, 0; ...
     0, 0, 0, 0, 0, 0, 0; ...
     0, 0, 0, 0, 0, 0, 0; ...
     0, 0, 0, 0, 0, 0, 0];
B = [0, 0, 0, 0; 0, 0, 0, 0; 0, 0, 0, 0; ...
     1, 0, 0, 0; 0, 1, 0, 0; 0, 0, 1, 0; 0, 0, 0, 1];
%% Simulation preliminary configuration
dt = 0.001; % Numerical computation step
tSpan = 0:dt:15; % Simulation time span
x1 = 1; % First and second vehicle inter-vehicle position error
x2 = -1; % Second and third vehicle inter-vehicle position error
x3 = 1; % Third and fourth vehicle inter-vehicle position error
v1 = 1; % First vehicle velocity error
v2 = -1; % Second vehicle velocity error
v3 = 1; % Third vehicle velocity error
v4 = -1; % Fourth vehicle velocity error
stt = [x1; x2; x3; v1; v2; v3; v4]; % Cooperative longitudinal
    state
% Record states in simulation
sttAll = zeros(length(stt), length(tSpan)); k = 0;

%% Design the gain matrix iteratively
lambdaE = -ones(1,7); % Expected eigenvalues
```

```
sttK1 = DesignGainMatrixSub(A, B(:,1), lambdaE);
A2 = A-B(:,1)*sttK1';
sttK2 = DesignGainMatrixSub(A2, B(:,2), lambdaE);
A3 = A2-B(:,2)*sttK2';
sttK3 = DesignGainMatrixSub(A3, B(:,3), lambdaE);
A4 = A3-B(:,3)*sttK3';
sttK4 = DesignGainMatrixSub(A4, B(:,4), lambdaE);
sttK = [sttK1, sttK2, sttK3, sttK4];
fprintf('Initial gain matrix K: '); sttK'
Q = eye(7); R = 0.6*eye(4); % LQR cost matrices
Pinit = SolveLyapunovEquation(A-B*sttK', -Q-sttK*R*sttK');
[P, sttK] = SolveRiccatiEquationLQR(A, B, Q, R, sttK);
fprintf('Optimal gain matrix K: '); sttK'

k = 0; figure(1), set(gcf, 'Position', [100, 0, 1100, 800]);
%% Simulation of cooperative longitudinal control
for t = tSpan
    %% Multiple-input-multiple-output control method
    accU = -sttK'*stt; % Full-state feedback control of LQR

    %% Cooperative longitudinal dynamics
    x1 = x1 + (v1 - v2)*dt;
    x2 = x2 + (v2 - v3)*dt;
    x3 = x3 + (v3 - v4)*dt;
    v1 = v1 + accU(1)*dt;
    v2 = v2 + accU(2)*dt;
    v3 = v3 + accU(3)*dt;
    v4 = v4 + accU(4)*dt;
    stt = [x1; x2; x3; v1; v2; v3; v4];
    k = k+1; sttAll(:,k) = stt;
end
subplot(4,2,3), plot(tSpan, sttAll(1,:), 'LineWidth', 2);
ylabel('\Delta x_1 Error'); grid on;
subplot(4,2,5), plot(tSpan, sttAll(2,:), 'LineWidth', 2);
ylabel('\Delta x_2 Error'); grid on;
subplot(4,2,7), plot(tSpan, sttAll(3,:), 'LineWidth', 2);
xlabel('Time'); ylabel('\Delta x_3 Error'); grid on;
subplot(4,2,2), plot(tSpan, sttAll(4,:), 'LineWidth', 2);
ylabel('v_1 Error'); grid on;
subplot(4,2,4), plot(tSpan, sttAll(5,:), 'LineWidth', 2);
ylabel('v_2 Error'); grid on;
subplot(4,2,6), plot(tSpan, sttAll(6,:), 'LineWidth', 2);
ylabel('v_3 Error'); grid on;
subplot(4,2,8), plot(tSpan, sttAll(7,:), 'LineWidth', 2);
xlabel('Time'); ylabel('v_4 Error'); grid on;
```

7.3 Model Predictive Control

7.3.1 System Model Simplification for Optimal Control

As presented in Sect. 7.2, given a control system that adopts generic state-space modelling described by (6.2):

$$\frac{\mathrm{d}}{\mathrm{d}t}\boldsymbol{x} = f(\boldsymbol{x}, \boldsymbol{u}).$$

We can treat the state function $\boldsymbol{x}_{:}$ as a functional in terms of the control input function $\boldsymbol{u}_{:}$. Once certain control cost functional $c(\boldsymbol{x}_{:}, \boldsymbol{u}_{:})$ is defined, the methodology of *optimal control* is formalized in (7.3):

$$\boldsymbol{u}_{:} = \arg\min_{\boldsymbol{u}_{:}} c(\boldsymbol{x}_{:}, \boldsymbol{u}_{:}) \equiv \arg\min_{\boldsymbol{u}_{:}} c(\boldsymbol{u}_{:}),$$

where the control cost functional $c(\boldsymbol{x}_{:}, \boldsymbol{u}_{:})$ can be treated as a functional implicitly in terms of $\boldsymbol{u}_{:}$, and we may abuse the control cost functional notation to simply denote $c(\boldsymbol{u}_{:}) \equiv c(\boldsymbol{x}_{:}, \boldsymbol{u}_{:})$.

If the control system can fairly adopt linear state-space modelling described by (6.5) and if the control cost functional $c(\boldsymbol{x}_{:}, \boldsymbol{u}_{:})$ adopts the quadratic form described in (7.4), then the methodology of optimal control is instantiated as the *linear quadratic regulator* described by (7.5):

$$\boldsymbol{u}_{:} = \arg\min_{\boldsymbol{u}_{:}} \int_{0}^{\infty} \left(\boldsymbol{x}^{\mathrm{T}}\boldsymbol{Q}\boldsymbol{x} + \boldsymbol{u}^{\mathrm{T}}\boldsymbol{R}\boldsymbol{u}\right) \mathrm{d}t,$$

which can be solved analytically and its solution is right the optimal full-state feedback control method.

What if linear state space modelling cannot be adopted for the control system? In this case, it is normally difficult to solve (7.3) analytically and even numerically. Despite such difficulty, can we still take advantage of the spirit of optimal control in regulating intermediate state evolution via minimization of certain control cost?

Fortunately, the answer is *yes*, yet we need to follow the spirit of optimal control in adapted way. More specifically, we may simplify the original system model described by (6.2) to a new formalism as

$$\frac{\mathrm{d}}{\mathrm{d}t}\boldsymbol{x} = \bar{f}(\boldsymbol{x}, \boldsymbol{u}) \tag{7.27}$$

such that (7.27) can somehow approximate dynamics of the control system's state on one hand and that

$$\boldsymbol{u}_{:} = \arg\min_{\boldsymbol{u}_{:}} c(\boldsymbol{x}_{:}, \boldsymbol{u}_{:})|_{\frac{\mathrm{d}}{\mathrm{d}t}\boldsymbol{x}=\bar{f}(\boldsymbol{x},\boldsymbol{u})} \tag{7.28}$$

can be effectively solved in analytical or numerical way on the other hand.

There are usually two directions for simplification of (6.2) into (7.27). First, we may simplify the system model directly by approximating partial state dynamics. Second, we may simplify the system model indirectly by confining the control input function to tractable patterns. A practical application namely intelligent vehicle lateral (steering) control will be presented later for demonstration.

As only a simplified version of the original system model described by (6.2), the new system model described by (7.27) may not predict state evolution so accurately in the long run, yet (7.27) tends to fairly predict general tendency of state evolution. Consequently, the control input function $\boldsymbol{u}_{:}$ obtained by solving (7.28) may not be really optimal in the long run, yet the obtained $\boldsymbol{u}_{:}$ in short time tends to enable the state to evolve optimally, at least in the sense of general tendency of state evolution. In other words, the obtained $\boldsymbol{u}_{:}$ in short time tends to be reasonable, whereas the obtained $\boldsymbol{u}_{:}$ in the long run may not.

7.3.2 *Dynamical Optimal Control*

How can we harmonize inconsistency between the performance of the obtained control input function $\boldsymbol{u}_{:}$ in short time and that in the long run? For this concern, an idea is: At the first control period $t = 0$, based on current state feedback, solve the following optimization problem

$$\boldsymbol{u}_{0:\infty} = \arg\min_{\boldsymbol{u}_{0:\infty}} c(\boldsymbol{x}_{0:\infty}, \boldsymbol{u}_{0:\infty})|_{\frac{\mathrm{d}}{\mathrm{d}t}\boldsymbol{x}=\bar{f}(\boldsymbol{x},\boldsymbol{u})}$$

to obtain the optimal control input function $\boldsymbol{u}_{0:\infty}$. Adopt the first control input $\boldsymbol{u}_0$ from the obtained $\boldsymbol{u}_{0:\infty}$ but discard all remaining part of the obtained $\boldsymbol{u}_{0:\infty}$. Apply only the first control input $\boldsymbol{u}_0$ to the control system at the first control period. At next control period $t = \Delta t$, based on state feedback at the moment then, solve the following optimization problem

$$\boldsymbol{u}_{\Delta t:\infty} = \arg\min_{\boldsymbol{u}_{\Delta t:\infty}} c(\boldsymbol{x}_{\Delta t:\infty}, \boldsymbol{u}_{\Delta t:\infty})|_{\frac{\mathrm{d}}{\mathrm{d}t}\boldsymbol{x}=\bar{f}(\boldsymbol{x},\boldsymbol{u})}$$

to obtain the optimal control input function $\boldsymbol{u}_{\Delta t:\infty}$. Also adopt the first control input $\boldsymbol{u}_{\Delta t}$ but discard all remaining part of the obtained $\boldsymbol{u}_{\Delta t:\infty}$. Also apply only the first control input $\boldsymbol{u}_{\Delta t}$ to the control system. Further at next control period $t = 2\Delta t$, solve the following optimization problem

$$\boldsymbol{u}_{2\Delta t:\infty} = \arg\min_{\boldsymbol{u}_{2\Delta t:\infty}} c(\boldsymbol{x}_{2\Delta t:\infty}, \boldsymbol{u}_{2\Delta t:\infty})|_{\frac{\mathrm{d}}{\mathrm{d}t}\boldsymbol{x}=\bar{f}(\boldsymbol{x},\boldsymbol{u})}$$

to obtain the optimal control input function $\boldsymbol{u}_{2\Delta t:\infty}$. Also adopt the first control input $\boldsymbol{u}_{2\Delta t}$ only and apply it to the control system. Continue above process in similar way at control periods $t = 3\Delta t, 4\Delta t, 5\Delta t, \ldots$, and so on.

In one word, *this idea consists in performing optimal control iteratively in dynamical way*: At each control period t, based on current state feedback, solve the following optimization problem

$$\boldsymbol{u}_{t:\infty} = \arg\min_{\boldsymbol{u}_{t:\infty}} c(\boldsymbol{x}_{t:\infty}, \boldsymbol{u}_{t:\infty})|_{\frac{\mathrm{d}}{\mathrm{d}t}\boldsymbol{x}=\bar{f}(\boldsymbol{x},\boldsymbol{u})} \tag{7.29}$$

to obtain the optimal control input function $\boldsymbol{u}_{t:\infty}$. Adopt the current control input namely the first control input $\boldsymbol{u}_t$ from the obtained $\boldsymbol{u}_{t:\infty}$ but discard all remaining part of the obtained $\boldsymbol{u}_{t:\infty}$. Apply only $\boldsymbol{u}_t$ to the control system at current control period t. To distinguish from the original methodology of optimal control, above methodology of dynamical optimal control is called the **model predictive control** or simply **predictive control** (Xi 2013).

Model predictive control
Initialization:
Approximate the system model (6.2) by a fairly simplified version (7.27).
Iteration:
Retrieve state feedback at current control period t.
Solve (7.29) to obtain the optimal control input function $\boldsymbol{u}_{t:\infty}$.
Adopt the first control input $\boldsymbol{u}_t$ but discard all remaining part of $\boldsymbol{u}_{t:\infty}$.
Apply only $\boldsymbol{u}_t$ to the control system at t. Then $t \to t + \Delta t$.

Intelligent vehicle lateral control

Consider a practical application namely intelligent vehicle lateral (steering) control for demonstration of model predictive control. For example, consider intelligent vehicle parking control illustrated in Fig. 1.10 in Chap. 1, which aims at controlling the intelligent vehicle to move from certain initial pose to the destination pose in a parking slot. Low-speed dynamics of the intelligent vehicle state $\boldsymbol{x}$ can be modelled by a state differential equation as

$$\frac{\mathrm{d}}{\mathrm{d}t}\boldsymbol{x} \equiv \frac{\mathrm{d}}{\mathrm{d}t}\begin{bmatrix} x \\ y \\ \theta \\ \beta \\ v \end{bmatrix} = \begin{bmatrix} v\cos\theta \\ v\sin\theta \\ \frac{1}{L}v\tan\beta \\ \max\{\min\{\frac{1}{\tau_\beta}(\beta_\mathrm{I}-\beta), s_\mathrm{M}\}, -s_\mathrm{M}\} \\ \max\{\min\{\frac{1}{\tau_v}(v_\mathrm{I}-v), a_\mathrm{M}\}, -a_\mathrm{M}\} \end{bmatrix} \equiv f(\boldsymbol{x}, \boldsymbol{u}), \tag{7.30}$$

where the state $\boldsymbol{x}$ consists of the horizontal position x, the vertical position y, and the orientation or heading θ of the intelligent vehicle in a global reference, as well as the steering angle β and the velocity v of the intelligent vehicle. Besides, L denotes the vehicle wheel-base, τ_β denotes the time-constant of the steer controller, s_M denotes the maximum steering velocity, τ_v denotes the time-constant of the velocity controller, and a_M denotes the maximum vehicle acceleration (or deceleration). The control input $\boldsymbol{u}$ is the multiple-input of vehicle steering angle command β_I and vehicle velocity command v_I.

We decouple intelligent vehicle longitudinal control and intelligent vehicle lateral control. For longitudinal control, we may empirically follow the constrained proportional control method as

$$v_\mathrm{I} = \max\{\min\{-P\Delta x, v_\mathrm{M}\}, -v_\mathrm{M}\},$$

where Δx denotes the distance of the intelligent vehicle in front of or behind the expected parking slot pose, and v_M denotes the maximum vehicle velocity allowed during parking.

The more difficult part of intelligent vehicle parking control is lateral control and we focus on this part. The system model described by (7.30) is nonlinear and complicated. It is difficult to apply optimal control with (7.30) as

$$\boldsymbol{u}_: = \arg\min_{\boldsymbol{u}_:} c(\boldsymbol{x}_:, \boldsymbol{u}_:)|_{\frac{\mathrm{d}}{\mathrm{d}t}\boldsymbol{x}=f(\boldsymbol{x},\boldsymbol{u})}.$$

Instead, we resort to the methodology of model predictive control. For this purpose, we simplify (7.30) in two directions: First, we simplify (7.30) directly by neglecting both lateral and longitudinal transient dynamics and obtain

$$\frac{\mathrm{d}}{\mathrm{d}t}\begin{bmatrix} x \\ y \\ \theta \\ \beta \\ v \end{bmatrix} = \begin{bmatrix} v_\mathrm{I}\cos\theta \\ v_\mathrm{I}\sin\theta \\ \frac{1}{L}v_\mathrm{I}\tan\beta_\mathrm{I} \\ \beta_\mathrm{I} \\ v_\mathrm{I} \end{bmatrix}. \tag{7.31}$$

Given constant β_I and v_I, based on (7.31), we can conveniently derive the state evolution between any two time instants t and t' as

$$\begin{cases} \theta_{t'} - \theta_t = \frac{v_\mathrm{I}}{L}\tan\beta_\mathrm{I}(t'-t) \equiv \omega_\mathrm{I}(t'-t), \\ x_{t'} - x_t = \int_t^{t'} v_\mathrm{I}\cos\theta \mathrm{d}t = \frac{v_\mathrm{I}}{\omega_\mathrm{I}}(\sin\theta_{t'} - \sin\theta_t), \\ y_{t'} - y_t = \int_t^{t'} v_\mathrm{I}\sin\theta \mathrm{d}t = \frac{v_\mathrm{I}}{\omega_\mathrm{I}}(\cos\theta_t - \cos\theta_{t'}). \end{cases} \tag{7.32}$$

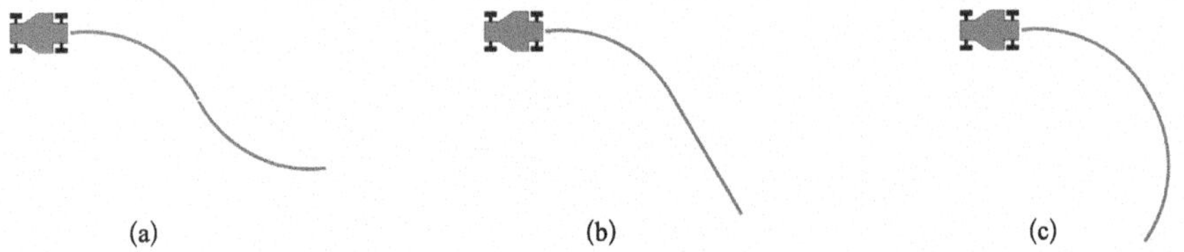

Fig. 7.3 Typical steering modes: (a) steering/back-steering mode; (b) steering/straight mode; (c) pure-steering mode

It is worth noting that when $\omega_{\mathrm{I}} \approx 0$, the last two equations in (7.32) are replaced by

$$\begin{cases} x_{t'} - x_t = v_{\mathrm{I}}(t' - t)\cos\frac{\theta_{t'}+\theta_t}{2}, \\ y_{t'} - y_t = v_{\mathrm{I}}(t' - t)\sin\frac{\theta_{t'}+\theta_t}{2}. \end{cases} \tag{7.33}$$

Second, we simplify (7.30) indirectly by confining the control input function to a special action pattern. More specifically, we set a predictive time span as

$$T_{\mathrm{P}} = \frac{\sqrt{(x - x_{\mathrm{E}})^2 + (y - y_{\mathrm{E}})^2}}{v_{\mathrm{I}}},$$

which heuristically represents the time roughly needed for the intelligent vehicle to move from its current pose to the expected parking slot pose. Besides, to avoid a too long predictive time span T_{P} that causes predictive ability of the simplified system model to deteriorate significantly, we set an upper limit for the predictive time span T_{P} as

$$T_{\mathrm{P}} = \min\left\{\frac{\sqrt{(x - x_{\mathrm{E}})^2 + (y - y_{\mathrm{E}})^2}}{v_{\mathrm{I}}}, T_{\mathrm{M}}\right\}. \tag{7.34}$$

Divide the predictive time horizon $[0, T_{\mathrm{P}}]$ into two even halves $[0, T_{\mathrm{P}}/2]$ and $[T_{\mathrm{P}}/2, T_{\mathrm{P}}]$. In the first half $[0, T_{\mathrm{P}}/2]$, set $\beta_{\mathrm{I}} = \beta_{\mathrm{I1}} \in [-\beta_{\mathrm{M}}, \beta_{\mathrm{M}}]$ where β_{M} denotes the maximum steering angle. Then in the second half $[T_{\mathrm{P}}/2, T_{\mathrm{P}}]$, set $\beta_{\mathrm{I}} = \beta_{\mathrm{I2}} \in \{-\beta_{\mathrm{I1}}, 0, \beta_{\mathrm{I1}}\}$. In other words, choices of the steering angle command β_{I2} for the second half $[T_{\mathrm{P}}/2, T_{\mathrm{P}}]$ depend on the choice of the steering angle command β_{I1} for the first half $[0, T_{\mathrm{P}}/2]$. Such pattern of the steering angle command pair $\{\beta_{\mathrm{I1}}, \beta_{\mathrm{I2}}\}$ is called the *double-action pattern*.

The double-action pattern $\{\beta_{\mathrm{I1}}, \beta_{\mathrm{I2}}\}$ has three modes namely the steering/back-steering mode $\{\beta_{\mathrm{I1}}, -\beta_{\mathrm{I1}}\}$, the steering/straight mode $\{\beta_{\mathrm{I1}}, 0\}$, and the pure-steering mode $\{\beta_{\mathrm{I1}}, \beta_{\mathrm{I1}}\}$ that correspond to three typical daily-life steering modes respectively, as illustrated in Fig. 7.3. The steering/straight mode $\{\beta_{\mathrm{I1}}, 0\}$ is reduced to the pure-straight mode if $\beta_{\mathrm{I1}} = 0$.

After above simplification, the control input function $\boldsymbol{u}_{t:t+T_{\mathrm{P}}}$ can be represented by the double-action pattern $\{\beta_{\mathrm{I1}}, \beta_{\mathrm{I2}}\}$. We can use (7.32) to conveniently predict the state at any time of the predictive time horizon $[0, T_{\mathrm{P}}]$, yet we only examine the predicted state at the end of the predictive time horizon and compare it with the expected parking slot pose. We have

$$\begin{aligned}
v_{t+T_P} &= v_I, \\
\beta_{t+T_P} &= \beta_{I2}, \\
\theta_{t+T_P/2} &= \theta_t + \omega_{I1}\frac{T_P}{2}, \\
\theta_{t+T_P} &= \theta_t + \omega_{I1}\frac{T_P}{2} + \omega_{I2}\frac{T_P}{2}, \\
x_{t+T_P} &= x_t + \frac{v_I}{\omega_{I1}}(\sin\theta_{t+T_P/2} - \sin\theta_t) + \frac{v_I}{\omega_{I2}}(\sin\theta_{t+T_P} - \sin\theta_{t+T_P/2}), \\
y_{t+T_P} &= y_t + \frac{v_I}{\omega_{I1}}(\cos\theta_t - \cos\theta_{t+T_P/2}) + \frac{v_I}{\omega_{I2}}(\cos\theta_{t+T_P/2} - \cos\theta_{t+T_P}),
\end{aligned}$$

compactly denoted as

$$\boldsymbol{x}_{t+T_P} = \bar{f}(\boldsymbol{x}_t, \boldsymbol{u}_{t:t+T_P}) \equiv \bar{f}(\boldsymbol{x}_t, \{\beta_{I1}, \beta_{I2}\}). \tag{7.35}$$

When $\omega_{I1} \approx 0$ or $\omega_{I2} \approx 0$, the terms associated with ω_{I1} or ω_{I2} in the last two equations above are replaced by corresponding terms as those in (7.33).

Define the control cost functional $c(\boldsymbol{x}_{t:t+T_P}, \boldsymbol{u}_{t:t+T_P})$ namely the control cost functional on the predictive time horizon as

$$\begin{aligned}
c(\boldsymbol{x}_{t:t+T_P}, \boldsymbol{u}_{t:t+T_P}) &\equiv c(\boldsymbol{x}_{t+T_P}, \{\beta_{I1}, \beta_{I2}\}) \\
&= (\boldsymbol{x}_{t+T_P} - \boldsymbol{x}_E)^T \boldsymbol{Q}(\boldsymbol{x}_{t+T_P} - \boldsymbol{x}_E) + R\beta_{I1}^2.
\end{aligned} \tag{7.36}$$

Then instantiate the methodology of model predictive control for intelligent vehicle lateral control as follows. At each control period t, based on current intelligent vehicle state feedback, solve the following optimization problem

$$\boldsymbol{u}_{t:t+T_P} = \arg\min_{\boldsymbol{u}_{t:t+T_P}} c(\boldsymbol{x}_{t:t+T_P}, \boldsymbol{u}_{t:t+T_P})|_{\boldsymbol{x}_{t+T_P} = \bar{f}(\boldsymbol{x}_t, \boldsymbol{u}_{t:t+T_P})}$$

namely

$$\{\beta_{I1}, \beta_{I2}\} = \arg\min_{\{\beta_{I1}, \beta_{I2}\}} c(\boldsymbol{x}_{t+T_P}, \{\beta_{I1}, \beta_{I2}\})|_{\boldsymbol{x}_{t+T_P} = \bar{f}(\boldsymbol{x}_t, \{\beta_{I1}, \beta_{I2}\})} \tag{7.37}$$

to obtain the optimal control input function $\boldsymbol{u}_{t:t+T_P}$ namely the optimal double-action pattern $\{\beta_{I1}, \beta_{I2}\}$. Adopt the first control input β_{I1} but discard β_{I2}. Apply only β_{I1} to the intelligent vehicle lateral control system at current control period t.

Model predictive control for intelligent vehicle lateral control

Initialization:

Approximate the system model (7.30) by a fairly simplified version (7.35).

Iteration:

Retrieve intelligent vehicle state feedback at current control period t.

Solve (7.37) to obtain the optimal double-action pattern $\{\beta_{I1}, \beta_{I2}\}$.

Adopt the first control input β_{I1} but discard β_{I2}.

Apply only β_{I1} to the intelligent vehicle lateral control system at t.

Then $t \to t + \Delta t$.

Matlab simulation code for complete demonstration of intelligent vehicle parking control (especially the part of intelligent vehicle lateral control) is given as follows.

IntelligentVehicleMPCPark.m

```
%% Intelligent vehicle parameters
vehL = 2; % Vehicle wheel-base
rotT = 0.2; % Steering time-constant
rotM = pi/2; % Maximum steering velocity
accT = 0.2; % Acceleration time-constant
accM = 4; % Maximum acceleration
%% Simulation preliminary configuration
dt = 0.02; % Numerical computation step
tSpan = 0:dt:9; % Simulation time span
SimConfig = [vehL, rotT, rotM, accT, accM, dt];
x = -1; y = -3; % Position (x, y)
a = 0; % Orientation angle
s = 0; % Lateral steering angle
v = 0; % Longitudinal velocity
stt = [x; y; a; s; v]; % Intelligent vehicle state
% Record states in simulation
sttAll = zeros(length(stt), length(tSpan)); k = 0;
sttE = [6; 0; pi; 0; 0]; % Expected intelligent vehicle state
parkX = [sttE(1)-vehL, sttE(1)+0.5*vehL, sttE(1)+0.5*vehL,
         sttE(1)-vehL];
parkY = [sttE(2)-vehL/2, sttE(2)-vehL/2, sttE(2)+vehL/2,
         sttE(2)+vehL/2];
%% Specify potential control input functions of steering
sC = (-pi/4:pi/60:pi/4)';
% Choices of double-action pattern (DAP)
sC = [[sC,0*sC]; [sC,sC]; [sC,-sC]];

%% Simulation of intelligent vehicle control
for t = tSpan
    %% Model predictive control method
    vIn = -5*(sttE(1)-stt(1)); vIn = max(min(vIn,2), -2);
    % Cost matrix for predicted state error
    Q = diag([1, 5, 1, 0.05, 0])^2;
    % Cost matrix for control input
    R = Q(4,4)*(1+1/max(abs(vIn), 0.01));
```

```
    costCMax = 10000;
    % Evaluate each control input function namely each choice of
     DAP
    for idx = 1:size(sC,1)
        sAng1 = sC(idx,1); sAng2 = sC(idx,2);
        if (abs(vIn)<0.001) break; end
        dT = min(sqrt((stt(1)-sttE(1))^2+(stt(2)
                 -sttE(2))^2)/abs(vIn), 1.6);
        % Prediction
        sttP = DynamicsIVforMPC([vehL, dT/2], stt, sAng1, vIn);
        % Prediction
        sttP = DynamicsIVforMPC([vehL, dT/2], sttP, sAng2, vIn);
        e = sttP-sttE;
        e(3) = min(mod(e(3),2*pi), 2*pi-mod(e(3),2*pi));
        costC = e'*Q*e + R*sAng1^2; % Control cost
        if (costC<costCMax) opt = idx; costCMax = costC; end
    end
    % Take only the current (first) control input element
    sIn = sC(opt,1);
    fprintf('[%f] Optimal steering of DAP: %f, %f\n', t,
            sIn, sC(opt,2));

    %% Intelligent vehicle dynamics
    stt = DynamicsIV(SimConfig, stt, sIn, vIn);
    k = k+1; sttAll(:,k) = stt;
    %% Intelligent vehicle visualization
    figure(1); clf, line(parkX, parkY, 'Color', 'r', 'LineWidth'
    , 3);
    hold on;
    DisplayIV(stt, vehL);
    axis equal; xlim([-6, 8]); ylim([-6, 6]); hold off; pause(dt
    );
end
```

The visualization code **DisplayIV.m**, the intelligent vehicle dynamics code **DynamicsIV.m** that corresponds to (7.30), and the model predictive control oriented intelligent vehicle dynamics code **DynamicsIVforMPC.m** that corresponds to (7.35) are given as follows.

DisplayIV.m

```
%% Intelligent vehicle visualization
function DisplayIV(stt, vehL)
    sttC = num2cell(stt); [x, y, a, s, v] = sttC{:};
    bX = [1, 3/4, 1/2, 1/4, 1/4, 0, 0, 1/4, 1/4, 1/2, 3/4, 1]*
    vehL;
    bY = [1/6, 1/6, 1/3, 1/3, 1/6, 1/6,-1/6,-1/6,-1/3,-1/3,
          -1/6,-1/6]*vehL;
    rX = [1/8, 5/24, 5/24, 0, 0, 1/12, 1/12, 0, 0, 5/24,
          5/24, 1/8]*vehL;
    rY = [1/4, 1/4, 1/3, 1/3, 1/4, 1/4,-1/4,-1/4,-1/3,-1/3,
          -1/4,-1/4]*vehL;
    fX = vehL - rX; fXm = mean(fX); fY = rY;
```

```
    R1 = [cos(a), -sin(a); sin(a), cos(a)];
    R2 = [cos(s), -sin(s); sin(s), cos(s)];
    bVeh = R1*[bX; bY] + repmat([x; y], 1, length(bX));
    rVeh = R1*[rX; rY] + repmat([x; y], 1, length(rX));
    fVeh = R2*[fX-fXm; fY]; fVeh(1,:) = fVeh(1,:) + fXm;
    fVeh = R1*fVeh + repmat([x; y], 1, length(fX));
    patch(rVeh(1,:), rVeh(2,:), 'k'); % Display the rear wheel
    patch(fVeh(1,:), fVeh(2,:), 'k'); % Display the front wheel
    patch(bVeh(1,:), bVeh(2,:), 'r'); % Display the vehicle body
end
```

DynamicsIV.m

```
%% Intelligent vehicle dynamics
function stt = DynamicsIV(SimConfig, sttIn, sIn, vIn)
    if (nargin<4) vIn = 2; end
    SC = num2cell(SimConfig);
    [vehL, rotT, rotM, accT, accM, dt] = SC{:};
    sttC = num2cell(sttIn); [x, y, a, s, v] = sttC{:};
    %% State evolution
    ds = (1-exp(-dt/rotT))*(sIn-s);
    sOld = s; s = s + max(min(ds, rotM*dt), -rotM*dt);
    sm = (s+sOld)/2;
    dv = (1-exp(-dt/accT))*(vIn-v);
    vOld = v; v = v + max(min(dv, accM*dt), -accM*dt);
    vm = (v+vOld)/2;
    aOld = a; a = a + vm*dt*tan(sm)/vehL; am = (a+aOld)/2;
    x = x + vm*dt*cos(am);
    y = y + vm*dt*sin(am);
    stt = [x; y; a; s; v];
end
```

DynamicsIVforMPC.m

```
%% Intelligent vehicle dynamics for model predictive control
function stt = DynamicsIVforMPC(ConfigMPC, sttIn, sIn, vIn)
    if (nargin<4) vIn = 2; end
    SC = num2cell(ConfigMPC); [vehL, dT] = SC{:};
    sttC = num2cell(sttIn); [x, y, a, s, v] = sttC{:};
    %% State evolution
    s = sIn; v = vIn;
    w = v*tan(s)/vehL;
    aOld = a; a = a + w*dT; am = (a+aOld)/2;
    if (abs(w) < 0.0001)
        x = x + v*dT*cos(am);
        y = y + v*dT*sin(am);
    else
        x = x + v*(sin(a)-sin(aOld))/w;
        y = y + v*(cos(aOld)-cos(a))/w;
    end
    stt = [x; y; a; s; v];
end
```

Fig. 7.4 Model predictive control for intelligent vehicle parking in various scenarios

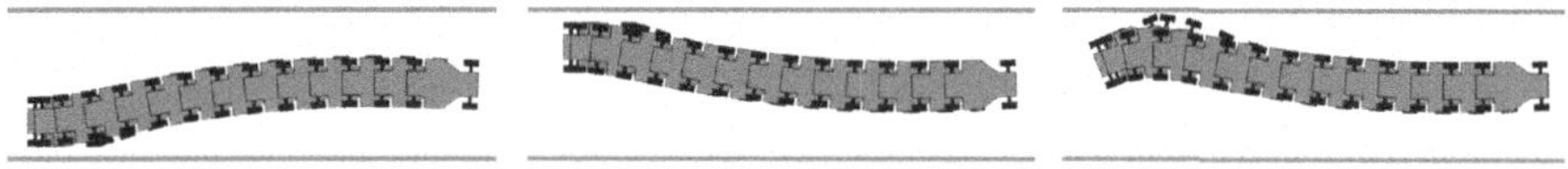

Fig. 7.5 Model predictive control for intelligent vehicle lane keeping in various scenarios

The performance of model predictive control for intelligent vehicle parking is demonstrated in Fig. 7.4.

Not only for intelligent vehicle parking, above intelligent vehicle lateral control method based on model predictive control can be conveniently adapted for intelligent vehicle lane keeping, as demonstrated in Fig. 7.5. For this purpose, the only point of adaptation in the intelligent vehicle lateral control method is that instead of fixing the expected intelligent vehicle state, we need to set the expected intelligent vehicle state dynamically according to the relative lane pose at certain preview distance in front of the intelligent vehicle.

It is worth noting that model predictive control is not the only solution for intelligent vehicle lateral control. Still taking intelligent vehicle parking as an example, we may follow the spirit of sliding mode control and design certain sliding mode for the intelligent vehicle state to evolve towards the expected parking slot pose.[6]

7.3.3 Note

Compared with almost all other control methods that we have presented throughout this book by so far, the model predictive control method (or methodology) is different in the sense that *it does not require linear system modelling*. In other words, the model predictive control method is generally applicable to not only linear but also nonlinear control systems, if realized properly. This characteristic of the model predictive control method is already demonstrated in the concrete application example of intelligent vehicle parking control, where the system model described by (7.30) is apparently nonlinear. In fact, even the simplified version of the system model, namely that described by (7.35) which is adopted for model predictive control purpose, is also nonlinear.

It is worth noting again that in practical applications, linearity is ideal whereas nonlinearity is ubiquitous, which implies that there is no strict linear control systems

[6] In the context of intelligent vehicle navigation, such instantiation of the spirit of sliding mode control may also be treated as a kind of *motion planning* (LaValle 2006).

as theoretically specified. Normally, what we refer to as linear control systems are those for which linear system modelling can be fairly adopted in their operation space.

Although linear system modelling can be fairly adopted in many practical applications, nonlinear control systems are more general because any control system is after all nonlinear by nature. It is worth developing control theory for nonlinear control systems directly.

However, in terms of both analysis and controller design, nonlinear control systems incur much more difficulties to control science than linear control systems. Taking the demonstrated example of intelligent vehicle parking as an example, theoretical questions arise naturally: In what circumstances will the model predictive control based intelligent vehicle lateral control method work? How to prove it works? We believe that readers will find it indeed difficult to answer these questions after reflection.

Detailed presentation of advanced control theory on nonlinear control systems is beyond the scope of this book. In Sect. 7.4, we will briefly present some fundamentals of nonlinear control systems.

7.4 Nonlinear Systems and Lyapunov Control

Stability is the first important concern for control systems. We have presented a number of criteria for analysing stability of a linear system. On the other hand, how to analyse stability of a nonlinear system? A nonlinear control system that adopts generic state-space modelling described by (6.2)[7] is

$$\frac{\mathrm{d}}{\mathrm{d}t}\boldsymbol{x} = f(\boldsymbol{x}, \boldsymbol{u}).$$

As already clarified in Sect. 6.2 in Chap. 6, the control input $\boldsymbol{u}$ is normally generated according to feedback of the state $\boldsymbol{x}$. Once the feedback control law $\boldsymbol{u} = g(\boldsymbol{x})$ is determined, dynamics of the closed-loop feedback control system is equivalent to dynamics of a self-evolutionary system:

$$\frac{\mathrm{d}}{\mathrm{d}t}\boldsymbol{x} = f(\boldsymbol{x}, g(\boldsymbol{x})) \equiv f_{\mathrm{c}}(\boldsymbol{x})$$

So we first consider stability of a nonlinear self-evolutionary system that adopts state-space modelling described by (6.8). For presentation convenience, suppose that the equilibrium or stable state of the nonlinear self-evolutionary system is $\boldsymbol{x}_{\mathrm{E}} = \boldsymbol{0}$ by default.

[7] Explicit expression of "nonlinear" is only to highlight consideration of systems that may not adopt linear system modelling, but by no means to imply that we exclude consideration of linear systems. So in this section, we treat linear systems as a special case of the so-called nonlinear systems.

7.4.1 *Stability of Nonlinear Self-Evolutionary Systems*

Readers may expect certain general stability criterion for nonlinear self-evolutionary systems as that for linear self-evolutionary systems. Unfortunately, we do not have such a general stability criterion. Instead, we have a strategy that may guide us to analyse stability of a nonlinear self-evolutionary system in *ad hoc* way. The strategy consists in finding a scalar function $V(\boldsymbol{x})$ in terms of the state $\boldsymbol{x}$ such that

- The scalar function $V(\boldsymbol{x})$ is *positive definite*.
- The derivative $\frac{\mathrm{d}}{\mathrm{d}t}V(\boldsymbol{x})$ is *negative definite* or at least *negative semi-definite*.[8]

The strategy is called the **Lyapunov strategy** (Liapounoff 1907). The scalar function $V(\boldsymbol{x})$ that satisfies above two conditions is called a **Lyapunov function**. In one word, the Lyapunov strategy consists in finding a Lyapunov function for the nonlinear self-evolutionary system. If we does find a Lyapunov function for the nonlinear self-evolutionary system, then we can conclude that the nonlinear self-evolutionary system is stable.

For example, consider a linear self-evolutionary system that adopts linear state-space modelling as

$$\frac{\mathrm{d}}{\mathrm{d}t}\boldsymbol{x} = \boldsymbol{A}_{\mathrm{c}}\boldsymbol{x},$$

where the state transition matrix $\boldsymbol{A}_{\mathrm{c}}$ has eigenvalues all with negative real part. The linear self-evolutionary system is stable according to the *stability criterion* presented in Sect. 6.2 in Chap. 6. Here, we apply the Lyapunov strategy to verify its stability. Consider the following Lyapunov equation

$$\boldsymbol{P}\boldsymbol{A}_{\mathrm{c}} + \boldsymbol{A}_{\mathrm{c}}^{\mathrm{T}}\boldsymbol{P} = -\boldsymbol{I}.$$

According to the *Lyapunov criterion II* presented in Sect. 7.2, above Lyapunov equation has a unique positive definite solution of $\boldsymbol{P}$. Then define the Lyapunov function as

$$V \equiv V(\boldsymbol{x}) = \boldsymbol{x}^{\mathrm{T}}\boldsymbol{P}\boldsymbol{x},$$

which is positive definite. Besides, we can verify

$$\frac{\mathrm{d}}{\mathrm{d}t}V = \boldsymbol{x}^{\mathrm{T}}\boldsymbol{P}\frac{\mathrm{d}\boldsymbol{x}}{\mathrm{d}t} + \frac{\mathrm{d}\boldsymbol{x}^{\mathrm{T}}}{\mathrm{d}t}\boldsymbol{P}\boldsymbol{x} = \boldsymbol{x}^{\mathrm{T}}(\boldsymbol{P}\boldsymbol{A}_{\mathrm{c}} + \boldsymbol{A}_{\mathrm{c}}^{\mathrm{T}}\boldsymbol{P})\boldsymbol{x} = -\boldsymbol{x}^{\mathrm{T}}\boldsymbol{x} < 0.$$

So the defined function $V(\boldsymbol{x})$ is indeed a Lyapunov function for the linear self-evolutionary system whose stability is then verified.

---

[8] If the derivative $\frac{\mathrm{d}}{\mathrm{d}t}V(\boldsymbol{x})$ is negative semi-definite, some further care is required, namely to verify that the stable state is the only fixed point or attractor for $\frac{\mathrm{d}}{\mathrm{d}t}V(\boldsymbol{x}) \equiv 0$. In other words, $\frac{\mathrm{d}}{\mathrm{d}t}V(\boldsymbol{x}) = 0$ may momentarily hold at some states other than the stable state, but will not continue to hold as the state evolves on.

For another example, consider a nonlinear self-evolutionary system that adopts state-space modelling as

$$\frac{\mathrm{d}}{\mathrm{d}t}\boldsymbol{x} = -(\boldsymbol{x}^{\mathrm{T}}\boldsymbol{x})\boldsymbol{x}.$$

For the nonlinear self-evolutionary system, we can find a Lyapunov function as $V(\boldsymbol{x}) = \boldsymbol{x}^{\mathrm{T}}\boldsymbol{x}$ because such $V(\boldsymbol{x})$ is positive definite and its derivative

$$\frac{\mathrm{d}}{\mathrm{d}t}V = \boldsymbol{x}^{\mathrm{T}}\frac{\mathrm{d}\boldsymbol{x}}{\mathrm{d}t} + \frac{\mathrm{d}\boldsymbol{x}^{\mathrm{T}}}{\mathrm{d}t}\boldsymbol{x} = \boldsymbol{x}^{\mathrm{T}}(-\boldsymbol{x}^{\mathrm{T}}\boldsymbol{x})\boldsymbol{x} + \boldsymbol{x}^{\mathrm{T}}(-\boldsymbol{x}^{\mathrm{T}}\boldsymbol{x})\boldsymbol{x} = -2(\boldsymbol{x}^{\mathrm{T}}\boldsymbol{x})^2 < 0.$$

So the nonlinear self-evolutionary system is stable.

7.4.2 Lyapunov Control

We may take advantage of the Lyapunov strategy to design a control law that stabilizes the nonlinear control system, if we can define a proper Lyapunov function. A nonlinear control system that adopts generic state-space modelling described by (6.2) is

$$\frac{\mathrm{d}}{\mathrm{d}t}\boldsymbol{x} = f(\boldsymbol{x}, \boldsymbol{u}).$$

The idea is to tentatively define certain positive definite scalar function $V(\boldsymbol{x})$ in terms of the state $\boldsymbol{x}$. Compute its derivative $\frac{\mathrm{d}}{\mathrm{d}t}V(\boldsymbol{x})$, which would normally be a scalar function in terms of the state $\boldsymbol{x}$ and the state derivative $\frac{\mathrm{d}x}{\mathrm{d}t}$, denoted as $W\left(\boldsymbol{x}, \frac{\mathrm{d}x}{\mathrm{d}t}\right)$. Substitute (6.2) into $W\left(\boldsymbol{x}, \frac{\mathrm{d}x}{\mathrm{d}t}\right)$ and obtain

$$\frac{\mathrm{d}}{\mathrm{d}t}V(\boldsymbol{x}) = W\left(\boldsymbol{x}, \frac{\mathrm{d}x}{\mathrm{d}t}\right) = W(\boldsymbol{x}, f(\boldsymbol{x}, \boldsymbol{u})).$$

Then we try to design a feedback control law $\boldsymbol{u} = g(\boldsymbol{x})$ such that

$$\frac{\mathrm{d}}{\mathrm{d}t}V(\boldsymbol{x}) = W(\boldsymbol{x}, f(\boldsymbol{x}, \boldsymbol{u})) = W(\boldsymbol{x}, f(\boldsymbol{x}, g(\boldsymbol{x}))) < 0 \text{ or } \leqslant 0 \tag{7.38}$$

namely the derivative $\frac{\mathrm{d}}{\mathrm{d}t}V(\boldsymbol{x})$ is negative definite or negative semi-definite if the feedback control law $\boldsymbol{u} = g(\boldsymbol{x})$ is adopted. If we does define a proper positive definite scalar function $V(\boldsymbol{x})$ and accordingly we does design a feedback control law $\boldsymbol{u} = g(\boldsymbol{x})$ that enables $\frac{\mathrm{d}}{\mathrm{d}t}V(\boldsymbol{x})$ to satisfy (7.38), then we can conclude that the defined positive definite scalar function $V(\boldsymbol{x})$ is a Lyapunov function for the closed-loop feedback control system if the designed feedback control law $\boldsymbol{u} = g(\boldsymbol{x})$ is adopted. Therefore, the designed feedback control law $\boldsymbol{u} = g(\boldsymbol{x})$ is a control law that can stabilize the nonlinear control system. Above strategy of designing the feedback control law $\boldsymbol{u} = g(\boldsymbol{x})$ is called the **Lyapunov control strategy**.

To facilitate understanding of the Lyapunov control strategy, we may consider a nonlinear control system example as follows.

$$\frac{\mathrm{d}}{\mathrm{d}t}\begin{bmatrix} x \\ y \end{bmatrix} = \begin{bmatrix} x \sin y \\ y \end{bmatrix} + \begin{bmatrix} 0 \\ 1 \end{bmatrix} u, \tag{7.39}$$

where $\boldsymbol{x} \equiv [x \quad y]^{\mathrm{T}}$ denotes the system state and u denotes the control input. For (7.39), tentatively define a positive definite scalar function $V(\boldsymbol{x})$ as

$$V(\boldsymbol{x}) = \frac{\boldsymbol{x}^{\mathrm{T}}\boldsymbol{x}}{2} = \frac{x^2 + y^2}{2}.$$

Compute its derivative $\frac{\mathrm{d}}{\mathrm{d}t} V(\boldsymbol{x})$ as

$$\frac{\mathrm{d}}{\mathrm{d}t} V(\boldsymbol{x}) = x\frac{\mathrm{d}x}{\mathrm{d}t} + y\frac{\mathrm{d}y}{\mathrm{d}t}.$$

Substitute (7.39) into above equation and obtain

$$\frac{\mathrm{d}}{\mathrm{d}t} V(\boldsymbol{x}) = x(x \sin y) + y(y + u) = x^2 \sin y + y^2 + yu.$$

After observation, we know that if we set the feedback control law of u as

$$u = -x^2\frac{\sin y}{y} - 2y \text{ or } - x^2 - 2y \text{ if } y \approx 0 \tag{7.40}$$

$$\text{(note that } \lim_{y\to 0} \frac{\sin y}{y} = 1),$$

then we have

$$\frac{\mathrm{d}}{\mathrm{d}t} V(\boldsymbol{x}) = -y^2 \leqslant 0.$$

Besides, for $\frac{\mathrm{d}}{\mathrm{d}t} V(\boldsymbol{x}) = -y^2 \equiv 0$, we must have $y \equiv 0$ and $u = -x^2 - 2y \equiv 0$ that implies $x \equiv 0$ as well. So the stable state is the only fixed point or attractor for $\frac{\mathrm{d}}{\mathrm{d}t} V(\boldsymbol{x}) = 0$. Then we conclude that the feedback control law of u described in (7.40) can stabilize the nonlinear control system described by (7.39).

Matlab simulation code for demonstration of the nonlinear control system example is given as follows. The simulation result is demonstrated in Fig. 7.6, where the black curve represents the convergence trajectory of the nonlinear control system's state.

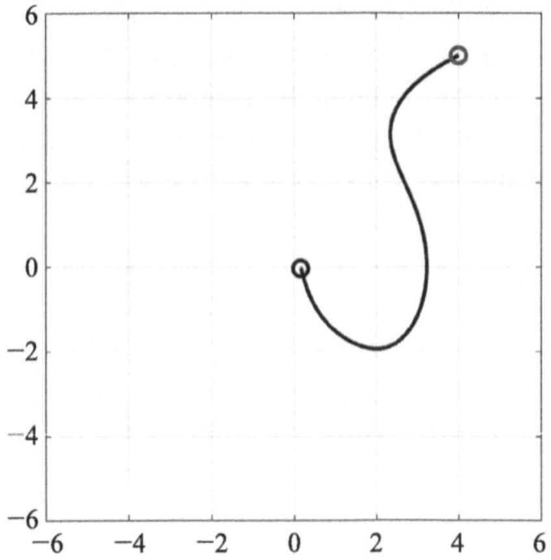

Fig. 7.6 Convergence of the nonlinear control system

LyapunovExample1.m

```
%% Simulation preliminary configuration
dt = 0.001; % Numerical computation step
tSpan = 0:dt:10; % Simulation time span
x = 4; y = 5; % Nonlinear system state
% Record states in simulation
sttAll = zeros(2, length(tSpan)); k = 0;
xExpected = 0; yExpected = 0; % Expected equilibrium status

%% Simulation of nonlinear system control
for t = tSpan
    %% Control method
    if (abs(y)>0.00001)
        u = -x^2*sin(y)/y - 2*y; % Lyapunov control
    else
        u = -x^2 - 2*y;
    end

    %% Nonlinear system dynamics
    x = x + x*sin(y)*dt;
    y = y + (y+u)*dt;
    k = k+1; sttAll(:,k) = [x; y];
    %% Nonlinear system visualization
    if (rem(k,50) == 0)
        plot(x, y, 'ob', 'MarkerSize', 8, 'LineWidth', 2);
        axis equal; grid on; xlim([-6, 6]); ylim([-6, 6]);
        pause(10*dt);
    end
end
```

Readers can try above Matlab simulation code with an arbitrarily different initial state of the nonlinear control system and will observe that the state always converges, consistent with the theoretical analysis that the feedback control law of u described in (7.40) can stabilize the nonlinear control system described by (7.39).

7.4.3 A Thinking Journey

Single inverted pendulum control

We further consider another nonlinear control system example, namely the single inverted pendulum control system, as illustrated in Fig. 1.11 in Chap. 1. Dynamics of the single inverted pendulum control system's state $\boldsymbol{x}$ is modelled by a state differential equation as

$$\frac{\mathrm{d}}{\mathrm{d}t}\boldsymbol{x} \equiv \frac{\mathrm{d}}{\mathrm{d}t}\begin{bmatrix} \theta \\ \frac{\mathrm{d}\theta}{\mathrm{d}t} \\ x \\ \frac{\mathrm{d}x}{\mathrm{d}t} \end{bmatrix} = \begin{bmatrix} \frac{\mathrm{d}\theta}{\mathrm{d}t} \\ \frac{\sin\theta}{L}g - \frac{\cos\theta}{L}a \\ \frac{\mathrm{d}x}{\mathrm{d}t} \\ a \end{bmatrix} \equiv f(\boldsymbol{x}, a), \tag{7.41}$$

where the state $\boldsymbol{x} \equiv \begin{bmatrix} \theta & \frac{\mathrm{d}\theta}{\mathrm{d}t} & x & \frac{\mathrm{d}x}{\mathrm{d}t} \end{bmatrix}^{\mathrm{T}}$ consists of the inverted pendulum angle and angular velocity, the cart position and velocity. The control input is the single-input of cart acceleration a.

If the inverted pendulum angle θ is close to zero, then like how we obtain the linear system model described by (6.3) for the double inverted pendulum control system, we can also obtain a linear system model for the single inverted pendulum control system as

$$\frac{\mathrm{d}}{\mathrm{d}t}\boldsymbol{x} = \begin{bmatrix} 0 & 1 & 0 & 0 \\ \frac{g}{L} & 0 & 0 & 0 \\ 0 & 0 & 0 & 1 \\ 0 & 0 & 0 & 0 \end{bmatrix}\boldsymbol{x} + \begin{bmatrix} 0 \\ -\frac{1}{L} \\ 0 \\ 1 \end{bmatrix} a \equiv \boldsymbol{A}\boldsymbol{x} + \boldsymbol{B}a. \tag{7.42}$$

As presented in Chaps. 6 and 7, we already have effective methods for single inverted pendulum control if the linear system model described by (7.42) can be fairly adopted for the single inverted pendulum control system.

However, if the inverted pendulum angle θ is far away from zero, namely the inverted pendulum is in a status of almost fallen down as illustrated in Fig. 7.7, then linear state-space modelling described by (7.42) will be no longer appropriate for the single inverted pendulum control system. We may follow the spirit of *piecewise linear approximation* to model dynamics of the single inverted pendulum control system's state, yet we intend to demonstrate how to directly handle the nonlinear system dynamics modelled by (7.41) in the light of the Lyapunov control strategy.

Since the cart position x is purely a linear factor for dynamics of the single inverted pendulum control system's state, we focus on handling the inverted pendulum angle θ, the inverted pendulum angular velocity $\theta' \equiv \frac{\mathrm{d}\theta}{\mathrm{d}t}$, and the cart velocity $x' \equiv \frac{\mathrm{d}x}{\mathrm{d}t}$ via Lyapunov control. In other words, we apply the Lyapunov control strategy to control θ, θ', and x' to zero.[9] For this purpose, we should define a proper Lyapunov function.

---

[9] For a generic variable z, denote its first-order derivative $\frac{\mathrm{d}z}{\mathrm{d}t}$, second-order derivative $\frac{\mathrm{d}^2z}{\mathrm{d}t^2}, \cdots$ simply as $z', z'', \cdots$ respectively.

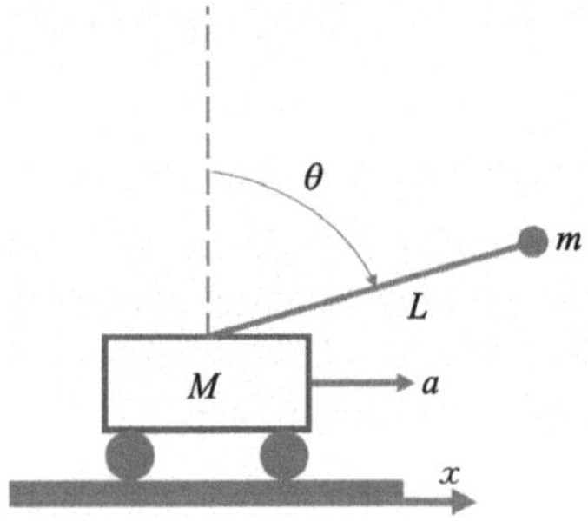

Fig. 7.7 Inverted pendulum control with large inverted pendulum angle θ

We may tentatively define certain candidate Lyapunov function namely certain positive definite scalar function in terms of θ, θ', and x'. The positive definite scalar functions $1 - \cos\theta$, θ'^2, and x'^2 seem to be natural choices for θ, θ', and x' respectively, so we combine them into a holistic positive definite scalar function as

$$V \equiv V(\theta, \theta', x') = 1 - \cos\theta + \theta'^2 + x'^2. \tag{7.43}$$

Compute its derivative $\frac{\mathrm{d}}{\mathrm{d}t} V(\boldsymbol{x})$ as

$$\frac{\mathrm{d}}{\mathrm{d}t} V(\boldsymbol{x}) = \sin\theta\theta' + 2\theta'\theta'' + 2x'x''.$$

Note following relationships conveyed by (7.41):

$$\begin{cases} \theta'' = \frac{\sin\theta}{L} g - \frac{\cos\theta}{L} a, \\ x'' = a. \end{cases} \tag{7.44}$$

Substitute (7.44) into $\frac{\mathrm{d}}{\mathrm{d}t} V(\boldsymbol{x})$ and obtain

$$\begin{aligned} \frac{\mathrm{d}}{\mathrm{d}t} V(\boldsymbol{x}) &= \sin\theta\theta' + 2\theta' \left(\frac{\sin\theta}{L} g - \frac{\cos\theta}{L} a \right) + 2x'a \\ &= \sin\theta\theta' \left(1 + 2\frac{g}{L}\right) + 2\left(x' - \frac{\cos\theta}{L}\theta'\right) a. \end{aligned}$$

If we set the feedback control law of a as

$$a = \frac{1}{2\left(x' - \frac{\cos\theta}{L}\theta'\right)} \left[-\sin\theta\theta' \left(1 + 2\frac{g}{L}\right) - y^2 \right], \tag{7.45}$$

then we have

$$\frac{\mathrm{d}}{\mathrm{d}t} V(\boldsymbol{x}) = -y^2 \leqslant 0.$$

According to the Lyapunov control strategy, it seems that the feedback control law of a described in (7.45) can stabilize the single inverted pendulum.

Is the single inverted pendulum control problem really solved via Lyapunov control as analysed above? Readers may try the feedback control law of a described in (7.45) with Matlab simulation code (given below in this section) and will soon realize that some problem exists for such feedback control law. What is the problem?

In fact, the feedback control law of a described in (7.45) is *ill-conditioned* because the denominator $2\left(x' - \frac{\cos\theta}{L}\theta'\right)$ becomes zero as θ, θ', and x' are zero. Besides, even when θ, θ', and x' are not zero, the denominator $2\left(x' - \frac{\cos\theta}{L}\theta'\right)$ may still by chance approach zero. In other words, (7.45) gives a *naive* feedback control law of a which is impractical.

Readers may be confused: For the nonlinear control system example described by (7.39), the Lyapunov control strategy based feedback control law described in (7.40) seems to have the same problem that the denominator of the control law becomes zero when the system state approaches the equilibrium state. Then why does the feedback control law described in (7.40) work? The reason is that as $y \to 0$, the feedback control law described in (7.40) just happens to have a finite limit thanks to $\lim_{y\to 0}\frac{\sin y}{y} = 1$ and is *well-conditioned* in all circumstances. This is purely a desirable coincidence.

However, the feedback control law of a described in (7.45) does not enjoy such desirable coincidence. Generally for nonlinear control systems, we cannot expect feedback control laws derived according to casually-defined candidate Lyapunov functions to commonly enjoy above desirable coincidence. If the candidate Lyapunov function is naive as that described in (7.43), the derived feedback control law also tends to be naive and impractical. Therefore, *we need to carefully define a candidate Lyapunov function of high quality*.

But how? Unfortunately, there is neither theory nor rule of thumb to let us know how. Defining a proper Lyapunov function together with its associated feedback control law is rather a kind of *arts* than techniques. It can be a long process of the *observation-trial-analysis* loop full of creativeness. In order to guide readers to better feel such "artistic" process, we demonstrate the "thinking journey" of defining a proper Lyapunov function for single inverted pendulum control, instead of just putting forward the "destination" namely the Lyapunov function abruptly.

Thinking journey for single inverted pendulum control

Now we start the "thinking journey". Although the candidate Lyapunov function described in (7.43) is naive, its composing terms $1 - \cos\theta$, θ'^2, and x'^2 do seem to be natural choices for θ, θ', and x' respectively. Perhaps we need to combine them in a more sophisticated way to form a new candidate Lyapunov function.

Consider the derivatives of the three composing functions $1 - \cos\theta$, θ'^2, and x'^2:

$$
\begin{aligned}
(1-\cos\theta)' &= \sin\theta\theta', \\
(\theta'^2)' &= 2\theta'\theta'' = 2\theta'\left(\frac{\sin\theta}{L}g - \frac{\cos\theta}{L}a\right), \\
(x'^2)' &= 2x'x'' = 2x'a.
\end{aligned}
\tag{7.46}
$$

As shown in (7.46), among the three functions, the function $1-\cos\theta$ is the only one whose derivative lacks the term of the control input a. In other words, the function $1-\cos\theta$ is the only one that lacks flexibility to adapt its derivative value by the control input a. Based on this observation, we intend to keep the less flexible composing function $1-\cos\theta$ in a simple form in the new candidate Lyapunov function while leaving more sophisticated way of integration to the other two composing functions that enjoy more adaptive flexibility. So we keep $1-\cos\theta$ simply as $A(1-\cos\theta)$ in the new candidate Lyapunov function, where A is a scaling factor. For the moment, we tentatively have

$$
V \equiv V(\theta, \theta', x') = A(1-\cos\theta) + \cdots
$$

The derivative of the function $1-\cos\theta$ is only in terms of θ and θ', yet the function $1-\cos\theta$ itself lacks flexibility to adapt such derivative. To adapt the derivative of the function $1-\cos\theta$, we need to resort to terms associated with θ' and x'. The term associated with θ' is apparently more suitable for such purpose, because its derivative is right related to θ and θ' without entangling a third factor such as x'. So we further integrate the term associated with θ' into the new candidate Lyapunov function.

Instead of naively integrating θ'^2 as in (7.43), we may consider a more general quadratic form as $B(\theta)\theta'^2 + B_1(\theta)\theta' + B_2(\theta)$ which enjoys much better flexibility to coordinate with $A(1-\cos\theta)$. On the other hand, such general quadratic form seems to be too complicated. For a balance between flexibility and complexity, we intend to adopt a compromised quadratic version namely $B(\theta)\theta'^2$ only (B should be positive definite)—After trials latter, if we fail in defining a proper Lyapunov function due to simplicity of $B(\theta)\theta'^2$, we can return to pick up the general quadratic form $B(\theta)\theta'^2 + B_1(\theta)\theta' + B_2(\theta)$ for new rounds of trials. If $B(\theta)\theta'^2 + B_1(\theta)\theta' + B_2(\theta)$ is still inappropriate, we may then try even more general forms for further trials. But for the moment, we try a moderate form as $B(\theta)\theta'^2$, and we tentatively have

$$
V = A(1-\cos\theta) + B(\theta)\theta'^2 + \cdots
$$

If we only consider stabilization of θ and θ', then we have (denote $B \equiv B(\theta)$):

$$\frac{\mathrm{d}}{\mathrm{d}t}[A(1-\cos\theta)+B\theta'^2] = A\sin\theta\theta' + 2B\theta'\theta'' + \frac{\mathrm{d}B}{\mathrm{d}\theta}\theta'^3$$
$$= A\sin\theta\theta' + 2B\theta'\left(\frac{\sin\theta}{L}g - \frac{\cos\theta}{L}a\right) + \frac{\mathrm{d}B}{\mathrm{d}\theta}\theta'^3$$
$$= -2\frac{\cos\theta}{L}\theta'\left[Ba - \frac{L}{2\cos\theta}\left(\frac{2g}{L}B\sin\theta + A\sin\theta + \frac{\mathrm{d}B}{\mathrm{d}\theta}\theta'^2\right)\right].$$

Since B should be positive definite, we can have a well-conditioned formula of a for any function y such that

$$Ba - D = y, \tag{7.47}$$

where

$$D = \frac{L}{2\cos\theta}\left(\frac{2g}{L}B\sin\theta + A\sin\theta + \frac{\mathrm{d}B}{\mathrm{d}\theta}\theta'^2\right).$$

Let $y = \cos\theta\theta'$. Then the corresponding formula of a obtained by solving (7.47) will be a well-conditioned feedback control law that can stabilize θ and θ'. By so far, integration of $A(1-\cos\theta)$ and $B(\theta)\theta'^2$ seems to be promising, because a well-conditioned equation $Ba - D = y$ in terms of a can be obtained (only if B is positive definite).

We move forward to integrate the last term associated with x' into the new candidate Lyapunov function, and we may consider a general form as $C(\theta,\theta',x')^2$ (denote $C \equiv C(\theta,\theta',x')$). The general form of C is abstract for the moment, but we will try to gradually specify details of C for it to coordinate with $A(1-\cos\theta)$ and $B(\theta)\theta'^2$. We tentatively have

$$V = A(1-\cos\theta) + B(\theta)\theta'^2 + C(\theta,\theta',x')^2. \tag{7.48}$$

Note that

$$\frac{\mathrm{d}}{\mathrm{d}t}[A(1-\cos\theta)+B\theta'^2] = -2\frac{\cos\theta}{L}\theta'(Ba - D) \tag{7.49}$$

$$\frac{\mathrm{d}}{\mathrm{d}t}(C^2) = 2C\left(\frac{\partial C}{\partial\theta}\theta' + \frac{\partial C}{\partial\theta'}\theta'' + \frac{\partial C}{\partial x'}x''\right) \tag{7.50}$$
$$= 2C\left[\frac{\partial C}{\partial\theta}\theta' + \frac{\partial C}{\partial\theta'}\left(\frac{\sin\theta}{L}g - \frac{\cos\theta}{L}a\right) + \frac{\partial C}{\partial x'}a\right]$$
$$= -2C\left[\left(\frac{\partial C}{\partial\theta'}\frac{\cos\theta}{L} - \frac{\partial C}{\partial x'}\right)a - \frac{\partial C}{\partial\theta}\theta' - \frac{\partial C}{\partial\theta'}\frac{\sin\theta}{L}g\right]$$
$$\equiv -2C(C_1 a - C_2),$$

where

$$C_1 \equiv \frac{\partial C}{\partial \theta'}\frac{\cos\theta}{L} - \frac{\partial C}{\partial x'}, \quad C_2 \equiv \frac{\partial C}{\partial \theta}\theta' + \frac{\partial C}{\partial \theta'}\frac{\sin\theta}{L}g.$$

By comparing (7.49) and (7.50), we tentatively hope

$$C_1 a - C_2 \propto Ba - D \text{ i.e. } k(C_1 a - C_2) \equiv Ba - D \tag{7.51}$$

for two reasons: First, as already clarified, $Ba - D$ is a well-conditioned expression. Second, if (7.51) does hold, we would have a concise form of $\frac{\mathrm{d}}{\mathrm{d}t}V$ as

$$\frac{\mathrm{d}}{\mathrm{d}t}V = -2\left(\frac{\cos\theta}{L}\theta' + \frac{C}{k}\right)(Ba - D), \tag{7.52}$$

which facilitates design of the feedback control law of a via the Lyapunov control strategy. We will see whether the hope of (7.51) can indeed be satisfied.[10]

For (7.51) to hold, we require

$$B = kC_1 = k\left(\frac{\partial C}{\partial \theta'}\frac{\cos\theta}{L} - \frac{\partial C}{\partial x'}\right) \tag{7.53}$$

and

$$D = \frac{L}{2\cos\theta}\left(\frac{2g}{L}B\sin\theta + A\sin\theta + \frac{\mathrm{d}B}{\mathrm{d}\theta}\theta'^2\right) = kC_2 = k\left(\frac{\partial C}{\partial \theta}\theta' + \frac{\partial C}{\partial \theta'}\frac{\sin\theta}{L}g\right). \tag{7.54}$$

We tentatively separate terms with and without θ' in (7.54) and hope

$$\frac{L}{2\cos\theta}\left(\frac{2g}{L}B + A\right)\sin\theta = k\frac{\partial C}{\partial \theta'}\frac{\sin\theta}{L}g \Leftarrow \frac{L^2}{2\cos\theta}\left(\frac{2g}{L}B + A\right) = k\frac{\partial C}{\partial \theta'}g, \tag{7.55}$$

$$\frac{L}{2\cos\theta}\frac{\mathrm{d}B}{\mathrm{d}\theta}\theta'^2 = k\frac{\partial C}{\partial \theta}\theta' \Leftarrow \frac{L}{2\cos\theta}\frac{\mathrm{d}B}{\mathrm{d}\theta}\theta' = k\frac{\partial C}{\partial \theta}. \tag{7.56}$$

Substitute (7.53) into (7.55) and obtain

$$\frac{L^2}{2\cos\theta}\left[\frac{2g}{L}k\left(\frac{\partial C}{\partial \theta'}\frac{\cos\theta}{L} - \frac{\partial C}{\partial x'}\right) + A\right] = k\frac{\partial C}{\partial \theta'}g \iff k = \frac{AL}{2g\frac{\partial C}{\partial x'}}. \tag{7.57}$$

For convenience, we tentatively suppose

[10] A Chinese proverb says: "Where there is a hope, there is a way". The philosophical wisdom reflected in this proverb is especially applicable to the "artistic" process of defining a proper Lyapunov function together with its associated feedback control law, and even generally applicable to research.

$$\frac{\partial C}{\partial x'} = 1. \tag{7.58}$$

Then (7.57) is reduced to

$$k = \frac{AL}{2g} \tag{7.59}$$

and (7.53) is reduced to

$$B = \frac{AL}{2g}\left(\frac{\partial C}{\partial \theta'}\frac{\cos\theta}{L} - 1\right). \tag{7.60}$$

So we have

$$\frac{\partial C}{\partial \theta'} = \frac{L}{\cos\theta}\left(1 + \frac{2g}{AL}B(\theta)\right) \equiv \bar{C}(\theta). \tag{7.61}$$

From (7.58) and (7.61), we can know that C is of the form

$$C = \bar{C}(\theta)\theta' + x', \tag{7.62}$$

where terms irrelevant to θ, θ', and x' are omitted.

Substitute (7.59), (7.60), and (7.62) into (7.56) and obtain

$$\begin{aligned}
&\frac{L}{2\cos\theta}\frac{\mathrm{d}}{\mathrm{d}\theta}\left[\frac{AL}{2g}\left(\bar{C}\frac{\cos\theta}{L} - 1\right)\right]\theta' = \frac{AL}{2g}\frac{\mathrm{d}\bar{C}}{\mathrm{d}\theta}\theta' \\
\Longleftrightarrow\ &\frac{1}{2\cos\theta}\frac{\mathrm{d}}{\mathrm{d}\theta}(\bar{C}\cos\theta)\theta' = \frac{\mathrm{d}\bar{C}}{\mathrm{d}\theta}\theta' \\
\Longleftrightarrow\ &\left(\frac{\mathrm{d}\bar{C}}{\mathrm{d}\theta}\cos\theta + \bar{C}\sin\theta\right)\theta' = 0 \Longleftrightarrow \frac{\mathrm{d}\bar{C}\cos\theta + \bar{C}\sin\theta d\theta}{\cos^2\theta}\theta' = 0 \\
\Longleftrightarrow\ &d\left(\frac{\bar{C}}{\cos\theta}\right)\theta' = 0,
\end{aligned}$$

which has a non-trivial solution as

$$\bar{C} = L_0\cos\theta \quad (L_0 \text{ is a constant}).$$

Substitute it back into (7.62) and (7.60) and obtain

$$C = L_0\cos\theta\theta' + x', \tag{7.63}$$

$$B = \frac{AL}{2g}\left(\frac{L_0}{L}\cos^2\theta - 1\right). \tag{7.64}$$

Then substitute (7.63) and (7.64) back into (7.48) and obtain

$$V \equiv V(\theta, \theta', x') = A(1 - \cos\theta) + \frac{AL}{2g}\left(\frac{L_0}{L}\cos^2\theta - 1\right)\theta'^2 + (L_0 \cos\theta\theta' + x')^2. \tag{7.65}$$

After a long "thinking journey", we have finally defined the new candidate Lyapunov function, which is specified in (7.65). The new candidate Lyapunov function seems to be a proper Lyapunov function, because in retrospect, all our hopes and tentative assumptions are consistent with each other without causing contradiction. Despite the promising analysis, it is still worth verifying directly whether the new candidate Lyapunov function is indeed a proper Lyapunov function.

$$\begin{aligned}
\frac{\mathrm{d}}{\mathrm{d}t}V &= -\frac{A}{g}\cos\theta\theta'\left[\left(\frac{L_0}{L}\cos^2\theta - 1\right)a - \left(g\frac{L_0}{L}\cos\theta - L_0\theta'^2\right)\sin\theta\right] \\
&\quad + 2(L_0\cos\theta\theta' + x')\left[-\left(\frac{L_0}{L}\cos^2\theta - 1\right)a\right. \\
&\quad \left. + \left(g\frac{L_0}{L}\cos\theta - L_0\theta'^2\right)\sin\theta\right] \\
&= -\left[\left(\frac{A}{g} + 2L_0\right)\cos\theta\theta' + 2x'\right]\left[\left(\frac{L_0}{L}\cos^2\theta - 1\right)a\right. \\
&\quad \left. - \left(g\frac{L_0}{L}\cos\theta - L_0\theta'^2\right)\sin\theta\right] \\
&\equiv -E(\theta, \theta', x')[B_0(\theta)a - D_0(\theta, \theta')],
\end{aligned} \tag{7.66}$$

where

$$\begin{aligned}
B_0(\theta) &\equiv \frac{L_0}{L}\cos^2\theta - 1, \\
D_0(\theta, \theta') &\equiv \left(g\frac{L_0}{L}\cos\theta - L_0\theta'^2\right)\sin\theta, \\
E(\theta, \theta', x') &\equiv \left(\frac{A}{g} + 2L_0\right)\cos\theta\theta' + 2x'.
\end{aligned}$$

If we set the feedback control law of a as

$$a = \frac{1}{B_0(\theta)}[D_0(\theta, \theta') + KE(\theta, \theta', x')] \quad (K > 0), \tag{7.67}$$

then we have

$$\frac{\mathrm{d}}{\mathrm{d}t}V = -KE(\theta, \theta', x')^2 \leqslant 0.$$

We can easily choose L_0 to guarantee the positive definiteness of $B_0(\theta) \equiv \frac{L_0}{L}\cos^2\theta - 1$, so the feedback control law of a described in (7.67) is well-conditioned. According to the Lyapunov control strategy, the feedback control law of a described in (7.67) can stabilize the inverted pendulum angle θ, the inverted pendulum angular velocity $\theta' \equiv \frac{d\theta}{dt}$, and the cart velocity $x' \equiv \frac{dx}{dt}$.

After stabilization of θ, θ', and x', the cart position x may still be far away from the equilibrium position, yet then the single inverted pendulum control system can fairly adopt linear state-space modelling. For remaining control of the single inverted pendulum after stabilization of θ, θ', and x', we can resort to the sliding mode control method or the linear quadratic regulator control method presented in Sects. 7.1 and 7.2 respectively.

Matlab simulation code for demonstrating single inverted pendulum control based on the Lyapunov control strategy is given as follows. In simulation, let $m = 1$, $L = 1$, and $g = 10$. For the Lyapunov control strategy based feedback control law of a, set the relevant parameters A, K, and L_0 respectively as $A = 2g^2$, $K = 0.5$, and $L_0 = L\left(1 + \frac{1}{\cos^2\theta}\right)$.

SingleInvertedPendulumLyapunov1.m

```
%% Single inverted pendulum parameters
m1 = 1; L1 = 1; g = 10;
%% Simulation preliminary configuration
dt = 0.001; % Numerical computation step
tSpan = 0:dt:20; % Simulation time span
x = 20; dx = 0; % Cart position and its velocity
% Inverted pendulum angle theta and its angular velocity
y = 0.4*pi; dy = 0;
stt = [y; dy; x; dx]; % Single inverted pendulum state
% Record states in simulation
sttAll = zeros(length(stt), length(tSpan)); k = 0;
xExpected = 0; yExpected = 0; % Expected equilibrium status
SimConfig = [m1, L1, g, dt];

%% Design the optimal LQR gain matrix (linear quadratic
    regulator)
A = [0, 1, 0, 0; g/L1, 0, 0, 0; 0, 0, 0, 1; 0, 0, 0, 0];
B = [0; -1/L1; 0; 1];
lambdaE = [-4;-4;-4;-4]; % Expected eigenvalues
sttK = DesignGainMatrixSub(A, B, lambdaE);
cLyapunov=1; fprintf('Start Lyapunov control\n');

%% Simulation of single inverted pendulum control
for t = tSpan
    %% Control method
    if (y^2+dy^2+dx^2 > 0.1 && cLyapunov==1) % Lyapunov control
        L0 = L1*(1+1/cos(y)^2); cA = 2*g^2; cK = 0.5;
        B0 = (L0/L1)*cos(y)^2 - 1;
        D0 = (g*L0*cos(y)/L1 - L0*dy^2)*sin(y);
        E = (cA/g+2*L0)*cos(y)*dy + 2*dx;
        acc = (D0+cK*E)/B0;
    else % Switch to sliding mode control
```

```
        if (1==cLyapunov)
            cLyapunov = 0; sttE = [0; 0; x; 0];
            fprintf('Switch to sliding mode control\n');
        end
        % Sliding mode design
        if (sttE(3)>0) sttE(3) = max(sttE(3) - 8*dt, 0);
        else sttE(3) = min(sttE(3) + 8*dt, 0); end
        acc = -sttK'*(stt-sttE); % Full-state feedback control
    end

    %% Single inverted pendulum dynamics
    stt = DynamicsSIP(SimConfig, stt, acc);
    sttC = num2cell(stt); [y, dy, x, dx] = sttC{:};
    if (abs(y)>=pi/2) fprintf('Control failure!\n'); break; end
    k = k+1; sttAll(:,k) = stt;
    %% Single inverted pendulum visualization
    if (rem(k,20) == 0)
        DisplaySIP(x, y, L1); pause(20*dt);
    end
end
```

SingleInvertedPendulumLyapunov2.m

```
%% Single inverted pendulum parameters
m1 = 1; L1 = 1; g = 10;
%% Simulation preliminary configuration
dt = 0.001; % Numerical computation step
tSpan = 0:dt:20; % Simulation time span
x = 20; dx = 0; % Cart position and its velocity
% Inverted pendulum angle theta and its angular velocity
y = 0.4*pi; dy = 0;
stt = [y; dy; x; dx]; % Single inverted pendulum state
% Record states in simulation
sttAll = zeros(length(stt), length(tSpan)); k = 0;
xExpected = 0; yExpected = 0; % Expected equilibrium status
SimConfig = [m1, L1, g, dt];

%% Design the optimal LQR gain matrix (linear quadratic
    regulator)
A = [0, 1, 0, 0; g/L1, 0, 0, 0; 0, 0, 0, 1; 0, 0, 0, 0];
B = [0; -1/L1; 0; 1];
lambdaE = [-4;-4;-4;-4]; % Expected eigenvalues
sttK = DesignGainMatrixSub(A, B, lambdaE);
Q = eye(4); R = 6; % LQR cost matrices
[P, sttK] = SolveRiccatiEquationLQR(A, B, Q, R, sttK);
cLyapunov=1; fprintf('Start Lyapunov control\n');

%% Simulation of single inverted pendulum control
for t = tSpan
    %% Control method
    if (y^2+dy^2+dx^2 > 0.1 && cLyapunov==1) % Lyapunov control
        L0 = L1*(1+1/cos(y)^2); cA = 2*g^2; cK = 0.5;
        B0 = (L0/L1)*cos(y)^2 - 1;
```

```
        D0 = (g*L0*cos(y)/L1 - L0*dy^2)*sin(y);
        E = (cA/g+2*L0)*cos(y)*dy + 2*dx;
        acc = (D0+cK*E)/B0;
    else % Switch to LQR optimal control
        if (1==cLyapunov)
            cLyapunov = 0; fprintf('Switch to LQR control\n');
        end
        acc = -sttK'*stt; % Full-state feedback control of LQR
    end

    %% Single inverted pendulum dynamics
    stt = DynamicsSIP(SimConfig, stt, acc);
    sttC = num2cell(stt); [y, dy, x, dx] = sttC{:};
    if (abs(y)>=pi/2) fprintf('Control failure!\n'); break; end
    k = k+1; sttAll(:,k) = stt;
    %% Single inverted pendulum visualization
    if (rem(k,20) == 0)
        DisplaySIP(x, y, L1); pause(20*dt);
    end
end
```

The visualization code **DisplaySIP.m** and the single inverted pendulum dynamics code **DynamicsSIP.m** in above code are given as follows.

DisplaySIP.m

```
%% Single inverted pendulum visualization
function DisplaySIP(x, y1, L1)
    Mx = x; My = 0;
    m1x = Mx + L1*sin(y1); m1y = My + L1*cos(y1);
    figure(1); cla; hold on;
    plot(Mx, My, 'or', 'MarkerSize', 16, 'LineWidth', 2);
    plot(m1x, m1y, 'ob', 'MarkerSize', 8, 'LineWidth', 2);
    line([Mx,m1x], [My,m1y], 'Color', 'b', 'LineWidth', 2);
    axis equal; grid on; ylim([0, 3]);
    if (abs(x)<2) xlim([-2,2]); else xlim([x-2, x+2]); end; hold
     off;
end
```

DynamicsSIP.m

```
%% Single inverted pendulum dynamics
function stt = DynamicsSIP(SimConfig, sttIn, acc, acc1)
    if (nargin<4) acc1 = 0; end
    SC = num2cell(SimConfig); [m1, L1, g, dt] = SC{:};
    sttC = num2cell(sttIn); [y, dy, x, dx] = sttC{:};
    %% State evolution
    x = x + dx*dt;
    dx = dx + acc*dt;
    yold = y; y = y + dy*dt;
    dy = dy + (sin(yold)*g/L1 - cos(yold)*acc/L1 + acc1)*dt;
    stt = [y; dy; x; dx];
end
```

7.5 Adaptive Control

7.5.1 Adaptive Modelling

In practical applications, we usually adopt a basic assumption that the control system is *time-invariant*. The time-invariance assumption has also been adopted throughout this book by so far. Just like linearity is ideal whereas nonlinearity is ubiquitous, time-invariance is also ideal whereas time-variance is ubiquitous. Normally, what we refer to as time-invariant control systems are those for which time-invariant system modelling can be fairly adopted in their operation space.

Although time-invariant system modelling can be fairly adopted in many practical applications, time-variant control systems are more general because any control system is after all time-variant by nature. Theoretically speaking, instead of adopting an ideal time-invariant system model established off-line, a control system had better possess a mechanism to adapt its system model via on-line estimation or identification so that the adaptive system model would better match its dynamics during its real-time operation, and had better also generate its control law with the adaptive system model taken into account in real time. This is the basic spirit of **adaptive control** (Astrom and Wittenmark 2008), which is illustrated in Fig. 7.8.

The basic spirit of adaptive control can be naturally incorporated into other control methods that have been presented. Taking model predictive control as an example, we can naturally perform on-line adaptation of the simplified system model function $\bar{f}(\boldsymbol{x}, \boldsymbol{u})$ described in (7.29), while the model predictive control method itself is performed in the same way as if the simplified system model function $\bar{f}(\boldsymbol{x}, \boldsymbol{u})$ is time-invariant.

7.5.2 Adaptive Linear State-Space Modelling Based Control

Given a nonlinear control system, even supposing it is time-invariant, we may sometimes still follow the basic spirit of adaptive control and handle the nonlinear control system as a linear control system with time-variant system model parameters. A non-

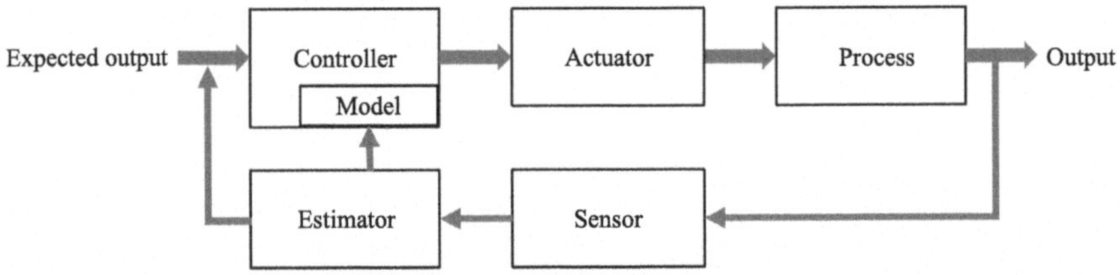

Fig. 7.8 Adaptive closed-loop feedback control system

linear control system that adopts generic state-space modelling described by (6.2) is given as

$$\frac{\mathrm{d}}{\mathrm{d}t}\boldsymbol{x} = f(\boldsymbol{x}, \boldsymbol{u}).$$

The idea is to approximate the system model function $f(\boldsymbol{x}, \boldsymbol{u})$, if possible, by an adaptive linear form with time-variant parameters (usually in terms of the state $\boldsymbol{x}$):

$$f(\boldsymbol{x}, \boldsymbol{u}) \approx \boldsymbol{A}(\boldsymbol{x})\boldsymbol{x} + \boldsymbol{B}(\boldsymbol{x})\boldsymbol{u}.$$

Then the nonlinear control system is approximately modelled by

$$\frac{\mathrm{d}}{\mathrm{d}t}\boldsymbol{x} = \boldsymbol{A}(\boldsymbol{x})\boldsymbol{x} + \boldsymbol{B}(\boldsymbol{x})\boldsymbol{u}, \tag{7.68}$$

which gives the generic formalism of **adaptive linear state-space modelling**. In (7.68), both the state transition matrix $\boldsymbol{A}(\boldsymbol{x})$ and the control input matrix $\boldsymbol{B}(\boldsymbol{x})$ are time-variant. The linear system model described by (6.5) can be regarded as a special case of the adaptive linear system model described by (7.68). In other words, if we set the state transition matrix $\boldsymbol{A}(\boldsymbol{x})$ and the control input matrix $\boldsymbol{B}(\boldsymbol{x})$ described in (7.68) to constant matrices $\boldsymbol{A}$ and $\boldsymbol{B}$ respectively, then (7.68) will be reduced to (6.5).

Compared with the linear system model described by (6.5), the adaptive linear system model described by (7.68) enjoys the advantage of being able to better match dynamics of the control system's state during real-time operation. Compared with the original nonlinear system model described by (6.2), the adaptive linear system model described by (7.68) enjoys the obvious advantage of control convenience, especially note that a variety of mature control methods exist for linear control systems.

Based on the adaptive linear system model described by (7.68), adaptive control can be naturally performed. For example, consider the full-state feedback control method described by (6.27):

$$\boldsymbol{u} = -\boldsymbol{K}^{\mathrm{T}}\boldsymbol{x}.$$

In the spirit of adaptive control, we no longer fix the gain matrix $\boldsymbol{K}$, but allow the gain matrix $\boldsymbol{K}$ to be adapted according to the time-variant state transition matrix $\boldsymbol{A}(\boldsymbol{x})$ and the time-variant control input matrix $\boldsymbol{B}(\boldsymbol{x})$. The adaptive full-state feedback control method can be described by

$$\boldsymbol{u} = -\boldsymbol{K}^{\mathrm{T}}_{\boldsymbol{A}(\boldsymbol{x}),\boldsymbol{B}(\boldsymbol{x})}\boldsymbol{x}, \tag{7.69}$$

where the gain matrix $\boldsymbol{K}_{\boldsymbol{A}(\boldsymbol{x}),\boldsymbol{B}(\boldsymbol{x})}$ is recomputed according to $\boldsymbol{A}(\boldsymbol{x})$ and $\boldsymbol{B}(\boldsymbol{x})$ regularly.

Sometimes, such kind of adaptive linear state-space modelling based control strategy may facilitate handling of nonlinear control problems. For demonstration of this point, we still resort to the example of single inverted pendulum control presented in Sect. 7.4.

Single inverted pendulum control

Dynamics of the single inverted pendulum control system's state $\boldsymbol{x}$ is modelled by the state differential equation described in (7.41):

$$\frac{\mathrm{d}}{\mathrm{d}t}\boldsymbol{x} \equiv \frac{\mathrm{d}}{\mathrm{d}t}\begin{bmatrix} \theta \\ \frac{\mathrm{d}\theta}{\mathrm{d}t} \\ x \\ \frac{\mathrm{d}x}{\mathrm{d}t} \end{bmatrix} = \begin{bmatrix} \frac{\mathrm{d}\theta}{\mathrm{d}t} \\ \frac{\sin\theta}{L}g - \frac{\cos\theta}{L}a \\ \frac{\mathrm{d}x}{\mathrm{d}t} \\ a \end{bmatrix},$$

where the state $\boldsymbol{x} \equiv \begin{bmatrix} \theta & \frac{\mathrm{d}\theta}{\mathrm{d}t} & x & \frac{\mathrm{d}x}{\mathrm{d}t} \end{bmatrix}^{\mathrm{T}}$ consists of the inverted pendulum angle and angular velocity, the cart position and velocity. The control input is the single-input of cart acceleration a. Transform (7.41) into an adaptive linear formalism as

$$\frac{\mathrm{d}}{\mathrm{d}t}\boldsymbol{x} = \boldsymbol{A}(\boldsymbol{x})\boldsymbol{x} + \boldsymbol{B}(\boldsymbol{x})a, \tag{7.70}$$

where

$$\boldsymbol{A}(\boldsymbol{x}) \equiv \begin{bmatrix} 0 & 1\,0\,0 \\ \frac{g}{L}\frac{\sin\theta}{\theta} & 0\,0\,0 \\ 0 & 0\,0\,1 \\ 0 & 0\,0\,0 \end{bmatrix}, \quad \boldsymbol{B}(\boldsymbol{x}) \equiv \begin{bmatrix} 0 \\ -\frac{\cos\theta}{L} \\ 0 \\ 1 \end{bmatrix}.$$

If the inverted pendulum angle θ is close to zero, then $\frac{\sin\theta}{\theta} \approx 1$ and $\cos\theta \approx 1$, so we can fairly approximate the time-variant state transition matrix $\boldsymbol{A}(\boldsymbol{x})$ and the time-variant control input matrix $\boldsymbol{B}(\boldsymbol{x})$ as

$$\boldsymbol{A}(\boldsymbol{x}) \approx \begin{bmatrix} 0\,1\,0\,0 \\ \frac{g}{L}\,0\,0\,0 \\ 0\,0\,0\,1 \\ 0\,0\,0\,0 \end{bmatrix}, \quad \boldsymbol{B}(\boldsymbol{x}) \approx \begin{bmatrix} 0 \\ -\frac{1}{L} \\ 0 \\ 1 \end{bmatrix}.$$

Then (7.70) is reduced to (7.42). However, if the inverted pendulum angle θ is far away from zero, (7.42) will be inappropriate for single inverted pendulum control.

In contrast, (7.70) is generally applicable regardless of whether the inverted pendulum angle θ is close to zero or not. We apply adaptive full-state feedback control based on the adaptive linear system model described by (7.70). Since the cart position x is purely a linear factor for dynamics of the state $\boldsymbol{x}$, we focus on handling the partial state $\boldsymbol{x}_{\mathrm{P}} \equiv \begin{bmatrix} \theta & \frac{\mathrm{d}\theta}{\mathrm{d}t} & \frac{\mathrm{d}x}{\mathrm{d}t} \end{bmatrix}^{\mathrm{T}}$ that consists of the inverted pendulum angle θ, the inverted pendulum angular velocity $\frac{\mathrm{d}\theta}{\mathrm{d}t}$, and the cart velocity $\frac{\mathrm{d}x}{\mathrm{d}t}$. Control of the partial state $\boldsymbol{x}_{\mathrm{P}}$ is the difficult nonlinear part of single inverted pendulum control—Upon stabilization of the partial state $\boldsymbol{x}_{\mathrm{P}}$, sliding mode control based on the easier linear system model described by (7.42) can be conveniently performed to converge the entire state $\boldsymbol{x}$ to the expected equilibrium state finally, as demonstrated in Sect. 7.4.

Extract the sub-model associated with the partial state $\boldsymbol{x}_\mathrm{P}$ from the adaptive linear system model described by (7.70) and obtain

$$\frac{\mathrm{d}}{\mathrm{d}t}\boldsymbol{x}_\mathrm{P} = \boldsymbol{A}_\mathrm{P}(\boldsymbol{x})\boldsymbol{x} + \boldsymbol{B}_\mathrm{P}(\boldsymbol{x})a, \tag{7.71}$$

where

$$\boldsymbol{A}_\mathrm{P}(\boldsymbol{x}) \equiv \begin{bmatrix} 0 & 1\,0 \\ \frac{g}{L}\frac{\sin\theta}{\theta} & 0\,0 \\ 0 & 0\,0 \end{bmatrix}, \quad \boldsymbol{B}_\mathrm{P}(\boldsymbol{x}) \equiv \begin{bmatrix} 0 \\ -\frac{\cos\theta}{L} \\ 1 \end{bmatrix}.$$

Apply the adaptive full-state feedback control method as

$$a = -\boldsymbol{K}^\mathrm{T}_{\boldsymbol{A}_\mathrm{P}(\boldsymbol{x}),\boldsymbol{B}_\mathrm{P}(\boldsymbol{x})}\boldsymbol{x}_\mathrm{P}, \tag{7.72}$$

where the gain matrix $\boldsymbol{K}_{\boldsymbol{A}_\mathrm{P}(\boldsymbol{x}),\boldsymbol{B}_\mathrm{P}(\boldsymbol{x})}$ is recomputed according to $\boldsymbol{A}_\mathrm{P}(\boldsymbol{x})$ and $\boldsymbol{B}_\mathrm{P}(\boldsymbol{x})$ regularly. Matlab simulation code for demonstrating adaptive full-state feedback control of the partial state $\boldsymbol{x}_\mathrm{P}$ is given as follows.

SingleInvertedPendulumAL.m

```
%% Single inverted pendulum parameters
m1 = 1; L1 = 1; g = 10;
%% Simulation preliminary configuration
dt = 0.001; % Numerical computation step
tSpan = 0:dt:3; % Simulation time span
x = 20; dx = 0; % Cart position and its velocity
% Inverted pendulum angle theta and its angular velocity
y = 0.4*pi; dy = 0;
stt = [y; dy; x; dx]; % Single inverted pendulum state
% Record states in simulation
sttAll = zeros(length(stt), length(tSpan)); k = 0;
xExpected = 0; yExpected = 0; % Expected equilibrium status
SimConfig = [m1, L1, g, dt];
%% Control initialization
lambdaE = [-4;-4;-4;-4]; % Expected eigenvalues
iStt = [1 2 4]; % Indices of partial state control
fprintf('Start adaptive control\n');

%% Simulation of single inverted pendulum control
for t = tSpan
    %% Control method
    if (abs(y)<0.1) A21 = g/L1; else A21 = (g/L1)*sin(y)/y; end
    A = [0, 1, 0, 0; A21, 0, 0, 0; 0, 0, 0, 1; 0, 0, 0, 0];
    % Adaptive linear state-space modelling
    B = [0; -cos(y)/L1; 0; 1];
    sttK = DesignGainMatrixSub(A(iStt,iStt), B(iStt),
    lambdaE(iStt));
```

```
    acc = -sttK'*stt(iStt);

    %% Single inverted pendulum dynamics
    stt = DynamicsSIP(SimConfig, stt, acc);
    sttC = num2cell(stt); [y, dy, x, dx] = sttC{:};
    if (abs(y)>=pi/2) fprintf('Control failure!\n'); break; end
    k = k+1; sttAll(:,k) = stt;
    %% Single inverted pendulum visualization
    if (rem(k,20) == 0)
        DisplaySIP(x, y, L1); pause(dt);
    end
end
```

After trials with the Matlab simulation code, readers will find that the adaptive linear system model based full-state feedback control method can effectively stabilize the partial state $\boldsymbol{x}_{\mathrm{P}}$, as demonstrated in Fig. 7.9.

Recall that in Sect. 7.4, we arrived at the Lyapunov strategy based control law described in (7.67) after a long "thinking journey". In contrast, we arrived at the adaptive full-state feedback control law described in (7.72) simply by incorporating the adaptive linear system model described by (7.71) into the full-state feedback control method. Readers can see the obvious control convenience brought by adaptive linear state-space modelling based control.

In above simulation, the gain matrix $\boldsymbol{K}_{\boldsymbol{A}_{\mathrm{P}}(\boldsymbol{x}),\boldsymbol{B}_{\mathrm{P}}(\boldsymbol{x})}$ is recomputed according to $\boldsymbol{A}_{\mathrm{P}}(\boldsymbol{x})$ and $\boldsymbol{B}_{\mathrm{P}}(\boldsymbol{x})$ in each control period. In fact, a more convenient and efficient realization of the adaptive control method is to establish a look-up table of gain matrices by dividing the state space into *piecewise linear operation subspaces* and pre-computing a gain matrix for each linear operation subspace.

For example, we may divide the single inverted pendulum state space into three piecewise linear operation subspaces according to the inverted pendulum angle θ, respectively $|\theta| \leqslant \pi/6$, $\pi/6 < |\theta| \leqslant \pi/3$, and $\pi/3 < |\theta| \leqslant 0.45\pi$ (suppose that 0.45π is the maximum allowable inverted pendulum angle for normal operation). For the first linear operation subspace, set $\theta = 0$ in the adaptive linear system model described by (7.71) and compute its associated gain matrix $\boldsymbol{K}_1$. For the second linear operation subspace, set $\theta = \pi/4$ in the adaptive linear system model described by (7.71) and compute its associated gain matrix $\boldsymbol{K}_2$. For the third linear operation subspace, set $\theta = 0.45\pi$ in the adaptive linear system model described by (7.71) and

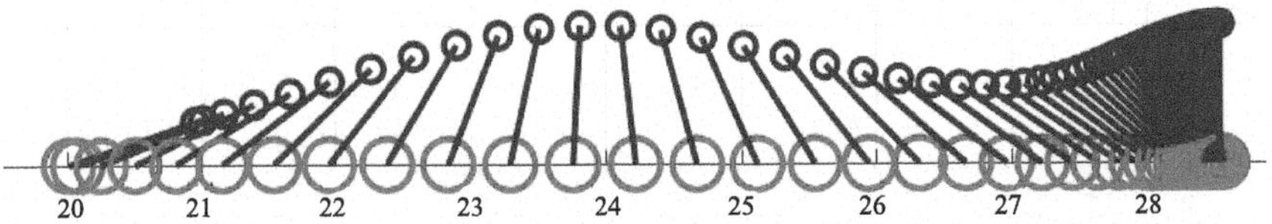

Fig. 7.9 Adaptive full-state feedback control of partial inverted pendulum state

compute its associated gain matrix $\boldsymbol{K}_3$. During adaptive control for stabilization of the partial state $\boldsymbol{x}_\mathrm{P}$, the controller can simply check the inverted pendulum angle θ and select a corresponding gain matrix among $\{\boldsymbol{K}_1, \boldsymbol{K}_2, \boldsymbol{K}_3\}$ to use.

Matlab simulation code for demonstrating such convenient and efficient realization of adaptive full-state feedback control of the partial state $\boldsymbol{x}_\mathrm{P}$ is given as follows—Once the partial state $\boldsymbol{x}_\mathrm{P}$ is stabilized, sliding mode control based on the easier linear system model described by (7.42) is performed to converge the entire state $\boldsymbol{x}$ to the expected equilibrium state finally.

SingleInvertedPendulumAL2.m

```
%% Single inverted pendulum parameters
m1 = 1; L1 = 1; g = 10;
%% Simulation preliminary configuration
dt = 0.001; % Numerical computation step
tSpan = 0:dt:10; % Simulation time span
x = 20; dx = 0; % Cart position and its velocity
% Inverted pendulum angle theta and its angular velocity
y = 0.4*pi; dy = 0;
stt = [y; dy; x; dx]; % Single inverted pendulum state
% Record states in simulation
sttAll = zeros(length(stt), length(tSpan)); k = 0;
xExpected = 0; yExpected = 0; % Expected equilibrium status
SimConfig = [m1, L1, g, dt];
%% Control initialization
lambdaE = [-4;-4;-4;-4]; % Expected eigenvalues
A = [0, 1, 0, 0; g/L1, 0, 0, 0; 0, 0, 0, 1; 0, 0, 0, 0];
B = [0; -1/L1; 0; 1];
sttK0 = DesignGainMatrixSub(A, B, lambdaE);
iStt = [1 2 4]; % Indices of partial state control
sttK1 = DesignGainMatrixSub(A(iStt,iStt), B(iStt),
    lambdaE(iStt));
yA = pi/4; A(2,1) = (g/L1)*sin(yA)/yA; B(2,1) = -cos(yA)/L1;
sttK2 = DesignGainMatrixSub(A(iStt,iStt), B(iStt),
    lambdaE(iStt));
yA = 0.45*pi; A(2,1) = (g/L1)*sin(yA)/yA; B(2,1) = -cos(yA)/L1;
sttK3 = DesignGainMatrixSub(A(iStt,iStt), B(iStt),
    lambdaE(iStt));
cAdaptive=1; fprintf('Start adaptive control\n');

%% Simulation of single inverted pendulum control
for t = tSpan
    %% Control method
    if (y^2+dy^2+dx^2 > 0.1 && cAdaptive==1) % Adaptive control
        if (abs(y)<pi/6) sttK = sttK1;
        elseif (abs(y)<pi/3) sttK = sttK2;
        % Adaptive linear state-space modelling
        else sttK = sttK3; end
        acc = -sttK'*stt(iStt);
    else % Switch to sliding mode control
        if (1==cAdaptive)
            cAdaptive = 0; sttE = [0; 0; x; 0]; sttK = sttK0;
            fprintf('Switch to sliding mode control\n');
        end
```

```
        % Sliding mode design
        if (sttE(3)>0) sttE(3) = max(sttE(3) - 8*dt, 0);
        else sttE(3) = min(sttE(3) + 8*dt, 0); end
        acc = -sttK'*(stt-sttE); % Full-state feedback control
    end

    %% Single inverted pendulum dynamics
    stt = DynamicsSIP(SimConfig, stt, acc);
    sttC = num2cell(stt); [y, dy, x, dx] = sttC{:};
    if (abs(y)>=pi/2) fprintf('Control failure!\n'); break; end
    k = k+1; sttAll(:,k) = stt;
    %% Single inverted pendulum visualization
    if (rem(k,20) == 0)
        DisplaySIP(x, y, L1); pause(dt);
    end
end
```

References

B. Anderson, J. Moore. *Optimal control: linear quadratic methods*. Englewood Cliffs, N.J.: Prentice Hall, 1990.

K. Astrom, B. Wittenmark. *Adaptive control*. Dover Publications, Inc. N.Y., 2008.

R.A. Horn, C.R. Johnson. *Matrix analysis*. Cambridge University Press, 2012.

R.A. Horn, C.R. Johnson. *Topics in Matrix Analysis*. Cambridge University Press, 1991.

D. Kleinman. On an iterative technique for riccati equation computations. *IEEE Transactions on Automatic Control*, 13(1):114 – 115, 1968.

S. LaValle. *Planning algorithms*. Cambridge university press, 2006.

A. Liapounoff. *Problème général de la stabilité du mouvement (traduit du russe)*. Princeton University Press, 1907.

V. Utkin. *Sliding modes in control and optimization*. Springer-Verlag Berlin, 1992.

Y. Xi. *Predictive control (in Chinese)*. National Defense Industry Press, 2013.

The manufacturer's authorised representative in the EU is Springer
Nature Customer Service Centre GmbH, Europaplatz 3, 69115 Heidelberg,
Germany. If you have any concerns regarding our products, please
contact ProductSafety@springernature.com

Printed and bound by CPI Group (UK) Ltd, Croydon, CR0 4YY
01/10/2026
02225925-0013